Handbook of Radar Scattering Statistics for Terrain

DISCLAIMER OF WARRANTY

The technical descriptions, procedures, and computer programs in this book have been developed with the greatest of care and they have been useful to the author in a broad range of applications; however, they are provided as is, without warranty of any kind. Artech House, Inc. and the author and editors of the book titled *Handbook of Radar Scattering Statistics for Terrain* make no warranties, expressed or implied, that the equations, programs, and procedures in this book or its associated software are free of error, or are consistent with any particular standard of merchantability, or will meet your requirements for any particular application. They should not be relied upon for solving a problem whose incorrect solution could result in injury to a person or loss of property. Any use of the programs or procedures in such a manner is at the user's own risk. The editors, author, and publisher disclaim all liability for direct, incidental, or consequent damages resulting from use of the programs or procedures in this book or the associated software.

Accompanying software available at: **http://radarstatistics.eecs.umich.edu**

User code: **$RadStat723!**

For a complete listing of titles in the
Artech House Remote Sensing Library,
turn to the back of this book.

Handbook of Radar Scattering Statistics for Terrain

Fawwaz T. Ulaby
The Radiation Laboratory

M. Craig Dobson
The University of Michigan

with updated
Python and MATLAB® software
by José Luis Álvarez-Pérez

**ARTECH
HOUSE**

BOSTON | LONDON
artechhouse.com

Library of Congress Cataloging-in-Publication Data
A catalog record for this book is available from the U.S. Library of Congress.

British Library Cataloguing in Publication Data
A catalogue record for this book is available from the British Library.

Cover design by John Gomes

ISBN 13: 978-1-63081-701-5

© 2019 ARTECH HOUSE
685 Canton Street
Norwood, MA 02062

10 9 8 7 6 5 4 3 2 1

*To Jean, Marina,
and Tanja*

Contents

Preface xiii

PART I. THEORY OF RADAR SCATTERING STATISTICS 1

1. INTRODUCTION 1
    1-1   Radar Clutter Statistics 1
    1-2   Causes of Radar Signal Variations 2
    1-3   Scattering Coefficient Data Base 4
    1-4   Scope of This Handbook 9

2. THE RADAR EQUATION 11
    2-1   Scattering from a Point Target 11
    2-2   Scattering from a Distributed Target 14
            2-2.1   Narrow-Beam Scatterometer 16
            2-2.2   Imaging Radar 16
    2-3   Statistical Properties of a Distributed Target 18
            2-3.1   Homogeneous Medium 21
            2-3.2   Inhomogeneous Random Medium 22
            2-3.3   Statistically Homogeneous Medium 24
            2-3.4   Statistically Uniform Target 25
    2-4   General Behavior of $(\sigma°)$ 25
            2-4.1   Angular Dependence of $(\sigma°)$ of a Random Surface 25
            2-4.2   Dependence of $(\sigma°)$ on Surface Roughness 26
            2-4.3   Dependence of $(\sigma°)$ on Surface Moisture 27
            2-4.4   Angular Dependence of $(\sigma°)$ for Vegetation 28
            2-4.5   Dependence of $(\sigma°)$ on Vegetation Biomass 30

2-4.6   Angular Dependence of ($\sigma°$) for Snow 31

2-4.7   Frequency Variation of ($\sigma°$) for Snow 33

3. STATISTICS OF SIGNAL FLUCTUATIONS 35

3-1   Rayleigh Fading Statistics 38

    3-1.1   Underlying Assumptions 38

    3-1.2   Output Voltage 41

    3-1.3   Interpretation 44

3-2   Multiple Independent Samples 44

    3-2.1   Linear Detection 44

    3-2.2   Square-Law Detection 48

3-3   Applicability of the Rayleigh Clutter Model 48

3-4   Non-Rayleigh Distributions 50

    3-4.1   Log-Normal Distribution 50

    3-4.2   Weibull Distribution 54

    3-4.3   Distribution for a Strong Point Target in Clutter Background 56

3-5   Clutter Model for Nonuniform Terrain 59

    3-5.1   Uniformly Distributed Background 60

    3-5.2   Gamma Distributed Background 63

    3-5.3   Log-Normal Distributed Background 63

4. SCINTILLATION REDUCTION BY SPATIAL AVERAGING AND FREQUENCY AGILITY 65

4-1   Spatial Averaging 67

    4-1.1   Discrete Samples 67

    4-1.2   Continuous Averaging in Azimuth 67

    4-1.3   Continuous Averaging in Range 69

4-2   Frequency Agility 69

    4-2.1   Decorrelation Bandwidth 70

    4-2.2   Continuous Averaging in Frequency 72

    4-2.3   Examples of Experimental Observations 74

PART II. DISTRIBUTION FUNCTION OF BACKSCATTERING COEFFICIENT 79

5. DATA QUALITY AND PRESENTATION FORMAT 79

5-1   Data-Quality Evaluation 79

5-2   Terrain Categories 80

    5-2.1   Barren and Sparsely Vegetated Land 83

    5-2.2   Vegetated Land 85

    5-2.3   Urban Land 86

    5-2.4   Snow 87

    5-2.5 Classification Summary 87

  5-3 Frequency Bands 88

    5-4 Statistical Distributions 88

    5-4.1 Statistical Distribution Tables 90

    5-4.2 Angular Plots 90

    5-4.3 Histograms 96

APPENDIX 5. A VARIABILITY IN THE MEASUREMENT OF RADAR
BACKSCATTER 99

    References 109

APPENDIX A. BACKSCATTERING DATA FOR SOIL AND ROCK
SURFACES 119

  A-1 Data Sources and Parameter Loadings 119

  A-2 L-Band Data 120

  A-3 S-Band Data 125

  A-4 C-Band Data 130

  A-5 X-Band Data 135

  A-6 Ku-Band Data 140

  A-7 Ka-Band Data 145

APPENDIX B. BACKSCATTERING DATA FOR TREES 147

  B-1 Data Sources and Parameter Loadings 147

  B-2 L-Band Data 148

  B-3 S-Band Data 151

  B-4 C-Band Data 152

  B-5 X-Band Data 154

  B-6 Ku-Band Data 159

  B-7 Ka-Band Data 164

  B-8 W-Band Data 165

APPENDIX C. BACKSCATTERING DATA FOR GRASSES 167

  C-1 Data Sources and Parameter Loadings 167

  C-2 L-Band Data 168

  C-3 S-Band Data 173

  C-4 C-Band Data 178

  C-5 X-Band Data 183

  C-6 Ku-Band Data 188

  C-7 Ka-Band Data 193

  C-8 W-Band Data 198

APPENDIX D. BACKSCATTERING DATA FOR SHRUBS 199

  D-1 Data Sources and Parameter Loadings 199

D-2   L-Band Data 200
D-3   S-Band Data 205
D-4   C-Band Data 210
D-5   X-Band Data 215
D-6   Ku-Band Data 220
D-7   Ka-Band Data 225

APPENDIX E. BACKSCATTERING DATA FOR SHORT VEGETATION .. 231
E-1 Data Sources and Parameter Loadings 231
E-2   L-Band Data 232
E-3   S-Band Data 237
E-4   C-Band Data 242
E-5   X-Band Data 247
E-6   Ku-Band Data 252
E-7   Ka-Band Data 257
E-8   W-Band Data 262

APPENDIX F. BACKSCATTERING DATA FOR ROAD SURFACES 263
F-1   Data Sources and Parameter Loadings 263
F-2   L-Band Data 264
F-3   X-Band Data 265
F-4   Ku-Band Data 270
F-5   Ka-Band Data 275
F-6   W-Band Data 280

APPENDIX G. BACKSCATTERING DATA FOR URBAN AREAS 281
G-1   Data Sources and Parameter Loadings 281
G-2   L-Band Data 282
G-3   S-Band Data 283
G-4   C-Band Data 284
G-5   X-Band Data 285

APPENDIX H. BACKSCATTERING DATA FOR DRY SNOW 287
H-1   Data Sources and Parameter Loadings 287
H-2   L-Band Data 288
H-3   S-Band Data 293
H-4   C-Band Data 298
H-5   X-Band Data 303
H-6   Ku-Band Data 308
H-7   Ka-Band Data 313
H-8   W-Band Data 318

APPENDIX I. BACKSCATTERING DATA FOR WET SNOW 323
    I-1   Data Sources and Parameter Loadings 323
    I-2   L-Band Data 324
    I-3   S-Band Data 329
    I-4   C-Band Data 334
    I-5   X-Band Data 339
    I-6   Ku-Band Data 344
    I-7   Ka-Band Data 349
    I-8   W-Band Data 354

PART III. COMPANION SOFTWARE BY JOSÉ LUIS ÁLVAREZ-PÉREZ .. 357

APPENDIX J.   PYTHON LIBRARY ACCOMPANYING THIS BOOK 357
    J-1   Introduction 357
    J-2   Scope and Objectives 358
    J-3   System Requirements 359
    J-4   Data Formats 359
    J-5   Data Parameters 360
    J-6   Importing Rsst 360
    J-7   The sdtData Class 361
    J-8   The modelData Class 365
    J-9   The histData Class 367
          J-9.1   KDE 370
          J-9.2   Gaussian Probability Distribution 371
          J-9.3   Weibull Probability Distribution 373
          J-9.4   Gamma Probability Distribution 374
          J-9.5   K-Distribution 375

APPENDIX K.   MATLAB® LIBRARY ACCOMPANYING THIS BOOK ... 377
    K-1   Introduction 377
    K-2   Scope and Objectives 378
    K-3   System Requirements 378
    K-4   Data Formats 378
    K-5   Data Parameters 378
    K-6   Databases And MATLAB Classes 378
    K-7   The scattdata Structure 379
    K-8   The sdtData Class 381
    K-9   The modelpar Structure 383
    K-10   The modelData Class 384
    K-11   The scatthis Structure 385

K-12   The histData Class 388
    K-12.1   KDE 390
    K-12.2   Gaussian Probability Distribution 391
    K-12.3   Weibull Probability Distribution 393
    K-12.4   Gamma Probability Distribution 394
    K-12.5   K-Distribution 395

APPENDIX L.   GRAPHICAL USER INTERFACE 397
  L-1   Introduction 397
  L-2   Data Products 398
    L-2.1   Statistical Distribution Table 398
    L-2.2   Angular Plots 399
    L-2.3   Histograms 399

ABOUT THE AUTHORS 401

INDEX 403

Preface

This is a new edition of the *Handbook of Radar Scattering Statistics for Terrain*, which was first published in 1989, exactly 30 years ago. The *Handbook* was relevant then and continues to be relevant today, primarily because it is based on the most comprehensive database of radar measurements ever assembled. The database contains calibrated multifrequency, multipolarization measurements of the radar backscattering coefficient for a wide range of terrain types and conditions. A special feature of the current edition is the companion software, which offers the user three different options for extracting data from the database: (1) Python-based software and commands, (2) MATLAB®-based software and commands, and (3) a graphical user interface that requires no experience with programming languages. The three software approaches are described in three dedicated appendixes at the end of this book.

A radar image of terrain usually exhibits two types of variations: tonal variations corresponding to the spatial variations of the physical properties of the entire image scene, and speckle-like pixel-to-pixel variations that permeate the entire image and give it a salt-and-pepper appearance. The total variation may be separated into a background component associated with the gross (or local average) properties of the scene and a fluctuating component associated with random variations at the local level. The visual representation of the fluctuating component is called image speckle.

The purpose of this book is twofold: To examine the statistical behavior of speckle and to provide ready access to statistical summaries and plots of radar backscatter from terrain. Accordingly, the book is organized into three parts. Part I focuses on the electromagnetic mechanisms that give rise to the speckle phenomenon and provides an overview of techniques used to reduce its significance.

Part II of this book is intended to support radar system design and signal processing applications. It provides probability density functions (PDFs) of the radar backscattering coefficient extracted from a comprehensive database comprised of a large number of data sources. The PDFs are organized by terrain category, microwave frequency band, incidence angle, and antenna polarization configuration.

Part III, new to this edition, includes software tools for extracting PDF data and other statistical parameters from the database.

The authors are grateful for the help and support of many individuals who contributed to the development of this book, most notably Saied Moezzi, Sebastian Lauer, Debbie Boulier, Richard Austin, Kyle McDonald, Michael Whitt, Brian Zuerndorfer, and Damien Farrell.

PART I

Chapter 1

Introduction

1-1 RADAR CLUTTER STATISTICS

The term *radar clutter* was introduced during World War II to denote unwanted radar echoes from extended targets such as rain, land, and water. When a downward-looking radar is used for the purpose of detection or tracking, the terrain background represents a source of interference because the energy backscattered from the terrain *clutters* the fidelity of the desired signal, namely that reflected by the airplane or object that the radar is intended to detect or track. In the 1950s and 1960s, *radar terrain clutter* encompassed all aspects of scattering from terrain surfaces [9]. This included theoretical models for electromagnetic interaction with surfaces and volumes, experimental measurements of the backscattering cross section per unit area for various types of terrain, and the statistics associated with the scattering process. Research in this field was conducted primarily in support of military applications. The advent of high resolution imaging radar in the 1950s has since led to the development of the new field of *radar remote sensing,* which has evolved into a major scientific discipline with wide-ranging civilian, military, and environmental applications [133, Chapter 1]. In many of these applications, the terrain-scattered radar wave, which previously was regarded as a source of interference, is itself the information carrier because the information sought is derived from knowledge of how waves scatter from rough surfaces and inhomogeneous media. Because of this wider scope of the intended applications of radar scattering, in today's literature the generic term *radar scattering from distributed targets* has become the standard term used (instead of terrain clutter) to refer to the topics previously classified as radar clutter.

From the standpoint of the detection and tracking applications, terrain scattering is viewed in statistical terms because the objective is to determine the degradation in the false alarm probability (associated with detecting the target) caused by interference of the radar return from the target with that contributed by the terrain background. Consequently, the term radar clutter has evolved into the more specific term *radar clutter statistics* and is used in today's literature to denote the probability density function (PDF) characterizing $\sigma°$, the backscattering cross section per unit area. As we shall see later, $\sigma°$ of terrain varies as a function of two sets of parameters: the sensor (or wave) parameters, namely the wavelength (or frequency), the incidence angle (defined here with respect to normal incidence), and the polarization configurations of the transmitting and receiving antennas, and the terrain parameters, which include its dielectric properties and geometrical characteristics. Hence, in general the PDF is both sensor- and terrain type-specific. Actually, the PDF is also terrain condition-specific because a forest, for example, undergoes changes in both its dielectric properties and geometry as a function of time.

A distributed target usually consists of a large number of randomly distributed scatterers. When a distributed target is illuminated by a coherent electromagnetic wave, the magnitude of the scattered signal is equal to the phasor sum of the returns from all of the scatterers illuminated by the incident beam. The backscattered signal is a random variable because the dielectric and geometrical properties of a terrain surface are random variables. Thus, two ground patches with the same statistical properties will produce backscattered signals with different magnitudes because the individual scatterers in the two patches have different locational arrangements. This variability in the magnitude of the backscattered signal is referred to as *signal fading* or *signal scintillation.* Its appearance in a radar image, which takes the form of a *speckle pattern,* complicates image interpretation and reduces the effectiveness of information extraction algorithms.

1-2 CAUSES OF RADAR SIGNAL VARIATIONS

The image segments shown in Figure 1.1 were extracted from the same strip of X-band radar imagery. The two segments, one of which corresponds to a corn field and the other to a forest parcel, have different average *tones,* exhibit significantly different *textures,* and both exhibit large pixel-to-pixel intensity variations. The average tone of an image is the average value of the image intensity for all pixels contained in that image. (Each image segment in Figure 1.1 is approximately 10^4 pixels.) This average tone is proportional to the average received power which, in turn, is directly proportional to the *backscattering coefficient,* $\sigma°$, of the imaged target. In other words, $\sigma°$ of a statistically homogeneous distributed target is, by definition, proportional to the *mean value* of the random process characterizing the intensity variations in the radar image.

Texture refers to the low spatial-frequency intensity variations of the image [139]; the corn field, being more spatially uniform than the forest parcel, exhibits the same type of random variations in all regions of the image, whereas the image

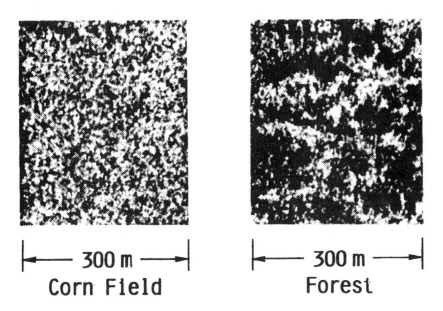

Figure 1.1 X-band SAR images of (a) a corn field and (b) a forested area. Note the textural differences between the two images.

of a "backscattering coefficient for a distributed target" is meaningful only for targets with uniform electromagnetic properties, then texture becomes the spatial variation of $\sigma°$ from one region of an image to another. In the case of the forest parcel, these variations are related to the spatial nonuniformity of tree density and height.

Unlike textural variations, which may or may not have specific directional properties and which are governed by the spatial variation of the target scattering properties relative to the dimensions of the radar resolution cell, the random variations that give the image its speckled appearance are due to phase-interference effects and are a characteristic feature of the scattering pattern for any distributed target (provided that the target satisfies certain conditions, as we shall discuss in Chapter 2). *Image speckle* is a visual manifestation of fading statistics.

Thus, there are three types of intensity variations that one may observe in a radar image: variations in average tone from one distributed target (such as a bare-soil field) to another (such as a forest parcel), textural variations from one region of a distributed target to another, and random fading variations at the pixel-to-pixel scale. These variations are governed by different processes and are characterized by different probability density functions (PDFs).

In some radar applications, these three types of variations are lumped together, treated as a single variation, and characterized as *terrain clutter*. To determine the statistics of the clutter random variable for a given terrain type or geographic area, the area is imaged and then a PDF of the received voltage or power is generated. Next, the data are tested against standard PDFs (such as the Gaussian, Rayleigh, and log-normal) to determine which fits best. Such an empirical approach may produce a statistical description appropriate to the imaged area, but it has some severe limitations. The empirically generated PDF is, in essence, a convolution of the three PDFs characterizing the three types of variations referred to previously. Hence, it is both target-specific and sensor-specific. It is target-specific in that it pertains to the specific mix of terrain categories and the specific conditions of those categories at the time the radar observations were made; most terrain surfaces exhibit dynamic variations with time of day, season, and weather history. The PDF is sensor-specific because one of the underlying variations (namely, that due to signal fading) is governed by the detection scheme used in the receiver (linear or square-law) and the type of filtering or smoothing technique employed in the signal processor. Filtering techniques, which are used to reduce fading variations, may involve spatial averaging or frequency averaging schemes and may be performed coherently or incoherently [35,42,86,100,158].

1-3 SCATTERING COEFFICIENT DATA BASE

By definition, a scatterometer is a calibrated radar system designed to measure the backscattering coefficient of a distributed target with good absolute accuracy. Whereas the primary objective of a *real aperture radar* (RAR) or a *synthetic aperture radar* (SAR) is to produce a high resolution image of the observed scene—often at the expense of measurement accuracy (lack of absolute calibration) and precision (uncertainty caused by signal fluctuation)—the primary objective of a scatterometer is to measure $\sigma°$ of a distributed target both accurately and precisely, and to do so over a wide range of the incidence angle θ (Figure 1.2), or its complement, the grazing angle ψ. For these reasons, a scatterometer usually is not designed to be an imaging system and relies on its antenna beam for spatial resolution.

Scattering coefficient measurements of terrain have been made with scatterometers mounted on a variety of platforms including trucks, towers, airplanes, helicopters, and satellites. The platforms provide different, but complementary information about terrain scattering statistics. Because of the relatively small footprint size of its antenna beam, a truck-mounted scatterometer is well suited for making observations of spatially uniform targets, such as an agricultural field or a flat ground surface with a uniform layer of snow cover, and the radar measurements can be augmented with detailed *ground truth information* about the terrain target. Airborne scatterometers can acquire much more data per unit time,

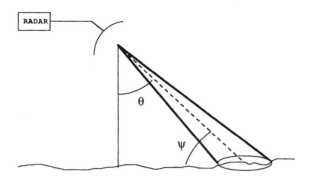

Figure 1.2 Radar illumination geometry, with θ defined as the angle of incidence and its complement ψ as the grazing angle.

but their spatial resolution is coarser. If the scatterometer observations are made from a height on the order of 500 m or less, the antenna footprint size on the ground is sufficiently small that it is still possible to measure $\sigma°$ of uniform targets, but if the airplane is flown at a much higher altitude the footprint size becomes large, thereby containing several terrain types, and the measurements will no longer be associated with uniform targets or conditions. There are a number of exceptions, of course, and these include heavily forested areas, certain desert regions, and water surfaces. For these terrain types, an airborne scatterometer system is better suited for the acquisition of scattering data than a ground-based system. Satellite-borne scatterometers, such as the four-beam wind-measurement system that flew on Seasat in 1978 [73], have resolution cells on the order of tens of square kilometers, which limits their use (with regard to terrain scattering statistics) to areas like the Amazon forest and the African Sahara.

One of the most extensive series of specialized scatterometer measurements was made during the 1950s and 1960s at Ohio State University [37,96,98,99] using a truck-mounted system (Figure 1.3) that operated at 10 GHz, 15.5 GHz, and 35 GHz. The like-polarized backscattering coefficient $\sigma°$ was measured as a function of incidence angle from 10° to 80° for many types of targets including agricultural crops, concrete and asphalt surfaces, and shallow snow cover. This set of measurements served as a standard reference of the backscattering coefficient for many years.

At the same time as the Ohio State University program, the Naval Research Laboratory (NRL) developed its own scattering-coefficient measurement program using an airborne four-frequency scatterometer covering the range from 0.428 GHz to 8.8 GHz. The system was flown extensively during the 1950s and 1960s and into the early 1970s, mostly over the sea but also over land [43,58,92]. The

Figure 1.3 Photo of the truck-mounted scatterometer system used by Ohio State University [37] in the 1950s and 1960s to measure the backscattering properties of distributed targets.

NRL system consisted of four pulsed radars with narrow pointable beams. By pointing the beams to different elevation angles, $\sigma°$ could be measured as a function of the incidence angle θ.

In the mid-1960s, NASA's Johnson Space Center developed a 13.3 GHz doppler scatterometer with a fan beam pointed along the flight path as illustrated in Figure 1.4. This arrangement allowed measurements to be made simultaneously over angles of incidence from 60° down to 5° behind the aircraft and again from 5° to 60° ahead of it. To measure the backscattering coefficient of a given terrain target at a given angle of incidence, the received signal was passed through a bandpass filter centered at the doppler frequency corresponding to that angle and then integrated over the time interval that the target was observed by the antenna beam along that direction. This system, and lower frequency versions that were added in later years, was flown on a C-130 aircraft to investigate the scattering properties of various types of land surfaces and sea surface conditions and covers (such as sea ice).

The backscattering measurements produced by the aforementioned programs were all made using independent systems at spot microwave frequencies. Because of differences in the systems and the calibrations used in the different bands, and because the frequencies were far apart from each other, it was not

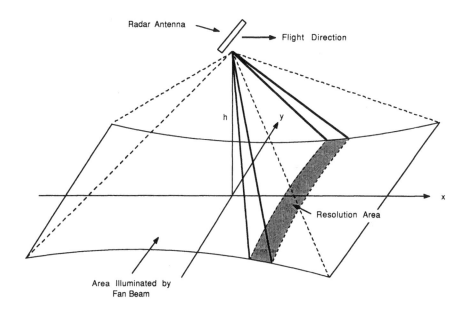

Figure 1.4 Illumination geometry of a downward-looking fan-beam doppler scatterometer. The antenna is long in the *y*-direction and short in the *x*-direction.

possible to ascertain the frequency response of the terrain with an acceptable degree of confidence. The need for continuous frequency coverage of terrain backscattering led to the development of radar spectrometers in the early 1970s at the University of Kansas [118]. A truck-mounted system was developed with continuous coverage from 1 to 18 GHz (utilizing separate antennas for the 1–8 GHz range and the 8–18 GHz range). Another system that provided spot coverage at 35 GHz was mounted on the same platform [128]. Between 1971 and 1984 these systems were used to acquire over 500,000 measurements of $\sigma°$ for a wide variety of terrain types and conditions. Moreover, each of these measurements was an average of a large number (typically ≥ 50) of individual measurements corresponding to different spatial locations of the distributed target under consideration.

One of the major conclusions reached on the basis of these spectral data is that $\sigma°$ of a given distributed target has a slowly varying response with frequency and does not exhibit resonant-like behavior in the microwave region. This is in stark contrast with the resonant spectra of reflectivity of certain terrain features (such as vegetation) at optical wavelengths. Consequently, interest in continuous spectral data subsided and was replaced with interest in making *polarimetric measurements* of $\sigma°$, wherein both the amplitude and phase of the scattered field are measured for all linear polarization combinations. This new interest, which was prompted by the development of polarimetric SAR systems in the 1980s

[148,157], led to the development of polarimetric scatterometers at the University of Michigan [140] at three microwave frequencies (1.6, 5, and 10 GHz) and at three millimeter-wave frequencies (35, 94, and 140 GHz). These systems (Figure 1.5) have been used to study the scattering properties of trees, snow-covered terrain, and other types of distributed targets [141–143,152].

Figure 1.5 Photo of the University of Michigan's truck-mounted millimeter-wave scatterometer-radiometer system. The system consists of a scatterometer and a radiometer at each of three frequencies: 35, 94, and 140 GHz.

In addition to the data bases generated in the major programs referred to previously, numerous other measurements of terrain scattering have been reported in the literature at both microwave and millimeter-wave frequencies. In order to extract meaningful statistics from these various data sets, it would be useful to integrate them into a master data base. Such a task is not as straightforward as it may seem, however, because the integration process must address questions related to differences between the systems used to acquire the data sets, particularly with regard to calibration accuracy. Otherwise, the combined data base will have built-in artifacts arising from calibration biases between the data sets comprising it. At the same time, it is not quite obvious how one would determine these biases without actually cross-calibrating the different systems against each other (or at least against one other system). In some cases, it may be possible to identify significant calibration biases between systems by comparing data measured by different systems for the same type of terrain target and condition (specific examples are given in Section 5-1).

Part II of this handbook provides distributions of $\sigma°$ generated from a master data base comprised of data reported in 124 references. The "data quality" criteria applied in evaluating individual data sets are also given.

1-4 SCOPE OF THIS HANDBOOK

This handbook is divided into two parts. Part I provides the user with a review of signal fading statistics for backscattering from distributed targets, an examination of techniques used to reduce the fading variation through spatial or frequency averaging, and an overview of the angular and spectral variations of the backscattering coefficient $\sigma°$ for various types and conditions of terrain surfaces. The contents of Part I lead us to Part II, which provides probability density functions (PDFs), in the form of both tables and graphs, for $\sigma°$ of terrain. These PDFs, which are based on a "carefully screened" set of *calibrated* radar backscattering measurements extracted from the open literature, are organized by terrain category (and condition, when possible), frequency band, incidence angle, and polarization configuration. For the category of *short vegetation,* for example, the data base is comprised of measured values of $\sigma°$ for various types of crops over a wide range of growth stages and conditions. The density of available data varies widely among categories and frequency bands because more data sources are available for certain categories than for others. Nevertheless, the data base used for generating the PDFs is the most complete data base of calibrated backscattering coefficient data generated to date.

In the process of assembling the data base, strict data-quality criteria were applied in the decision to include or exclude a certain data set. These included calibration accuracy, measurement precision, and category identification. Details of this process are given in Section 5-1. The category classification scheme used in this handbook, which is described in some detail in Section 5-2, divides terrain targets into specific terrain classes on the basis of ancillary target descriptions for the source data. These specific terrain classes are concatenated in the following generalized categories for presentation in Part II: (1) Soil and Rock Surfaces, (2) Trees, (3) Grasses, (4) Shrubs, (5) Short Vegetation, (6) Road, (7) Urban Areas, (8) Dry Snow, and (9) Wet Snow. Open water bodies such as lakes and rivers are not treated herein in either the liquid or frozen state (ice).

With few exceptions, most imaging radar systems flown to date have lacked absolute calibration with respect to a reference of known *radar cross section*

(RCS). Consequently, data sets obtained by different imaging radar systems could not be properly combined or easily compared with one another, even when the radars in question observed the same type of terrain and operated at the same wavelength, polarization configuration, and incidence angle range. Hence, all the data presented in Part II of this handbook are derived exclusively from measurements made by ground-based and airborne scatterometer systems.

Chapter 2
The Radar Equation

2-1 SCATTERING FROM A POINT TARGET

The radar equation relates the power intercepted by the radar receiving antenna to the characteristics of the radar system, including the power of its transmitter and the gain of its antenna, the location of the scattering target relative to the transmitting and receiving antennas, and the target's scattering properties. For *monostatic radar,* wherein the transmitting and receiving antennas are collocated, or the same antenna is used for both transmission and reception (Figure 2.1), the power received by the antenna, as a result of backscattering from a point target at a range R, is given by

$$P_r(\theta_a,\phi_a) = \frac{P_t G^2(\theta_a,\phi_a)\lambda^2}{(4\pi)^3 R^4} \cdot \sigma \tag{2.1}$$

where (θ_a,ϕ_a) defines the direction from the radar to the target, with θ_a measured relative to the boresight direction of the antenna, P_t is the transmitted power, λ is the wavelength, and σ is the radar cross section (RCS) of the target. The above equation pertains to a target whose physical dimensions are such that the solid angle it subtends (from where the radar is located) is much smaller than the solid angle of the radar beam. Such a target will be referred to as a *point target* even though it may have a complex geometry and nonuniform scattering properties.

The radar cross section σ, which has units of area, characterizes the scattering "strength" of the target in the backscattering direction in the form of an effective area. In general, σ of a given target is related to its shape and dielectric constant, the viewing geometry, and the wavelength λ and polarization directions of the incident and scattered waves. From the standpoint of electromagnetic scattering theory, the standard definition for σ is cast in terms of the ratio of the scattered power density S_p^s measured at a distance R from the scatterer to the power density S_q^i of an incident plane wave. Thus,

$$\sigma_{pq} = \lim_{R \to \infty} \left(4\pi R^2 \frac{S_p^s}{S_q^i} \right) \tag{2.2}$$

where the limit as $R \to \infty$ is included to denote that the observation point (Figure 2.2) is in the far-field region. The far-field power density S_p^s varies as R^{-2}, which renders σ_{pq} independent of R, as it should be.

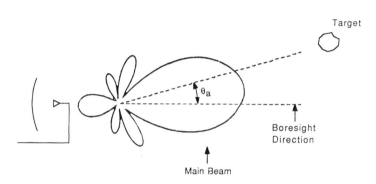

Figure 2.1 Radar antenna observing a target in a direction (θ_a, ϕ_a) relative to boresight.

The subscripts q and p denote the polarization directions of the incident and scattered waves, respectively. When the earth's surface is used as a reference plane, it is often convenient to use the orthogonal pair of linear polarizations known as *horizontal* and *vertical* to describe the direction of the electric field. For the configuration shown in Figure 2.2, the reference plane is the x-y plane, the propagation direction of the incident wave is $\hat{\mathbf{k}}_i$ and that of the scattered wave is $\hat{\mathbf{k}}_s$. The electric field vector of the incident wave can be expressed as

$$\mathbf{E}^i = E_h^i \, \hat{\mathbf{h}}_i + E_v^i \, \hat{\mathbf{v}}_i \tag{2.3}$$

where E_h and E_v are the horizontally polarized and vertically polarized components of $\mathbf{E}^i$, and $\hat{\mathbf{h}}_i$ and $\hat{\mathbf{v}}_i$ are unit vectors defined by

$$\hat{\mathbf{h}}_i = \frac{\hat{\mathbf{z}} \times \hat{\mathbf{k}}_i}{|\hat{\mathbf{z}} \times \hat{\mathbf{k}}_i|} = \hat{\mathbf{x}} \sin \phi_i - \hat{\mathbf{y}} \cos \phi_i \tag{2.4a}$$

and

$$\hat{\mathbf{v}}_i = \hat{\mathbf{k}}_i \times \hat{\mathbf{h}}_i = -\hat{\mathbf{x}} \cos \theta_i \cos \phi_i - \hat{\mathbf{y}} \cos \theta_i \sin \phi_i + \hat{\mathbf{z}} \sin \theta_i \tag{2.4b}$$

For the scattered field,

$$\mathbf{E}^s = E_h^s \, \hat{\mathbf{h}}_s + E_v^s \, \hat{\mathbf{v}}_s$$

the polarization vectors may be defined relative to $\hat{\mathbf{k}}_s$, the propagation direction of the scattering wave, or relative to $\hat{\mathbf{k}}_r = -\hat{\mathbf{k}}_s$, the unit vector pointing from the receive antenna towards the scatterer. The latter approach is more convenient in practice because the polarization vectors

$$\hat{\mathbf{h}}_s = \frac{\hat{\mathbf{z}} \times \hat{\mathbf{k}}_r}{|\hat{\mathbf{z}} \times \hat{\mathbf{k}}_s|} = \hat{\mathbf{x}} \sin \phi_s - \hat{\mathbf{y}} \cos \phi_s$$

and

$$\hat{\mathbf{v}}_s = \hat{\mathbf{k}}_r \times \hat{\mathbf{h}}_s = -\hat{\mathbf{x}} \cos \theta_s \cos \phi_s - \hat{\mathbf{y}} \cos \theta_s \sin \phi_s + \hat{\mathbf{z}} \sin \theta_s$$

assume the same form as $\hat{\mathbf{h}}_i$ and $\hat{\mathbf{v}}_i$ (in terms of θ_i and ϕ_i), and in the backscattering case ($\theta_i = \theta_s$ and $\phi_i = \phi_s$), the two sets ($\hat{\mathbf{h}}_i$, $\hat{\mathbf{v}}_i$) and ($\hat{\mathbf{h}}_s$, $\hat{\mathbf{v}}_s$) become identical.

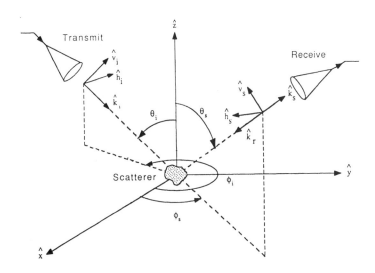

Figure 2.2 Scattering geometry.

Returning to (2.2), the hh-polarized and vh-polarized backscattering cross sections are given by

$$\sigma_{hh} = \lim_{R \to \infty} \left(4\pi R^2 \frac{|E_h^s|^2}{|E_h^i|^2} \right) \tag{2.5a}$$

$$\sigma_{vh} = \lim_{R \to \infty} \left(4\pi R^2 \frac{|E_v^s|^2}{|E_h^i|^2} \right) \tag{2.5b}$$

where we have used the relation $S_q = |E_q|^2/\eta$ for the power densities of both the incident and scattered waves. The quantity η is the intrinsic impedance of the medium, which in this case is air. Definitions similar to those given by (2.5) may be written for σ_{vv} and σ_{hv}. In the backscattering direction, the scattering process is reciprocal in character, which leads to the equality

$$\sigma_{vh} = \sigma_{hv} \tag{2.6}$$

The cross sections σ_{vv} and σ_{hh} are called the *like-polarized components*, whereas σ_{vh} and σ_{hv} are called the *cross-polarized* or *depolarized components* because they are generated as a result of a change in the direction of the electric field (of the incident wave) upon scattering by the target. In practice, σ_{hh} and σ_{vh} are measured as a pair by using an *h*-polarized transmitting antenna and a two-channel dual-polarized receiving antenna. If the transmitting antenna polarization is switched to v, the receive antenna will measure σ_{vv} and σ_{hv}.

2-2 SCATTERING FROM A DISTRIBUTED TARGET

The radar equation given by (2.1) for a point target may be extended to the distributed target shown in Figure 2.3 by integrating the backscattered power over the illuminated area A:

$$P_{rp}(\theta) = \iint\limits_{A} \frac{P_{tq} G^2(\theta_a, \phi_a) \lambda^2}{(4\pi)^3 R^4} \cdot \sigma_{pq}^\circ \, dA \tag{2.7}$$

where θ is the *incidence angle* of the boresight direction of the antenna, relative to normal incidence. Sometimes radar scattering data are reported in terms of the *grazing angle* $\psi = \pi/2 - \theta$. The polarization indices p and q are included in the above expression to show the connection between the q-polarized transmitted power, the p-polarized received power, and the pq-polarized *backscattering cross section per unit area* σ_{pq}°, which is defined as the backscattering cross section of a distributed target of horizontal area A, normalized with respect to A,

$$\sigma_{pq}^\circ = \sigma_{pq}/A \tag{2.8}$$

Often, σ° is referred to as the *backscattering coefficient*, for short.

Some authors describe the backscattering coefficient in terms of σ per unit area perpendicular to the antenna beam, rather than per unit area on the ground (Figure 2.4), and denote it γ (see Cosgriff *et al.* [37]). Thus,

$$\gamma_{pq} = \sigma_{pq}/A'$$
$$= \sigma_{pq}/(A \cos \theta)$$
$$= \sigma^{\circ}_{pq}/\cos \theta \qquad\qquad (2.9)$$

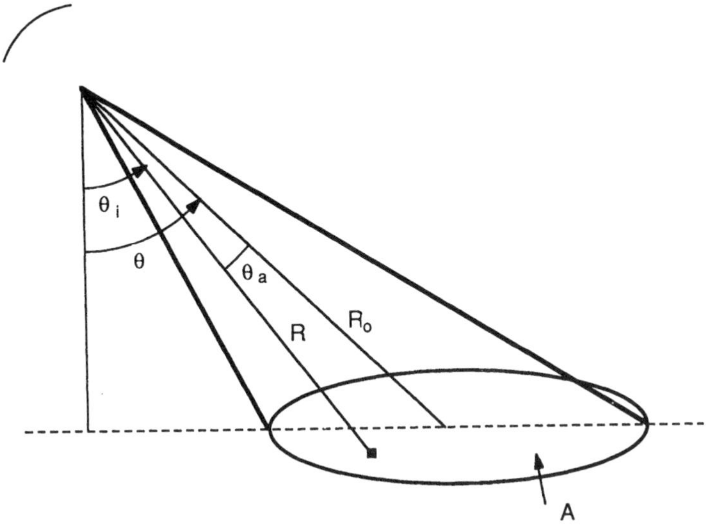

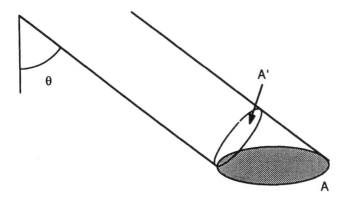

Figure 2.3 Illumination geometry for a distributed target.

Figure 2.4 The ground illuminated area A is related to the beam area A' by $A = A'/\cos \theta$.

2-2.1 Narrow-Beam Scatterometer

If the distributed target has uniform properties across the area illuminated by the antenna beam, and if the beam is sufficiently narrow that $\sigma°(\theta_i)$ at the local angle of incidence θ_i may be regarded as constant over the angular extent of the beam, then (2.7) may be simplified to

$$
P_{rp}(\theta) = \frac{P_{tq}\lambda^2}{(4\pi)^3} \sigma°_{pq}(\theta) \iint\limits_{A_i} \frac{G^2(\theta_a, \phi_a)\,\mathrm{d}A}{R^4}
$$

$$
= \frac{P_{tq}\lambda^2}{(4\pi)^3} \sigma°_{pq}(\theta) \cdot \mathrm{I} \tag{2.10}
$$

where

$$
\mathrm{I} = \iint\limits_{A} \frac{G^2(\theta_a, \phi_a)\,\mathrm{d}A}{R^4} \tag{2.11}
$$

is called the *illumination integral*. Often, an additional approximation is made, namely that $R \approx R_0$ over the solid angle of the antenna beam, in which case R can be taken outside the integral in (2.11), and the antenna pattern is replaced with an equivalent pattern of gain G_0 and effective width β. These approximations lead to

$$
\mathrm{I} \approx \frac{G_0^2}{R_0^4} A \tag{2.12}
$$

and

$$
P_{rp}(\theta) \approx \left[\frac{P_{tq}\lambda^2 G_0^2 A}{(4\pi)^3 R_0^4} \right] \cdot \sigma°_{pq}(\theta) \tag{2.13}
$$

where A is the area illuminated by the equivalent beam of the antenna. We should note that because the integration in (2.11) involves G^2, rather than just G, β is the effective beamwidth of the product pattern G^2. For a symmetrical Gaussian pattern $G(\theta_a)$, the effective beamwidth is approximately equal to its half-power (3 dB) beamwidth $\beta_{1/2}$. The half-power beamwidth of $G^2(\theta_a)$ is $\beta = \beta_{1/2}/\sqrt{2}$.

2-2.2 Imaging Radar

The assumptions made in deriving (2.13) are well suited to the imaging radar case because the dimensions of the illuminated ground cell are very small relative to the distance R between the radar and the cell. Figures 2.5 and 2.6 show the imaging geometries associated with the real aperture and synthetic aperture radars when used in the side-looking mode. The real-aperture radar uses an antenna that is wide horizontally and narrow vertically. The horizontal beamwidth β_h is typically $1°$ or less, and the azimuth resolution is

$$r_a = \beta_h R, \tag{2.14}$$

where R is the slant range. Resolution in the range or across-track dimension is achieved by transmitting a narrow pulse of length τ. The image is produced by recording the returns from successive pulses as the aircraft moves past the area being covered. The ground-range resolution is independent of R and is given by

$$r_\rho = \frac{c\tau}{2 \sin \theta} \tag{2.15}$$

and the area of the illuminated cell is

$$A = r_a\, r_\rho = \frac{\beta_h R c \tau}{2 \sin \theta} \tag{2.16}$$

Inserting (2.16) in (2.13) leads to the following form of the radar equation for real aperture radar:

$$P_{rp}(\theta) = \left[\frac{P_{tq}\lambda^2 G^2(\theta)\beta_h c\tau}{2\,(4\pi)^3 R^3 \sin \theta}\right] \cdot \sigma^{\circ}_{pq}(\theta) \tag{2.17}$$

where $G(\theta)$ is the antenna gain in the direction of the ground cell under consideration.

The antenna used in a synthetic aperture radar system is usually shorter in the along-track (azimuth) direction, but successive pulses are processed together to "focus" the azimuth beam of the antenna down to a very narrow beam corresponding to a very long synthetic array. Theoretically, the cell dimension in the along-track direction can be made independent of R and as small as

$$r_a = l/2 \tag{2.18}$$

where l is the along-track length of the antenna. When this is achieved, the SAR is said to be *fully focused*. In the across-track direction, resolution is achieved in the same way as with the RAR, so (2.15) applies. For a fully focused SAR, the radar equation becomes

$$P_{rp}(\theta) = \left[\frac{P_{tq}\lambda^2 G^2(\theta) l c \tau}{4(4\pi)^3 R^4 \sin\theta} \right] \cdot \sigma^\circ_{pq}(\theta) \tag{2.19}$$

In summary, the power received due to backscatter from an illuminated area that subtends a small solid angle as viewed from the radar is directly proportional to the backscattering coefficient σ°, regardless of the specific type of radar used. Thus, the radar equations for the narrow-beam scatterometer, the real aperture radar, and the synthetic aperture radar can all be written in the compact form

$$P_r(\theta) = K\sigma^\circ(\theta) \tag{2.20}$$

where K may be regarded as a system constant representing the quantity inside the square bracket in (2.13), (2.17), or (2.19). The polarization indices have been suppressed for the sake of simplicity.

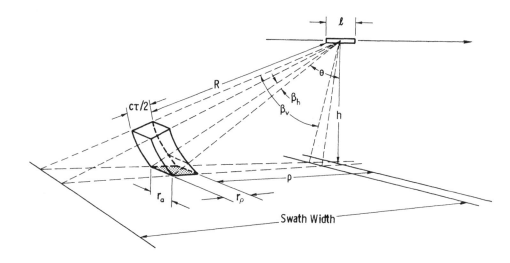

Figure 2.5 Real-aperture radar (RAR) geometry and resolved surface area.

2-3 STATISTICAL PROPERTIES OF A DISTRIBUTED TARGET

If a flat surface is illuminated by a radar wave at oblique incidence, the wave is specularly reflected in accordance with Snell's law and no energy is scattered toward the radar. Most terrain surfaces are not flat, however, and, consequently, they scatter the incident energy in many directions, including the backscattering direction. The directional pattern of the energy scattered by a rough surface is governed by the dielectric constant, the degree of roughness of the surface, and

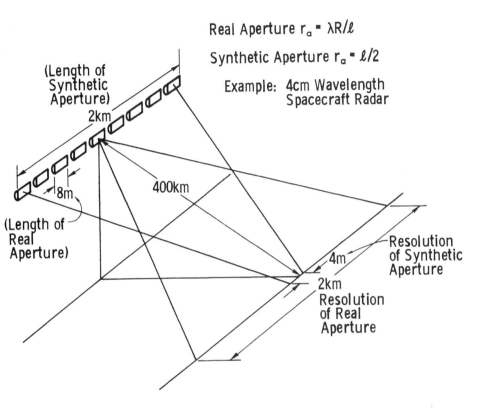

Figure 2.6 Illustration of advantage of SAR over RAR in a space application.

the parameters of the incident wave (incidence angle, wavelength, and polarization direction). Backscattering from terrain can also occur as a result of volume scattering in inhomogeneous media, such as vegetation canopies and snow cover.

When illuminated by a radar beam, a distributed terrain target—be it a bareground surface or a ground surface covered with a layer of another material—is usually modeled as an ensemble of *scattering centers* randomly distributed in spatial position over the illuminated area (or volume). One of the simplest models used to characterize a rough surface is the random facet model. The rough surface is approximated through a series of small planar facets, each tangential to the actual surface (Figure 2.7a). The facet model treats the scattering from the assemblage of such facets by taking into account both their reradiation patterns and the distribution of their slopes. The center of each facet represents a scattering

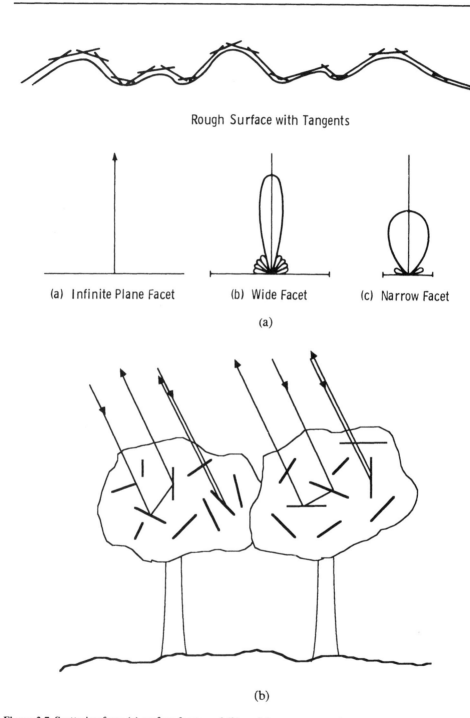

Rough Surface with Tangents

(a) Infinite Plane Facet   (b) Wide Facet   (c) Narrow Facet

(a)

(b)

Figure 2.7 Scattering from (a) surface facets and (b) an inhomogeneous volume.

center. Because the transmitted wave is a narrow-band coherent signal, the received signal is the *phasor sum* of the signals reradiated by all of the scattering centers located in the illuminated area. Thus, the amplitude of the received radar signal is governed not only by the strengths of the reradiated signals, but also by their locations. This means that the received signal is a random variable governed by a statistical process related to the statistical properties of the scattering centers. A similar scenario can be developed for radar backscattering from leaves and branches in a vegetation canopy (Figure 2.7b) or from ice crystals in a snow layer.

In the next chapter, we shall examine the statistical character of the received radar signal in some detail. Before we do so, however, we shall define some of the key terms used to characterize the statistical properties of a distributed target.

2-3.1 Homogeneous Medium

An object is electromagnetically *homogeneous* if its constitutive parameters, the complex dielectric constant ϵ and magnetic permeability μ, are constant everywhere within its volume and on its surface. Strictly speaking, no material object can be perfectly homogeneous because all objects are composed of molecules and atoms with specific spatial arrangements. However, an object may be considered spatially homogeneous at a given scale if its ϵ and μ do not vary when sampled at that scale. An object is homogeneous at a 1-cm scale, for example, if its ϵ (and similarly its μ) assumes the same value when averaged over any 1 cm^3 section of the object. For most natural materials, $\mu = \mu_0$, the permeability of free space. Hence, the spatial variation of ϵ alone determines the *degree* of inhomogeneity of natural objects.

The same criteria used above to characterize the homogeneous scale of an object can also be applied to characterize natural media, such as vegetation canopies and snow cover. In terms of electromagnetic wave propagation in a medium, the medium may be regarded as homogeneous if its ϵ is spatially homogeneous at a scale much smaller than the wavelength λ of the propagating wave, where λ is the wavelength measured in the medium; $\lambda = \lambda_0/n_a$ where λ_0 is the free-space wavelength and $n_a = \sqrt{\epsilon_a}$ is the average index of refraction of the medium. A layer of snow comprised of a connected lattice of ice crystals distributed in an air background would "appear" homogeneous to a wave propagating in the snow medium if $\lambda \geq 10$ cm because the ice crystals and the spacings between them typically are on the order of 1 mm or less. On the other hand, snow behaves like an inhomogeneous medium at centimeter and shorter wavelengths. By the same token, a forest canopy is an inhomogeneous medium even at wavelength scales of tens of meters.

2-3.2 Inhomogeneous Random Medium

Both snow and vegetation, as well as other natural media, often are treated as *random media*. A random medium is characterized by a dielectric constant $\epsilon(x, y, z)$ that varies randomly as a function of the spatial position (x, y, z). Over a given volume V of the random medium, we can express $\epsilon(x, y, z)$ as

$$\epsilon(x, y, z) = \epsilon_a + \epsilon_f(x, y, z) \tag{2.21}$$

where $\epsilon_a = \langle \epsilon(x, y, z) \rangle$ is the mean value of $\epsilon(x, y, z)$ over V and $\epsilon_f(x, y, z)$ is a zero-mean fluctuating component. That is, $\langle \epsilon_f(x, y, z) \rangle = 0$. The mean dielectric constant ϵ_a is related to the bulk dielectric constants of the background and inclusion materials comprising the medium and their respective volume fractions (and possibly the shapes and orientations of the inclusions relative to the direction of the electric field of the propagating wave). For a snow medium, the background is air and the inclusions are the ice particles. In vegetation media, the inclusions are the leaves, branches, *et cetera*.

The statistical properties of the fluctuating component $\epsilon_f(x, y, z)$ also are related to the dielectric constants of the inclusions (relative to that of the background), and to their shapes, orientations, and spatial distributions. The variance

$$s_\epsilon^2 = \langle \epsilon_f^2(x, y, z) \rangle \tag{2.22}$$

of ϵ_f is a measure of the degree of inhomogeneity of the random medium, and the correlation function relating the magnitude of ϵ_f at (x, y, z) to its magnitude at (x', y', z'),

$$R_\epsilon(x, y, z; x', y', z') = \langle \epsilon_f(x, y, z) \, \epsilon_f(x', y', z') \rangle \tag{2.23}$$

describes the correlation dimensions of ϵ_f. If ϵ_f is a stationary random variable, meaning that R_ϵ is a function of the distance between the points (x, y, z) and (x', y', z') rather than their specific locations, then

$$R_\epsilon(x, y, z; x', y', z') = R_\epsilon(\Delta x, \Delta y, \Delta z) \tag{2.24}$$

where $\Delta x = x' - x$, and similar definitions apply to Δy and Δz. As an example, let us consider the correlation function

$$R_\epsilon(\Delta x, \Delta y, \Delta z) = s_\epsilon^2 \exp\left[-\frac{(\Delta x)^2 + (\Delta y)^2}{l_\rho^2} \right] \exp\left[-\frac{|\Delta z|}{l_z} \right] \tag{2.25}$$

which is Gaussian in the x-y plane and exponential in the vertical direction. The parameters l_ρ and l_z are the correlation distances in the horizontal and vertical directions, respectively. A plot of $R_\epsilon(\Delta\rho, 0)$ *versus* $\Delta\rho$ is shown is Figure 2.8, where

$$\Delta\rho = [(\Delta x)^2 + (\Delta y)^2]^{1/2} \tag{2.26}$$

is the displacement in the *x-y* plane.

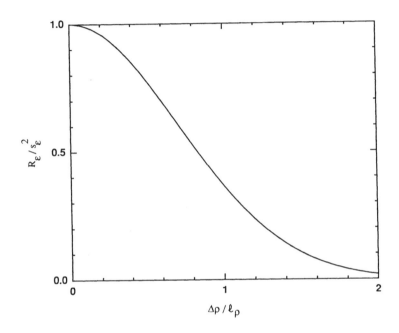

Figure 2.8 Plot of the normalized correlation function *versus* the normalized displacement.

Using profiles of the dielectric constant measured in the ρ- and z-directions for snow samples, Vallese and Kong [146] computed the correlation function $R_\epsilon(\Delta\rho, \Delta z)$ and found it to match the form given by (2.25) with $l_\rho = l_z \approx$ grain size of the sample. Figure 2.9 shows a photograph of one of their coarse-grain snow samples and a plot of the corresponding correlation function.

In contrast with the relatively simple correlation function associated with the dielectric fluctuations in a snow sample, the dielectric correlation function appropriate for a forest canopy is much more complicated. The complexity is related to the fact that the canopy volume contains several types of inclusions (leaves, branches, trunks, *et cetera*) with different dielectric properties, sizes, and shapes. To date, no "realistic" correlation function has been proposed for tree canopies.

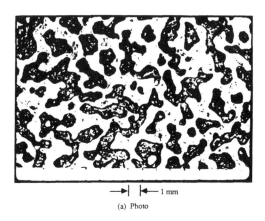

(a) Photo

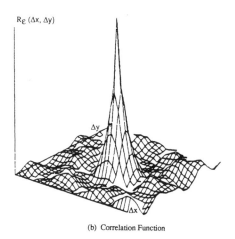

(b) Correlation Function

Figure 2.9 (a) Photo and (b) correlation function of a coarse-grain snow sample. The correlation lengths are $l_z = l_\rho = 0.3$ mm. (From [146].)

2-3.3 Statistically Homogeneous Medium

The statistics associated with $\epsilon(x, y, z)$ of a random medium, including ϵ_a, s_ϵ and R_ϵ, are specific to the test volume V over which $\epsilon(x, y, z)$ is defined. In order for these statistics to be valid, V has to be sufficiently large to provide adequate representation of the random process characterizing the variation of $\epsilon(x, y, z)$. For a medium characterized by a single correlation distance in a given direction, the dimension of V in that direction should be at least ten times longer than that correlation distance. Thus, for the snow-medium example, V should have dimensions of at least $10\ l_\rho$ in the x- and y-directions and $10\ l_z$ in the z-direction.

Now let us consider a distributed target with dimensions many times greater than those of the test volume V. Such a target is said to be *statistically homogeneous* if the statistics of $\epsilon(x, y, z)$ are the same for any V throughout the volume of that target. A similar definition applies to a two-dimensional rough surface with random height fluctuations.

2-3.4 Statistically Uniform Target

Some distributed targets may consist of two or more horizontal layers, each characterized by its own statistics. If the vertical structure that defines the thicknesses of the layers is uniform over a horizontal area A, and if, in addition, each layer is statistically homogeneous over A, then the distributed target is said to be *statistically uniform* across the area A. Similarly, a rough surface is considered statistically uniform over a given area if its average dielectric constant and its surface-height statistics are stationary over that area. An important consequence of statistical uniformity is that a distributed target exhibits the same average radar response for any segment of the area, provided the segment is large enough to contain a large number of scattering centers.

2-4 GENERAL BEHAVIOR OF $\sigma°$

This section is intended to provide the reader with an overview of the dependence of the backscattering coefficient $\sigma°$ of terrain on the microwave frequency f, the angle of incidence θ, the polarization configuration rt (which stands for the receiving-transmitting polarization configurations of the antenna), and some of the physical properties of the terrain target. The overview is presented in the form of examples of experimental observations with brief explanations of the underlying physics. No provision is made here for discussions of the theoretical models or detailed examinations of the behavior of $\sigma°$ for specific terrain surfaces. The reader interested in such discussions and examinations is referred to Ulaby *et al.* [138].

2-4.1 Angular Dependence of $\sigma°$ of a Random Surface

The variation of $\sigma°$ of a random surface with angle of incidence is diagrammed in Figure 2.10. The angular range includes a quasispecular region near normal incidence, a plateau region that typically extends between $\theta = 40°$ and $70°$, and a shadow region at angles close to grazing incidence. The variation of $\sigma°(\theta)$ in the shadow region is the least understood of the three regions.

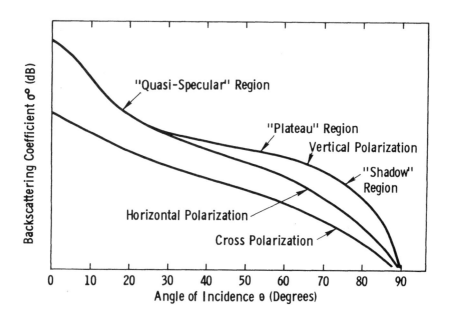

Figure 2.10 General characteristics of backscattering-coefficient variation with angle of incidence.

The curves shown in Figure 2.11 illustrate the angular response of $\sigma°$ for a relatively smooth asphalt surface. These curves, which are based on multifrequency, multipolarization measurements made by truck-mounted scatterometers, display the angular response of $\sigma°$ up to $\theta = 80°$. At all frequencies, the hh- and vv-polarized responses are identical between normal incidence and $\theta = 10°$, and then diverge as a function of θ, with $\sigma°_{vv}$ maintaining a higher level than $\sigma°_{hh}$. The cross-polarized scattering coefficient, $\sigma°_{hv}$, is always lower in level than the like-polarized components, and the difference in level is usually greatest at nadir.

2-4.2 Dependence of $\sigma°$ on Surface Roughness

Figure 2.12 shows five angular-response curves for $\sigma°$ based on measurements made at 1.1 GHz for five bare-soil fields with approximately the same soil-moisture content (and hence dielectric constant), but with rms surface heights ranging from 1.1 cm for the smoothest field to 4.1 cm for the roughest. The surfaces contained no observable large-scale periodic patterns. The effect of surface roughness on $\sigma°$ is clearly evident in these curves; for the smoothest surface, field 1, $\sigma°$ decreases rapidly with increasing angle of incidence, from about 18 dB at nadir to -27 dB at $\theta = 30°$; in contrast, the roughest surface, field 5, exhibits a variation of only 3 dB in $\sigma°$ between nadir and 30°.

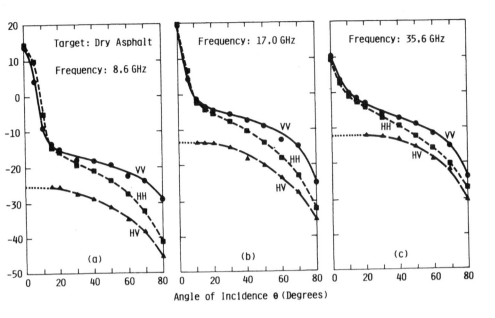

Figure 2.11 Angular response of the scattering coefficient of dry asphalt at (a) 8.6 GHz, (b) 17.0 GHz, and (c) 35.6 GHz. (From [138].)

Electromagnetically, the rms surface height s should be expressed in units of the wavelength λ. In theoretical models, the quantity ks (where $k = 2\pi/\lambda$) often is used for this purpose. For the above example $\lambda = 27.3$ cm (at 1.1 GHz), and therefore $ks = 0.25$ for field 1 and 0.94 for field 5. As a general guideline, a surface may be considered smooth if its $ks < 0.2$ and may be considered very rough if its $ks > 1.0$. We should note, however, that ks is not the only parameter determining the roughness of a surface; the correlation length l (or rather kl) should be considered also.

2-4.3 Dependence of $\sigma°$ on Surface Moisture

The backscattering coefficient of a random surface is governed by both the roughness parameters of that surface and its dielectric properties. In most natural materials, the dielectric constant is dominated by the water content of the material. This is because the microwave relative dielectric constant of liquid water is about an order of magnitude larger than that of dry materials, including dry soil, rocks, frozen snow, ice, and dry vegetation. Consequently, if the roughness of a surface remains unchanged, its $\sigma°$ increases with increasing moisture content (Figure 2.13) and may exhibit a dynamic range of about 10 dB between dry soil and very wet soil conditions.

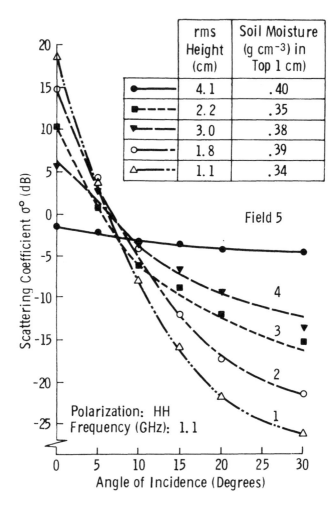

| | rms Height (cm) | Soil Moisture (g cm⁻³) in Top 1 cm |

Figure 2.12 Angular patterns of the scattering coefficient at 1.1 GHz for five bare-soil fields with different surface roughnesses. (From [126].)

2-4.4 Angular Dependence of $\sigma°$ for Vegetation

The results of backscattering measurements for a field of grass are given in Figure 2.14 to illustrate the behavior of $\sigma°$ for a volume-scattering medium. At the time of the experiment, the grass was green and approximately 80 cm in height, which

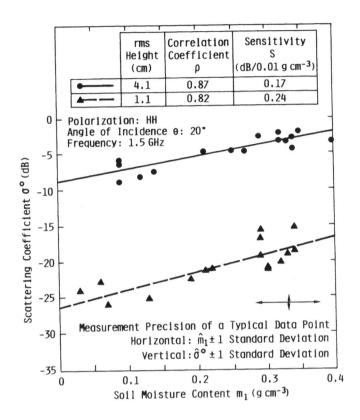

Figure 2.13 Backscattering response to soil moisture of a smooth surface and a rough surface for $\theta = 20°$ at $f = 1.5$ GHz. (From [138].)

means that the canopy attenuation at the frequencies under consideration would have been large enough to mask the backscattering contribution of the underlying soil surface, except perhaps at angles near normal incidence. The effect on $\sigma°$ of the underlying soil surface, which was relatively smooth, is evident at $\theta = 0°$ in Figure 2.14a. Overall, $\sigma°$ varies with θ approximately as $\cos \theta$ for all polarizations, and $\sigma°_{vv} \approx \sigma°_{hh}$.

If $\sigma°$ varies with θ as $\cos \theta$, then $\gamma(\theta) = \sigma°(\theta)/\cos \theta$ would exhibit a constant level with θ. Such an angular variation is typical of most observations made for dense tree canopies at X-band and higher frequencies. The example given in Figure 2.15 for deciduous trees at 8.6 GHz shows that $\gamma(\theta)$ is essentially constant (within 2 dB) for $0 \le \theta \le 60°$ for all linear polarization configurations.

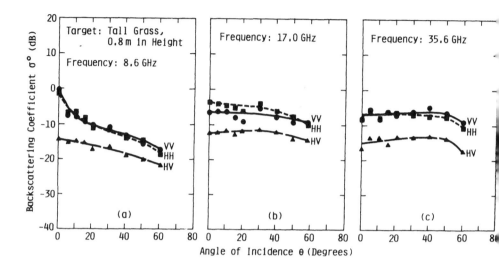

Figure 2.14 Angular response of the scattering coefficient of tall grass at (a) 8.6 GHz, (b) 17.0 GHz, and (c) 35.6 GHz.

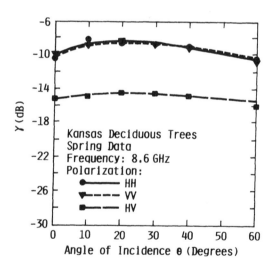

Figure 2.15 Angular variation of γ at 8.6 GHz as a function of θ for deciduous trees with leaves.

2-4.5 Dependence of $\sigma°$ on Vegetation Biomass

The dependence of $\sigma°$ on the type and density of vegetation biomass is illustrated by the examples shown in Figures 2.16 and 2.17. According to the data presented

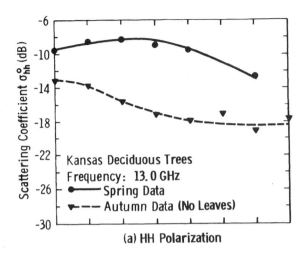

Figure 2.16 Angular variation of σ°_{hh} of trees measured at 13.0 GHz in the spring and autumn. (From [138].)

in Figure 2.16, deciduous trees in leaf exhibit a σ° level higher than that of defoliated trees by 2–8 dB, depending on angle. The purpose of the second example, Figure 2.17, is to show the variation of σ° as a function of the amount of biomass (per unit area) contained in the canopy. The biomass is represented here by the green *leaf area index* (LAI), which is the total single-sided surface area of all the leaves contained in the canopy over a unit area of ground. The data shown in Figure 2.17 for a field of sorghum indicate that σ° increases from -14 dB to about -8.5 dB as the LAI increases from ≈ 0 (corresponding to a vegetation-free soil surface) to about 2, and thereafter σ° maintains a constant level at higher values of LAI.

2-4.6 Angular Dependence of σ° for Snow

In regard to radar backscattering, a snow-covered surface is fundamentally similar to a vegetation-covered surface because in both cases the total backscatter is composed of surface scattering contributions from the underlying ground surface, volume scattering contributions from the snow or vegetation layer, and multiple scattering contributions involving both the ground surface and the layer above it. Penetration through a snow layer is strongly dependent upon the wavelength, the size of the snow crystals, and the snow liquid water content. Figure 2.18 illustrates the general angular and frequency behavior of σ° for a 58 cm deep dry snow layer over a soil surface. At 1.6 GHz, the snow layer is essentially transparent, and the return is primarily due to the soil surface. At higher frequencies, the snow

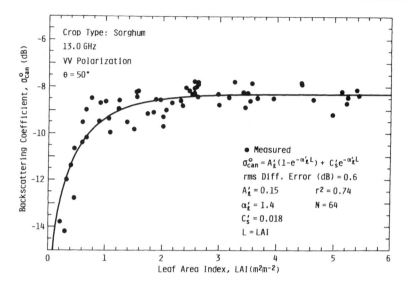

Figure 2.17 Variation of the measured backscattering coefficient of sorghum with LAI.

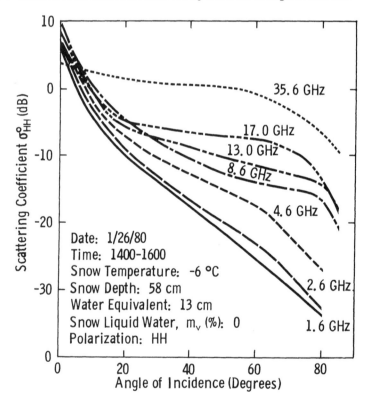

Figure 2.18 Angular patterns of measured backscattering coefficient of dry snow. (From [125].)

ayer plays a more important role, and at 35.6 GHz, the snow is the dominant source of backscattering.

2-4.7 Frequency Variation of $\sigma°$ for Snow

The spectral curves shown in Figure 2.19 illustrate the importance of the frequency f with regard to the sensitivity of $\sigma°$ to snow wetness m_v. For a snowpack 48 cm deep, a change in wetness from zero percent to 1.26% results in a drop in level of only 1 dB at 1 GHz, but as f increases, the two curves diverge, and the difference between them increases to a maximum of 15 dB at 35.6 GHz.

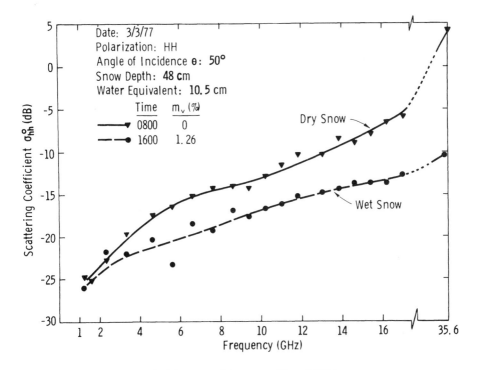

Figure 2.19 Spectral response of $\sigma°$ for wet and dry snow. (From [138].)

Chapter 3
Statistics of Signal Fluctuations

The polar pattern shown in Figure 3.1 represents the radar cross section σ for a B-26 aircraft plotted as a function of aspect angle. The data were obtained by mounting the aircraft on a turntable and then observing the backscattered signal measured by a 3-GHz radar as the turntable was rotated about its axis. Perhaps the most striking feature of the scattering pattern is the sensitivity of σ to aspect angle; σ changes by an order of magnitude (10 dB) for a change of a fraction of a degree in aspect angle.

This example is intended to illustrate the interference phenomenon associated with coherent scattering from a complex target. The aircraft may be considered to consist of a number of scattering elements, each characterized by a scattering pattern that reflects its particular shape and orientation. The composite backscatter in a given direction is the phasor sum of the waves backscattered from the individual elements, which includes the propagation phase delay to and from each element. The peaks in the pattern shown in Figure 3.1 are associated with constructive interference of the backscattered phasors and the nulls are associated with destructive interference.

Although an aircraft has a distinctly different geometry from that of a distributed terrain surface, the nature of the scattering process is fundamentally the same in both cases. The aircraft has a *deterministic scattering pattern* because it has a well-defined geometry, whereas a distributed target has a *random scattering pattern* (Figure 3.2) because its scattering elements are randomly distributed. Both patterns exhibit large fluctuations. The pattern shown in Figure 3.2b was generated from radar backscatter measurements made by a 35-GHz truck-mounted scatterometer as the truck was driven across an asphalt surface with the antenna beam pointing downward along the aft direction at an incidence angle of 40° relative to normal incidence (Figure 3.2a). The antennas were mounted atop a telescopic boom at a height of 10.3 m above the asphalt surface. The sampling rate was such that the footprints (on the asphalt surface) corresponding to adjacent samples were totally independent (no overlap). The vertical axis in Figure 3.2b represents

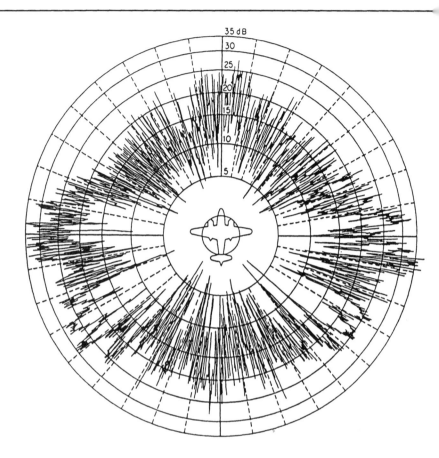

Figure 3.1 Measured azimuthal pattern of the radar backscatter from a B-26 airplane at 10-cm wavelength. (From Ridenour [103], courtesy McGraw-Hill Book Company).

$F = P/\overline{P}$, the ratio of the received power to the average value computed for all 1000 measurements, expressed in dB. The measured values of F extend over a range of 50.2 dB, and the standard deviation of F, s_F, is equal to 0.97, in close agreement with the *Rayleigh clutter model,* which predicts $s_F = 1$ (see Eq. (3.20)).

To characterize the fluctuation statistics associated with a terrain surface of uniform electromagnetic properties, the usual approach is to model the surface as an ensemble of independent, randomly located scatterers, all of comparable scattering strengths. Such a model leads to the result that the amplitude of the backscattered signal is Rayleigh-distributed (Section 3-1). If the return is dominated by backscatter from one or a few strong scatterers, the fading process is characterized by the *Rice distribution* (Section 3-4.3). Some experimental observations support the Rayleigh behavior [31,139,150] whereas others, particularly those measured for complex terrain categories, are in closer agreement with the *log-normal distribution,* the *Weibull distribution* [74,104,105,145,149], or other

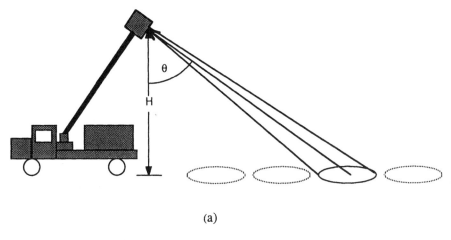

(a)

Variation of F(dB) with spatial position

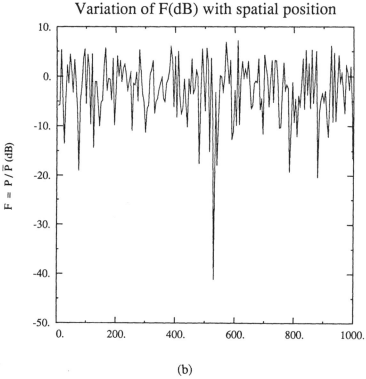

(b)

Figure 3.2 The sketch in (a) shows how the measurements (shown in (b)) of the backscatter from an asphalt surface were acquired. The incidence angle was 40°, the frequency 35 GHz, the platform height 10.4 m, and the polarization vv (From [143].)

more complicated distributions [72]. In this chapter we will review the properties of the Rayleigh and other clutter models and examine the statistics associated with scattering from nonhomogeneous terrain.

3-1 RAYLEIGH FADING STATISTICS

3-1.1 Underlying Assumptions

The Rayleigh clutter model used for describing radar scattering from an area-extended (distributed) target is essentially the same as the model used for random noise and is based on the same mathematical assumptions. A review of these assumptions will prove useful in later sections.

The sketch shown in Figure 3.3 depicts a radar beam illuminating an area A of an area-extended target. The illuminated area contains N_s point scatterers designated by the index $i = 1, 2, \ldots, N_s$. For simplicity, we shall confine our present discussion to the backscatter case. The field intensity at the input of the receiving antenna due to backscatter by the ith scatterer may be expressed as

$$E_i = K_i E_{i0} \exp\left[j(\omega t - 2kr_i + \theta_i)\right] \tag{3.1}$$

where E_{i0} is the scattering magnitude and θ_i is the scattering phase of the ith scatterer; r_i is the range from the antenna to the scatterer; $k = 2\pi/\lambda$ is the wavenumber; and K_i is a system constant that accounts for several radar system factors including propagation losses to and from the scatterer and antenna gain in the direction of the scatterer. The expression given by (3.1) may be abbreviated as

$$E_i = K_i E_{i0}\, e^{j\phi_i} \tag{3.2}$$

where

$$\phi_i = \omega t - 2kr_i + \theta_i \tag{3.3}$$

is the instantaneous phase of E_i.

Assumption 1: The scatterers are statistically independent. This assumption allows us to express the total instantaneous field due to the N_s scatterers contained in the area A as a simple sum,

$$E = \sum_{i=1}^{N_s} K_i E_{i0} e^{j\phi_i}$$

and it implies that interaction effects between adjacent scatterers may be ignored.

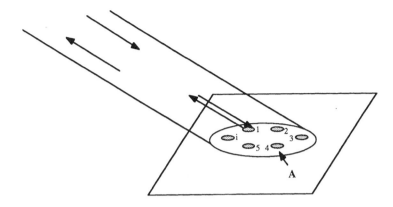

Figure 3.3 The illuminated area A contains N_s randomly distributed scatterers.

Assumption 2: The maximum range extent of the target $\Delta r = |r_i - r_j|_{max}$ is much smaller than the mean range to the target area A, and the antenna gain is uniform across A. This allows us to set $K_i = K$ for all i. For convenience, we shall set $K = 1$. Hence,

$$E = \sum_{i=1}^{N_s} E_{i0}e^{j\phi_i} \tag{3.4}$$

The total field E is a vector sum of N_s phasors. If we express these phasors graphically (Figure 3.4), with the first one starting at the origin and the successive ones each starting at the tip of the preceding one, the result is a vector from the origin to the tip of the last phasor. The length of this vector (which sometimes is called the voltage envelope) and its phase angle are denoted E_e and ϕ, respectively. Thus,

$$E = E_e e^{j\phi} \tag{3.5}$$

Assumption 3: N_s is a large number. This assumption allows us to use the central-limit theorem which, in turn, allows us to assume that the x- and y-components of E, E_x and E_y, are normally distributed. However, it can be shown through computer simulation that this condition can be satisfied (approximately) for N_s as small as 10. (The same conclusion was reported in Kerr [75] in the 1940s.)

Assumption 4: The scattering amplitude E_{i0} and the instantaneous phase ϕ_i are independent random variables. This condition is easily satisfied if E_{i0} is independent of the range r_i, which would be the case if the scatterers are randomly distributed in range.

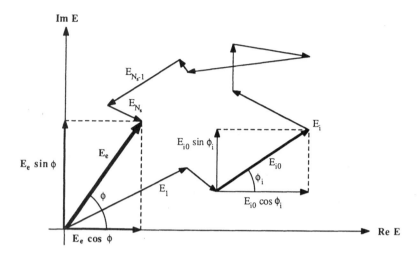

Figure 3.4 The vector E is the phasor sum of N_s fields.

Assumption 5: The phase ϕ_i is uniformly distributed over the range $[0, 2\pi]$. To satisfy this condition it is not only necessary that the scatterers be randomly distributed in range, but the maximum range extent of the target Δr must also be several wavelengths across.

Assumption 6: No one individual scatterer produces a field intensity of magnitude commensurate with the resultant field from all scatterers. In other words, the field E is not dominated by one (or a few) very strong scatterer(s). If this condition is not satisfied, the Rayleigh noise-like statistics do not apply and the statistics developed by Rice [13] for one or more large signals contained in a background of noise should be used instead (see Section 3-4.3).

Use of Assumptions 3–6 can be shown to lead to the following properties [135, p. 479]:

$$p(E_e) = \frac{E_e}{s^2} \exp\left(-E_e^2/2s^2\right), \quad E_e \geq 0$$

$$p(\phi) = \frac{1}{2\pi}, \quad 0 \leq \phi \leq 2\pi \tag{3.6}$$

$$\bar{E}_e = \left(\frac{\pi}{2}\right)^{1/2} s$$

where $p(E_e)$ and $p(\phi)$ denote the probability density functions (PDFs) of E_e and ϕ, respectively, $\overline{E}_e$ is the ensemble average (mean value) of E_e, and s is the standard deviation of both components of E_e, E_x and E_y. Equation (3.6) is known as the *Rayleigh PDF*.

The radar variable E_e represents the magnitude of the cumulative electric field at the receiving antenna prior to detection by the receiver. Most receivers use either linear detection or square-law detection to convert the input signal to an output voltage. As we will see in the next section, the output voltage of the linear receiver also is characterized by a Rayleigh PDF, but that of the square-law receiver is characterized by an exponential PDF. Both cases, however, belong to the Rayleigh clutter model because the underlying PDF, $p(E_e)$, is Rayleigh. Thus, the terms *Rayleigh fading* and *Rayleigh clutter* refer to the model and the assumptions leading to (3.6) and should not be confused with the form of the PDF of the detected voltage.

3-1.2 Output Voltage

Linear Detection

If the receiver uses a linear detector, its output voltage V is directly proportional to E_e; that is,

$$v = K_1 E_e$$
$$= K_1 \overline{E}_e \frac{E_e}{\overline{E}_e}$$
$$= K_1 K_2 (\sigma^\circ)^{1/2} f$$
$$= \overline{v} f$$

where K_1 is a system constant, K_2 relates the mean field $\overline{E}_e$ to the backscattering coefficient of the target σ° (actually, σ° is directly proportional to $\overline{E^2_e}$, but $\overline{E^2_e} = (4/\pi)\,\overline{E}^2_e$), and f is the normalized *fading random variable* given by

$$f = E_e/\overline{E}_e = v/\overline{v}$$

We should note that σ° is defined as the mean value of the backscatter cross section of the distributed target and f accounts for variations between observations of different combinations (or locations) of the scatterers comprising the target. In other words, in the absence of fading, all observations of the distributed target would give the same value for v regardless of spatial location.

Without loss of generality we may set $K_1 = 1$ and write down the following properties of the Rayleigh-distributed voltage:

$$p(v) = \frac{v}{s^2} \exp\left(-v^2/2s^2\right), \quad v \geq 0 \tag{3.7}$$

$$\bar{v} = \left(\frac{\pi}{2}\right)^{1/2} s \tag{3.8}$$

$$s_v^2 = \overline{v^2} - \bar{v}^2 = \left(2 - \frac{\pi}{2}\right) s \tag{3.9}$$

$$m_v = (2 \ln 2)^{1/2} s \tag{3.10}$$

$$W(v \leq v') = \int_0^{v'} p(v) \, dv, \quad v' \geq 0$$

$$= 1 - \exp\left(-v'^2/2s^2\right) \tag{3.11}$$

where $\bar{v}$, s_v^2, and m_v are the mean value, the variance, and the median value of v, respectively, and $W(v \leq v')$ is its *cumulative distribution*.

Sometimes it is convenient to use the PDF of v normalized with respect to its median m_v. Thus, for $v_n = v/m_v$, the PDF is given by

$$p(v_n) = 2 (\ln 2) \, v_n \exp\left[-(\ln 2) \, v_n^2\right], \quad v_n \geq 0 \tag{3.12}$$

The fading random variable f is equal to the voltage v normalized to its mean value $\bar{v}$. Using the relation $p(f) \, df = p(v) \, dv$, we obtain

$$p(f) = \frac{\pi f}{2} \exp\left(-\pi f^2/4\right), \quad f \geq 0 \tag{3.13}$$

$$\bar{f} = 1, \tag{3.14}$$

$$s_f = \frac{s_v}{\bar{v}}$$

$$= [(\overline{f^2} - \bar{f}^2)/\bar{f}^2]^{1/2} = \left(\frac{4}{\pi} - 1\right)^{1/2} = 0.523 \tag{3.15}$$

$$W(f \leq f') = 1 - \exp\left(-\pi f'^2/4\right), \quad f' \geq 0 \tag{3.16}$$

Because the output voltage v is a product of its mean value $\bar{v} = K_1 K_2 (\sigma^\circ)^{1/2}$ and the random variable f, the process is sometimes referred to as a *multiplicative noise model*.

Square-Law Detection

The voltage output of a square-law detector is directly proportional to the power of the input signal rather than to its field intensity E_e. Thus, within a proportionality constant, the output is represented by the input power P,

$$P = K_3 \overline{E_e^2} \frac{E_e^2}{\overline{E_e^2}}$$

$$= K_3 K_4 \, \sigma^\circ \, F$$

where

$$F = E_e^2/\overline{E_e^2} = P/\bar{P} \tag{3.17}$$

is the *normalized fading random variable* for power. Ignoring the constant $K_3 K_4$ and using (3.6) and the relation $p(E_e) \, dE_e = p(P)dP = p(F) \, dF$ leads to the exponential PDF [135, p. 480]:

$$p(P) = \frac{1}{\bar{P}} e^{-P/\bar{P}} \qquad\qquad P(F) = e^{-F} \tag{3.18}$$

with

$$\bar{P} = \sigma^\circ \qquad\qquad \bar{F} = 1 \tag{3.19}$$

$$\frac{S_P}{\bar{P}} = 1 \qquad\qquad S_F = 1 \tag{3.20}$$

and

$$W\,(P \le P') = 1 - e^{-P'/\bar{P}} \qquad W\,(F \le F') = 1 - e^{-F'} \tag{3.21}$$

3-1.3 Interpretation

What do these statistics tell us? To answer this question, we start by examining Figure 3.5(a) which shows plots of $p(f)$ and $p(F)$ for the Rayleigh and exponential distributions, respectively, and Figure 3.5(b) which shows the corresponding cumulative distributions. We observe that the range of fading associated with these distributions is very large. That is, if we take a single sample of the signal from a Rayleigh-distributed or exponentially distributed ensemble, we have very little chance of selecting a value close to the mean. Let us illustrate this with a specific example. According to the Rayleigh distribution in Figure 3.5(b), the value of f exceeded 5 percent of the time is 1.95 (relative to the mean) and that exceeded 95 percent of the time is 0.25. In decibels, these levels correspond to $+5.8$ dB and -11.9 dB, respectively. If we select a sample at random, the probability is 90 percent (95 to 5 range) that its value will be within the range extending from 11.9 dB below the mean to 5.8 dB above the mean. We may think of this as the 90-percent confidence interval associated with our measurement. The important point to note here is the fact that this interval (17.7 dB) is very large.

The situation is not much different when square-law detection is used; the 5 percent and 95 percent levels of the cumulative distribution for the exponential PDF are $+4.8$ and -12.9 dB, also totalling 17.7 dB. For the data presented earlier in Figure 3.2, the data points that fall within the -12.9 dB to $+4.8$ dB range constitute 90.8 percent of the total, which is in close agreement with the 90-percent figure predicted by the exponential distribution. Furthermore, the measured PDF of F, shown in Figure 3.6, exhibits a good fit to the exponential distribution.

3-2 MULTIPLE INDEPENDENT SAMPLES

To reduce the uncertainty of a radar measurement of the backscatter from a terrain surface, it is necessary to average many independent samples together. An easy way to increase the number of independent samples N contained in an estimate of the radar backscatter is through spatial averaging, which amounts to trading spatial resolution for improved radiometric resolution. Other ways to increase N are discussed in Chapter 4.

3-2.1 Linear Detection

If N randomly selected samples of a Rayleigh-distributed voltage v are averaged together, the average value v_N has the following properties:

$$v_N = \frac{1}{N} \sum_{i=1}^{N} v_i$$

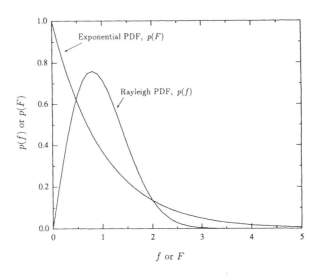

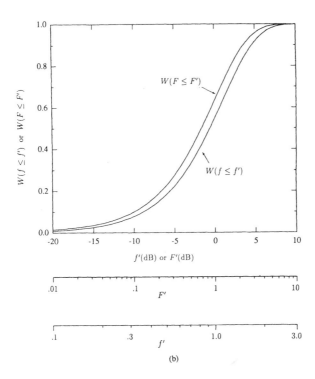

Figure 3.5 Plot of (a) probability density functions and (b) cumulative distributions for f and F.

$$= K_1 K_2 \, (\sigma^\circ)^{1/2} \left[\frac{1}{N} \sum_{i=1}^{N} f_i \right]$$

$$= K_1 K_2 \, (\sigma^\circ)^{1/2} \, f_N$$

where we define

$$f_N = \frac{1}{N} \sum_{i=1}^{N} f_i \tag{3.22}$$

as the fading random variable corresponding to the average of N independent samples. Its properties are

$$\bar{f}_N = 1$$

$$s_{f_N} = \frac{0.523}{\sqrt{N}} \tag{3.23}$$

and its PDF may be obtained by N-successive convolutions of the Rayleigh PDF (3.13). Plots of $p(f_N)$ are shown in Figure 3.7(a) for several values of N. As expected, as N increases, the distribution becomes more peaked and narrow (the standard deviation decreases as $N^{-1/2}$) and eventually approaches the Gaussian PDF.

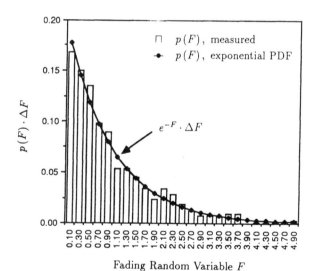

Figure 3.6 The measured PDF of the backscatter from asphalt (corresponding to the data in Fig. 3.2 (b)) is found to be in good agreement with the exponential PDF based on the Rayleigh fading model. The bin size $\Delta F = 0.2$. (From [143].)

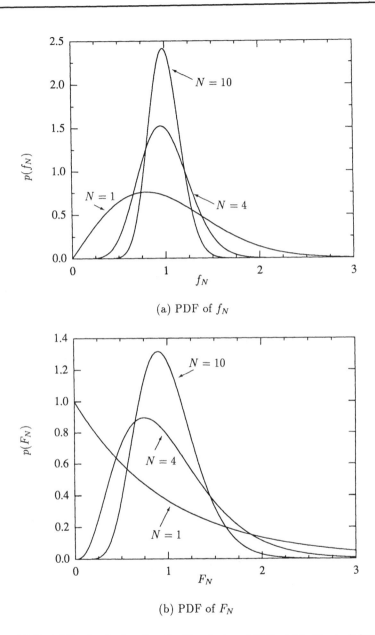

(a) PDF of f_N

(b) PDF of F_N

Figure 3.7 Probability density functions for $N = 1$, **4**, and **10** for (a) f_N (**linear deduction**) and (b) F_N (**square-law detection**).

3-2.2 Square-Law Detection

If the receiver uses square-law detection, the average received power due to N independent samples is

$$P_N = K_3 K_4 \, \sigma^\circ F_N \tag{3.24}$$

with

$$
\begin{aligned}
F_N &= \frac{P_N}{\overline{P}_N} \\
&= \frac{1}{N} \sum_{i=1}^{N} F_i
\end{aligned}
\tag{3.25}
$$

The mean value of F_N is 1, its standard deviation is

$$s_{F_N} = \frac{s_{P_N}}{\overline{P}_N} = \frac{1}{\sqrt{N}} \tag{3.26}$$

and its PDF is a χ^2 distribution with $2N$ degrees of freedom [138, p. 1914]:

$$p(F_N) = \frac{F_N^{N-1} N^N e^{-NF_N}}{(N-1)!}, \quad F_N \geq 0 \tag{3.27a}$$

The corresponding PDF for P_N is

$$p(P_N) = \frac{p(F_N)}{\overline{P}_N} \tag{3.27b}$$

Plots of $p(F_N)$ are shown in Figure 3.7(b) for various values of N, and Figure 3.8 shows plots of the 5% and 95% levels versus N. The 5% level is a plot of F_N' and represents the value of F_N' exceeded only 5% of the time; i.e., $W\,(F_N \geq F_N') = 0.05$. A similar definition applies to the 95% level.

3-3 APPLICABILITY OF THE RAYLEIGH CLUTTER MODEL

Does the Rayleigh fading model provide an appropriate approach for characterizing the statistics of radar backscatter from terrain? The answer is a qualified yes. If the assumptions underlying the Rayleigh fading model are reasonably satisfied, the available experimental evidence suggests that the Rayleigh model is quite applicable [31,139,143]. Terrain targets satisfying the Rayleigh assumptions include bare ground surfaces, agricultural fields, dense forest canopies, and snow-covered ground. In all cases the target must have stationary statistics, which requires that its "local-average" electromagnetic properties be uniform across the extent of the target.

Rayleigh fading is inapplicable for a sparse forest observed by a high resolution radar because the high spatial variations in tree density at the scale of the radar resolution violate the stationarity assumption. Thus, a very important parameter governing applicability of Rayleigh statistics to backscatter from terrain is the size of the radar resolution cell relative to the spatial frequency spectrum characterizing the scattering from the terrain target under consideration.

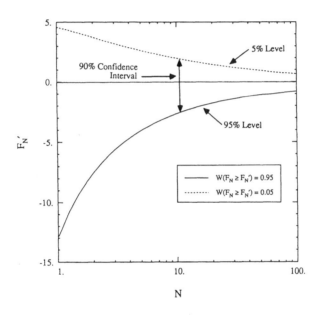

Figure 3.8 Plots of the 5% and 95% cumulative distribution levels *versus N*. The vertical spacing between these two curves is a measure of the "confidence interval" associated with a measurement of the radar backscatter.

An urban scene is another target class or condition for which Rayleigh statistics may not apply. If the resolution cell size is such that the backscatter is likely to be dominated by the return from one or a few strong scatterers, such as a building or a corner reflector formed by two intersecting flat surfaces, the Rayleigh clutter model is no longer applicable. Instead, other statistical models should be used, examples of which are given in the next section.

For nonuniform terrain composed of many distributed targets with different scattering properties, the spatial variability in radar backscatter is governed by both the statistics associated with fading and the statistics characterizing the distribution of the scattering coefficient of the scene. Clutter models for nonuniform terrain are discussed in Section 3-5.

3-4 NON-RAYLEIGH DISTRIBUTIONS

3-4.1 Log-Normal Distribution

The random variable F defined by (3.17) represents the backscattered power P normalized to its mean value $\bar{P}$. The cumulative distribution $W(F \leq F')$ shown in Figure 3.5(b) increases exponentially with F' and exceeds the level 0.999 for $F' > 7$ (or 8.5 dB). For some terrain targets, such as the sea surface, the cumulative distribution of $(P/\bar{P})$ is observed to depart markedly from the exponential distribution due to the presence of strong backscatter components with amplitudes many times larger than $\bar{P}$. In other words, the cumulative distribution $W(F \leq F')$ increases with F' at a rate slower than that of the exponential distribution. It has been proposed [104] to model such targets with the log-normal PDF because it has a long tail in the high probability region. The log-normal distribution is applicable when the backscattered power expressed in dB is Gaussian distributed. That is,

$$
\begin{aligned}
P_{dB} &= 10 \log P \\
&= 4.34 \ln P
\end{aligned}
\tag{3.28}
$$

and

$$
p(\ln P) = \frac{1}{\sqrt{2\pi}\, s} \exp\left[-(\ln P - \mu)^2/2s^2\right], \quad P \geq 0
\tag{3.29}
$$

where s and μ are the standard deviation and mean values of $\ln P$, respectively.

(a) Square-Law Detection

The corresponding PDF for P can be obtained by applying the relationship

$$
p(\ln P)\, d(\ln P) = p(P)\, dP
$$

which gives [1]

$$
p(P) = \frac{1}{\sqrt{2\pi}\, Ps} \exp\left[-(\ln(P/m_P))^2/2s^2\right], \quad P \geq 0
\tag{3.30}
$$

in which we used the relation

$$\mu = \overline{\ln P} = \ln m_P$$

where m_P is the median value of P. The standard deviation s of $\ln P$ is related to the mean and standard deviation of P, s_P and $\overline{P}$, by

$$s^2 = \ln [(s_P^2 + \overline{P}^2)/\overline{P}^2] \tag{3.31}$$

and the mean and the median are related by

$$\frac{\overline{P}}{m_P} = \exp (s^2/2) \tag{3.32}$$

If we normalize P with respect to its median value m_P by defining $P_n = P/m_P$, the PDF of P_n is given by

$$p(P_n) = \frac{1}{\sqrt{2\pi}\, P_n s} \exp [- (\ln P_n)^2/2s^2], \quad P_n \geq 0 \tag{3.33}$$

Figures 3.9 and 3.10 provide plots of the log-normal distribution for power for two values of s. A plot of the exponential distribution is also shown for comparison.

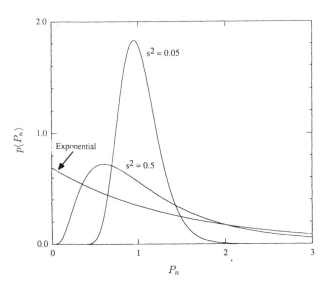

Figure 3.9 Plots of the log-normal PDF of P_n for two values of s^2. The exponential PDF is shown also for comparison.

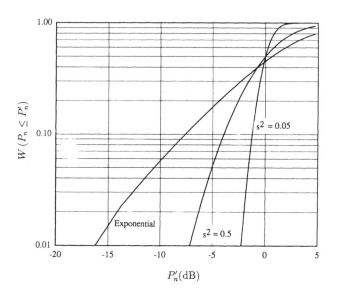

Figure 3.10 Plots of the cumulative distribution for the log-normal density of P_n.

(b) Linear Detection

The PDF that characterizes the statistics of the voltage amplitude at the output of a linear detector can be obtained through the transformation

$$v = \sqrt{P},$$

and the relationship $p(v)\, dv = p(P)\, dP$, and is given by

$$p(v) = \frac{2}{\sqrt{2\pi}\, sv} \exp\left[-2\left(\ln\left(v/m_v\right)\right)^2/s^2\right], \quad v \geq 0 \tag{3.34}$$

where m_v is the median value of v, and s retains the same meaning given earlier. The PDF of the normalized voltage $v_n = v/m_v$ is given by

$$p(v_n) = \frac{2}{\sqrt{2\pi}\, v_n s} \exp\left[-2\left(\ln v_n\right)^2/s^2\right], \quad v_n \geq 0 \tag{3.35}$$

Figures 3.11 and 3.12 provide plots of the log-normal distribution for the voltage v_n.

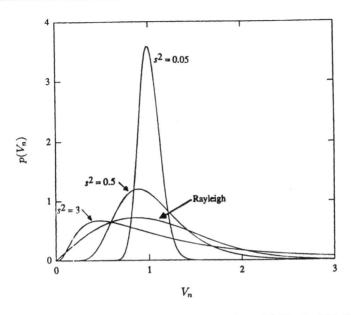

Figure 3.11 Plots of the log-normal PDF of V_n for various values of s^2. The Rayleigh PDF is also shown for comparison.

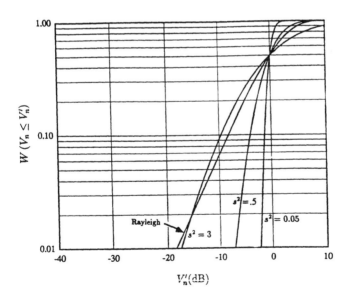

Figure 3.12 Plots of the cumulative distribution for the log-normal density of V_n.

3-4.2 Weibull Distribution

Whereas the Rayleigh model has been found useful for characterizing radar clutter when the clutter distribution encompasses a limited dynamic range, the log-normal distribution tends to overestimate the dynamic range. The Weibull model has been proposed as a solution that provides flexibility over a wide dynamic range because the Weibull distribution can be made to approach either the Rayleigh distribution or the log-normal distribution by appropriately adjusting its parameters [24,105].

(a) Linear Detection

The Weibull PDF for the normalized detected voltage amplitude $v_n = v/m_v$ is given by [105]

$$p(v_n) = \alpha(\ln 2)\, v_n^{\alpha-1} \exp\left[-(\ln 2)\, v_n^{\alpha}\right], \quad v_n \geq 0 \tag{3.36}$$

where m_v is the median value of v and α is a parameter that relates to the skewness of the distribution. The mean value $\bar{v}$, the standard deviation, s_v, and the cumulative distribution $W(v_n \leq v_n')$ are given by

$$\bar{v} = m_v(\ln 2)^{-1/\alpha}\, \Gamma(1 + 1/\alpha) \tag{3.37}$$

$$s_v = m_v\,(\ln 2)^{-1/\alpha}\, [\Gamma(1 + 2/\alpha) - \Gamma^2(1 + 1/\alpha)]^{1/2} \tag{3.38}$$

and

$$W(v_n \leq v_n') = 1 - \exp\left[-(\ln 2)(v_n')^{\alpha}\right] \tag{3.39}$$

where $\Gamma(\ )$ is the Gamma function. It can easily be shown that if we set $\alpha = 2$ in (3.36), it reduces to the Rayleigh PDF given by (3.12). Thus, the Rayleigh PDF is a member of the Weibull family. Figures 3.13 and 3.14 show plots of the Weibull distribution for various values of the *skew parameter* α.

(b) Square-Law Detection

The transformation $P = v^2$, when used with the relation $p(P)\, dP = p(v)\, dv$, leads to

$$p(P_n) = \beta\,(\ln 2)\, P_n^{\beta-1} \exp\left[-(\ln 2)\, P_n^{\beta}\right], \quad P_n \geq 0 \tag{3.40}$$

where $\beta = \alpha/2$, $P_n = P/m_P$, and $m_P = m_v^2$ is the median value of P. The PDF given by (3.40) is identical in form with the PDF for the voltage amplitude given by (3.36). Hence, the mean value $\overline{P}$, the standard deviation s_P and the cumulative distribution $W(P_n \leq P'_n)$ are given by expressions identical in form with those given by (3.37) through (3.39), upon replacing v with P and α with β.

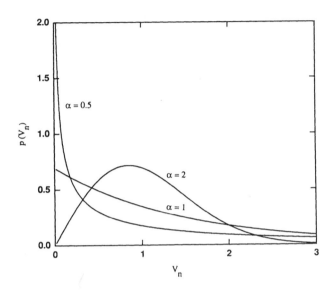

Figure 3.13 Plots of the Weibull PDF of V_n for various values of the skew parameter α. These also are plots of P_n if V_n and α are replaced with P_n and β, respectively.

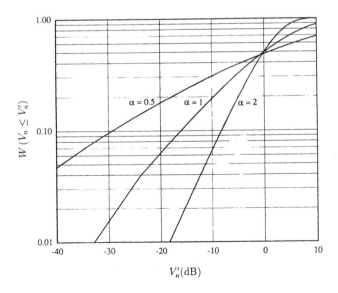

Figure 3.14 Plots of the cumulative distribution for the Weibull density of V_n.

3-4.3 Distribution for a Strong Point Target in Clutter Background

If a point target with constant, deterministic backscatter is present in a background comprised of a random distribution of scatterers, the total electric field at the receiving antenna is

$$E_t = E_0 + E \tag{3.41}$$

where E_0 is a steady field due to the point target and E is the field due to the background. The statistics of the detected voltage amplitude (linear detection) or input power (square-law detection) depend on the ratio of the backscatter cross section of the point target to the average backscatter cross section of the background, R, and on the statistics used to characterize the background return. As an example, we shall consider the case where the background is Rayleigh distributed. For situations involving a point target in a background characterized by log-normal or Weibull statistics, the reader is referred to Trunk and George [117] and Schleher [105], respectively.

(a) Linear Detection

The PDF that describes the statistics of the voltage amplitude due to backscattering from a Rayleigh-distributed extended target containing a constant-echo point target is known as the *Ricean distribution* and is given by [13]

$$p(v_t) = \frac{v}{s^2} \exp\left[-(v^2 + v_0^2)/2s^2\right] I_0 (v_o v/s^2), \quad v_t \geq 0 \tag{3.42}$$

where $v_t = v_0 + v$, v_0 and v are the voltage amplitudes due to the constant point target and the random background, respectively, and $I_0(\ )$ is the Bessel function of the first kind, zero order, with imaginary argument. The quantity s is the standard deviation of the x- and y-components of v.

The mean amplitude is given by

$$\bar{v}_t = s\sqrt{\pi/2}\left[(1 + v_1^2)I_0 (v_1^2/2) + v_1^2 I_1 (v_1^2/2)\right] \exp(-v_1^2/2) \tag{3.43}$$

where

$$v_1 = v_0/\sqrt{2}\, s \tag{3.44}$$

and $I_1(\ )$ is the Bessel function of the first kind, first order, with imaginary argument. The variance of v_t may be found from

$$s_{v_t}^2 = \overline{v_t^2} - \overline{v_t}^2 \tag{3.45}$$

and the relationship:

$$\overline{v_t^2} = 2s^2 + v_0^2$$
$$= \overline{P} + P_0 = \overline{P}(1 + R) \tag{3.46}$$

where $\overline{P} = 2s^2$ is the mean value of the power of the signal backscattered from the random background, P_0 is the power backscattered from the deterministic point target, and $R = P_0/\overline{P}$.

(b) Square-Law Detection

For the total input power,

$$P_t = P + P_0$$

the Rice PDF is given by

$$p(P_t) = \frac{(1 + R)}{\overline{P}_t} e^{-R} \exp\left[-\frac{P_t}{\overline{P}_t}(1 + R)\right]$$

$$\times I_0\left(2\left[R(1 + R)\left(\frac{P_t}{\overline{P}_t}\right)\right]^{1/2}\right), \quad P_t \geq 0 \tag{3.47}$$

the mean value of P_t is given by (3.46) and the normalized standard deviation is given by

$$\frac{s_{P_t}}{\overline{P}_t} = \frac{\sqrt{1 + 2R}}{1 + R} \tag{3.48}$$

For $R = 0$, (3.47) reduces to the exponential PDF (3.18).

If a radar system were to observe a tree canopy as a function of time, the backscattered signal would consist of the sum of a constant echo due to the trunks and other stationary objects and a fluctuating echo due to the leaves and branches as they move in the wind. This expectation led Goldstein (in Kerr [75], pp. 582–584) to use the Ricean PDF to model the statistics of the backscatter from heavily wooded terrain (Figure 3.15).

EXPERIMENTAL RESULTS

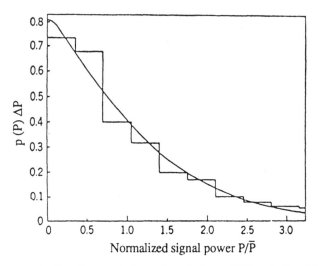

(a) Heavily wooded terrain, wind speed 25 mph, R = 0.8.

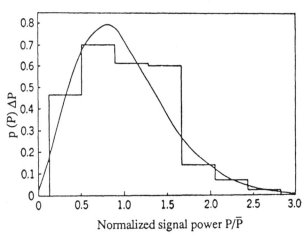

(b) Heavily wooded terrain, wind speed 10 mph, R = 5.2.

Figure 3.15 Plots of the measured PDFs of radar backscatter from heavily wooded terrain under two different windspeed conditions. Plots of the Ricean PDF also are shown, with R chosen in each case to obtain a good fit between theory and experiment.

3-5 CLUTTER MODEL FOR NONUNIFORM TERRAIN

Suppose we are interested in characterizing the statistics of a radar image of an agricultural scene, such as the one depicted in Figure 3.16. The scene consists of a large number of fields planted in different crops, and each field (on the image) may consist of any number of pixels. We shall denote the image intensity (or received power) of pixel j of field i by $P(i, j)$, and we shall assume that each field is statistically homogeneous (as defined in Section 2-3.3) over its spatial dimensions. The image intensity exhibits two types of variations: within-field pixel-to-pixel variations that are a consequence of the signal-fading process discussed in Section 3-2, and field-to-field variations of the field-average intensity corresponding to variations in the properties of the vegetation targets growing in the different fields. Because these two types of variations are governed by independent processes, they may be assumed to be statistically independent. Thus, the variation due to fading is characterized by the same statistics for all fields in the scene.

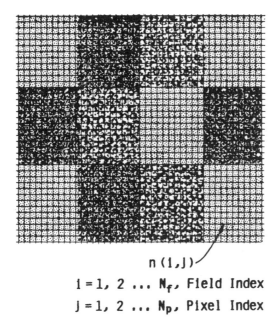

$$n\,(i, j)$$

$$i = 1,\ 2\ \dots\ N_f,\ \text{Field Index}$$

$$j = 1,\ 2\ \dots\ N_p,\ \text{Pixel Index}$$

Figure 3.16 Representation of an image consisting of rectangular fields with N_p pixels each.

For any field i, the random variable $P \triangleq P(i, j)$ is characterized by the conditional probability density $p_1(P|P_i)$ where P_i is the true mean value of P for field i. That is, P_i is the power that would be observed for all pixels in field i in the absence of fading, and it is directly proportional to the backscattering coefficient $\sigma°$ of that field. We shall refer to P_i as the *background power* level and to the distribution of P_i over the scene as the *background distribution*. If the background power P_i (which represents the field-to-field variability) is characterized by a background PDF $p_2(P_i)$, then the unconditional density for P is given by [85]

$$p(P) = \int_0^\infty p_1(P|P_i)p_2(P_i)\, dP_i, \quad P \geq 0 \tag{3.49}$$

This relationship, which may be generalized to any type of terrain, provides the means to characterize the statistics of the power P backscattered from a randomly located spot on the ground in terms of the fading PDF, $p_1(P|P_i)$, and the background PDF $p_2(P_i)$ characterizing the distribution of the backscattering coefficient $\sigma°$ of the scene.

For an N-look image governed by Rayleigh fading statistics, $p_1(P|P_i)$ is given by (3.27b):

$$p_1(P|P_i) = \frac{P^{N-1}N^N \exp(-NP/P_i)}{(N-1)!P_i^N}, \quad P \geq 0 \tag{3.50}$$

which leads to

$$p(P) = \int_0^\infty \frac{P^{N-1}N^N \exp(-NP/P_i)}{(N-1)!P_i^N}\, p_2(P_i)\, dP_i \tag{3.51}$$

The nth moment of P can be determined in terms of the nth moment of P_i [85] as

$$E(P^n) = \frac{(N+n-1)!}{(N-1)!N^n}\, E(P_i^n) \tag{3.52}$$

where $E(x)$ denotes the expected value of x.

Lewinski [85] used (3.51) to compute the unconditional density $p(P)$ for various densities $p_2(P_i)$ of the background level P_i. Some of his results are summarized in the following sections.

3-5.1 Uniformly Distributed Background

For a uniform density,

$$p_2(P_i) = \frac{1}{P_b - P_a}, \quad P_b \geq P_i \geq P_a \tag{3.53}$$

extending over the range $P_i = P_a$ to $P_i = P_b$, the *scene mean* is

$$\overline{P}_i = (P_b + P_a)/2 \tag{3.54}$$

and the dynamic range is $D = P_b/P_a$. Substitution of (3.53) into (3.51) leads to [85]:

$$p(P) = \frac{1}{(P_b - P_a)(N - 1)!} \left[\gamma\left(N - 1, \frac{NP}{P_a}\right) - \gamma\left(N - 1, \frac{NP}{P_b}\right) \right] \tag{3.55}$$

where γ is the *incomplete gamma function:*

$$\gamma(c, x) = \int_0^x t^{c-1} e^{-t} \, dt \tag{3.56}$$

Figure 3.17 shows normalized plots of (3.55) for $N = 1$ and two extreme values of D, namely $D = 1$ which represents a statistically homogeneous scene with constant background level, and $D = \infty$ representing a very nonhomogeneous scene. Plots for intermediate values of D are always bound by the plots shown in Figure 3.17.

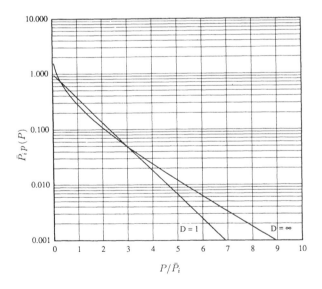

Figure 3.17 Normalized unconditional density $p(P)$ for a uniformly distributed background with mean $\overline{P}_i$ and dynamic range D, shown here for $D = 1$ and $D = \infty$.

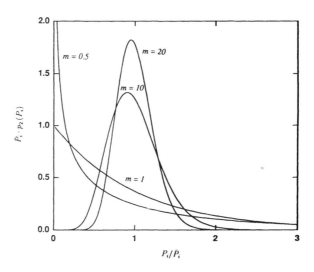

Figure 3.18 Plots of the normalized gamma PDF $p_2(P_i)$ for various values of m.

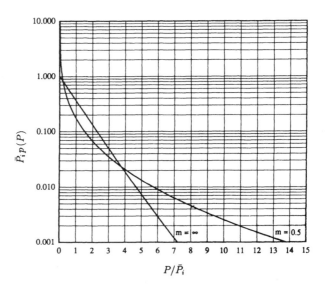

Figure 3.19 Normalized unconditional density $p(P)$ for a Gamma-distributed background with mean P_i and normalized variance m, shown here for $m = 0.5$, and $m = \infty$ (corresponding to a background with constant level).

3-5.2 Gamma Distributed Background

According to the experimental data given in Part II of this handbook, the backscattering coefficient $\sigma°$ of a given terrain category (such as crops, forests, snow-covered terrain, *et cetera*) is more likely to have a Gaussian- or Rayleigh-shaped PDF than a uniform one. The two-parameter gamma PDF is used in this regard because it provides a large family of functions to choose from. The gamma PDF is given by

$$p_2(P_i) = \frac{m^m P_i^{m-1} \exp\left(-mP_i/\overline{P}_i\right)}{\Gamma(m)\overline{P}_i^m}, \quad P_i \geq 0 \tag{3.57}$$

where $\overline{P}_i$ is the scene (or category) mean, and m is the inverse of the normalized variance of P_i:

$$m = \overline{P}_i^2/s^2 \tag{3.58}$$

with s being the variance of P_i. Plots of the normalized Gamma density are shown in Figure 3.18 for various values of m. Upon substituting (3.57) into (3.51) and evaluating the integral, the following result is obtained [85]:

$$p(P) = \frac{2\,(Nm)^{(N+m)/2}\,P^{(N+m-2)/2}}{(N-1)!\Gamma(m)\overline{P}_i^{(N+m)/2}}\,K_{m-N}\left(\sqrt{\frac{4\,mNP}{\overline{P}_i}}\right) \tag{3.59}$$

where $K_v(x)$ is the modified Bessel function of the second kind, of order v. Figure 3.19 presents plots of $\overline{P}_i p(P)$ for $N = 1$ and $m = 0.5$ and ∞, with the latter corresponding to a background with constant level ($P_i = \overline{P}_i$).

3-5.3 Log-Normal Distributed Background

Because the backscattering coefficient $\sigma°$ of terrain often exhibits a dynamic range of $100:1$ or greater, it is usually expressed in dB for convenience. Hence, the log-normal distribution is a natural candidate for characterizing the background power level P_i (which, as noted earlier, is directly proportional to $\sigma°$). Thus, if

$$P_i(dB) = 10 \log P_i = \sigma°(dB) + K$$
$$= 4.34 \ln P_i \tag{3.60}$$

has a uniform distribution with mean $\overline{P}_i$(dB) and standard deviation s (dB), then P_i has the log-normal density given by (see Section 3-4.1):

$$p_2(P_i) = \frac{1}{\sqrt{2\pi}\,sP_i} \exp\left[-(\ln P_i - \overline{\ln P_i})^2/2s^2\right], \quad P_i \geq 0 \tag{3.61}$$

where $\overline{\ln P_i}$ and s are the mean and standard deviation of $\ln P_i$. These two quantities are related to the mean and standard deviation of P_i (dB) by

$$\overline{\ln P_i} = \overline{P}_i \text{ (dB)}/4.34 \tag{3.62}$$

and

$$s = s \text{ (dB)}/4.34 \tag{3.63}$$

Insertion of (3.61) into (3.51) leads to the unconditional density:

$$p(P) = \int_0^\infty \frac{P^{N-1}N^N \exp(-NP/P_i) \exp\left[-(\ln P_i - \overline{\ln P_i})^2/2s^2\right]}{\sqrt{2\pi}\,(N-1)!\,sP_i^{N+1}}\, dP_i$$

$$\tag{3.64}$$

which has no closed-form solution. Sample plots based on numerical integration of (3.64) for $N = 1$ are shown in Figure 3.20.

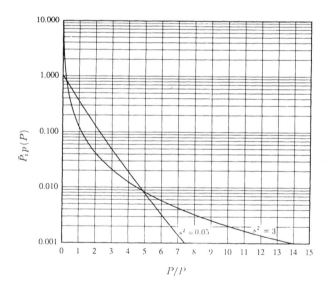

Figure 3.20 Normalized unconditional density $p(P)$ for a log-normal distributed background with mean $\overline{P}_i$ for two values of s, the standard deviation.

Chapter 4

Scintillation Reduction by Spatial Averaging
and Frequency Agility

According to Section 3-2, if a radar is used to measure the backscattering from a uniform, randomly distributed target with backscattering coefficient σ°, the voltage observed at the receiver output will be proportional to $(\sigma^\circ)^n$, with $n = \frac{1}{2}$ for linear detection and $n = 1$ for square-law detection. However, associated with the measurement process there will be a *multiplicative error* represented by the random variable f_N (for linear detection) or F_N (for square-law detection). These random variables both have means of 1 and standard deviations proportional to $N^{-1/2}$. Hence, the key to improving the precision of the measurement process is to make N as large as possible.

Fundamentally, increasing N is equivalent to trading off spatial resolution for improved radiometric resolution. This statement is true when discrete measurements (corresponding to discrete resolution cells) are averaged together after detection, as well as when the averaging process is an integral part of the detection process (as we shall discuss later).

Let us demonstrate the implications of some of the preceding statements with an example. When generating a SAR image, steps may be taken during the image formation process to reduce image speckle. The fading or speckle pattern of the resultant image can be reduced by processing N separate segments of the total SAR bandwidth and then averaging these separate *looks* together. This process, which is referred to as *mixed-integration* or *multiple-looking* [100], produces images with $N \geq 1$. This gain in the magnitude of N, however, is achieved at the expense of spatial resolution because the azimuth resolution of the SAR image is inversely proportional to the bandwidth.

The image shown in Figure 4.1(d), which is a fully focused SAR image with $N = 1$, has an image resolution capability corresponding to a ground resolution area approximately 1.5 m $\times$ 2.1 m. From the raw data of this image, three other images were generated, all having the same ground resolution of 6 m $\times$ 6 m. The

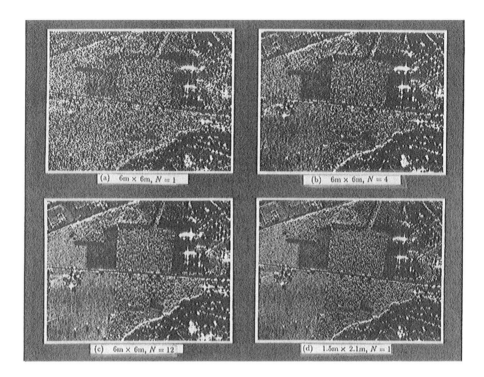

Figure 4.1 The four image set provides a visual illustration of the relationship between image speckle, the number of looks N, spatial resolution, and interpretability. (Courtesy of E. Kasischke, ERIM.)

image in Figure 4.1(a) was generated by subsampling the raw data in both the azimuth and range dimensions to degrade the resolution to 6 m × 6 m. Hence, no subsequent averaging was employed and N remained equal to 1. Partial and maximum possible averaging (or multi-looking) was used in the generation of the images in parts (b) and (c) of Figure 4.1, respectively.

The relationship between image speckle and the magnitude of N is quite *visible* in Figures 4.1(a)-(c). Despite the fact that the spatial resolution is the same for all three images, some features that are clearly distinguishable in Figure 4.1(c) are totally indistinguishable in Figure 4.1(a). An equally significant observation can be derived from a comparison of the 6 m × 6 m image with $N = 12$ (Figure 4.1(c)) with the 1.5 m × 2.1 m image with $N = 1$ (Figure 4.1(d)). From an *interpretability* point of view, the two images offer comparable *resolving power* despite the fact that the area of the resolution cell of the high-resolution image is 12 times smaller than that of the other image. Thus, the resolution quality of an image is governed by both the spatial resolution of the ground cell represented by an image pixel and the number of independent samples (looks) N "contained" in the measurement of the power backscattered from that ground cell.

4-1 SPATIAL AVERAGING

4-1.1 Discrete Samples

If N measurements corresponding to statistically independent non-overlapping footprints are averaged together, then the number of independent samples characterizing the average value is simply N. Statistical independence requires that the spacing between adjacent footprints be greater than the spatial correlation length of the random surface, L_s. Thus, reflections from two non-overlapping footprints on a very smooth surface are not considered independent because the correlation length of a smooth surface is very long (it is infinite for a specular surface). Conversely, for a random surface the returns from two footprints may be considered independent even when the footprints overlap, provided that the spacing between the centers of the two footprints is greater than a certain distance which we shall call the *fading decorrelation distance*, L_d. Expressions for L_d are given in succeeding sections for specific antenna pointing configurations. In all cases the condition $L_d > L_s$ must be satisfied in order for the samples to be statistically independent.

4-1.2 Continuous Averaging in Azimuth

Consider the antenna beam shown in Figure 4.2; the boresight direction is in the x-z plane, pointing at an angle θ, and the effective beamwidth (of the product-gain pattern) is β_y in the y-direction. The antenna is moving along the y-direction (azimuth) at a velocity u_y, the nominal range to the antenna footprint is R, and the nominal azimuth resolution (which is the width of the footprint in the y-direction) is

$$r_y = \beta_y R \qquad (4.1)$$

If the radar output voltage is recorded as a function of time as the beam traverses the ground surface at the velocity u_y, the beam performs a form of continuous averaging equivalent to low-pass filtering. From consideration of the time it takes to travel over a distance r_y and the doppler bandwidth of the signal backscattered from the illuminated cell, it has been shown [135, pp. 585–586] that the output voltage represents an average of N_y-equivalent discrete independent samples, and that N_y is given by the approximate expression

$$N_y \approx \frac{r_y}{(l_y/2)} \tag{4.2}$$

where l_y is the length of the antenna along the y-direction. The result, which is independent of u_y, may be interpreted as saying that the fading signal decorrelates whenever the antenna moves a distance $l_y/2$ in the y-direction, and therefore a resolution cell of width r_y contains $r_y/(l_y/2)$ independent samples. Thus, the fading decorrelation distance is simply

$$L_d \approx \frac{l_y}{2} \tag{4.3}$$

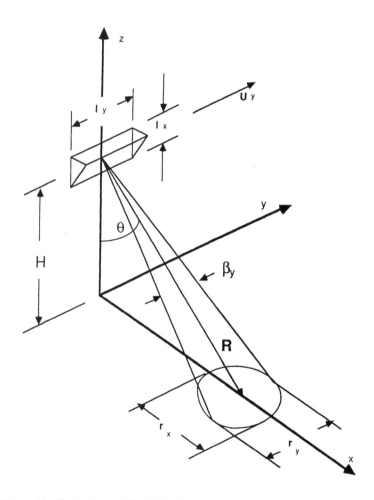

Figure 4.2 Antenna with effective beamwidth β_y illuminating a target at range R and incidence angle θ.

The result given in (4.2) applies equally to a pencil-beam scatterometer and to a fan-beam, side-looking, real-aperture radar. In the case of a side-looking fully focused synthetic-aperture radar, the doppler bandwidth is used to improve the azimuth resolution from $r_y = \beta_y R$ to the resolution $r_s = l_y/2$ corresponding to a synthetic aperture of length $l_s = \beta_y R$. Thus, for the fully focused SAR,

$$N_y = \frac{r_s}{(l_y/2)} = 1$$

Looked at another way, N_y represents, in general, the degradation in spatial resolution from the best achievable $(l_y/2)$ *down* to r_y. For the fully focused SAR, there is no degradation in spatial resolution, but N_y assumes the smallest value it can have, namely 1. For an RAR or a partially focused SAR, the spatial resolution is not the best, but N is usually larger than 1.

4-1.3 Continuous Averaging in Range

For a narrow pencil-beam scatterometer traveling in the x-direction, consideration of the time-bandwidth product leads to [135, pp. 792–793].

$$N_x \approx \frac{r_x}{L_d} \tag{4.4}$$

and

$$L_d \approx \left(\frac{l_e}{2}\right) \sec \theta \tag{4.5}$$

where r_x is the ground resolution in the x-direction, and l_e is the height of the antenna in the elevation plane.

4-2 FREQUENCY AGILITY

Amplitude scintillation of the signal backscattered from a distributed target can be reduced by using a frequency-agile radar. Frequency agility refers to changing the frequency of the signal carrier between one group of pulses and the next, between one pulse and the next one, or during a pulse. An example of the latter is the FM chirp technique wherein the carrier frequency is swept linearly across some bandwidth B over the *chirped-pulse* duration τ_c. Usually, the chirp technique is used instead of transmitting short pulses with very high peak power. The same energy in the short pulse is distributed over a much longer pulse τ_c at a lower peak power, while simultaneously modulating the carrier [135, pp. 517–525]. By demodulating (dechirping) the received pulse, its duration can be shortened to $\tau \geq 1/B$, thereby achieving high resolution in the range direction. If the chirped transmitted pulse is dechirped in the receiver (after reflection by the target) down to the narrowest pulse length possible, namely $\tau = 1/B$, no improvement is realized in terms of reducing the variability of the backscattered signal. If, on the other hand, the chirped transmitted pulse is not dechirped or is dechirped only partially, the frequency averaging provided by the "excess bandwidth" will serve to reduce the variability of the received signal. This section is intended to provide an overview of the relationships between the number of independent samples N and frequency averaging, and between signal variability and the bandwidth over which the signal is averaged.

4-2.1 Decorrelation Bandwidth

The criteria used to decide whether or not a pair of signals, v_1 and v_2, backscattered from two ground footprints, may be treated as statistically independent observations are based on the magnitude of the correlation coefficient between them, $\rho(v_1, v_2)$. If, on the average, ρ is smaller than some specified value, such as 0.2, the two observations may be regarded as statistically independent. Decorrelation is a consequence of differences in the instantaneous phases of the scatterers present in the observed cells. The phase of a given scatterer, as given by (3.3),

$$\phi_i = \omega t - 2 k r_i + \theta_i$$

$$= \omega t - \frac{4\pi}{c} \nu r_i + \theta_i \tag{4.6}$$

may be changed by altering the range r_i between the scatterer and the antenna, or by changing the wave frequency ν. Birkemeier and Wallace [20] derived an expression for the *correlation function* for two signals (one at frequency ν_1 and the other at frequency ν_2) scattered from the same randomly distributed target as a function of the illumination geometry and the frequency separation $\Delta\nu = \nu_2 - \nu_1$. They considered the case of a linear target of range extent r_x observed at an angle θ, as shown in Figure 4.3, and ignored the effects of the azimuth dimension on the correlation function. This is a reasonable first-order approximation if the azimuth resolution is comparable to or smaller than the range resolution r_x.

For a square-law receiver with output voltage proportional to the input power P, the *autocorrelation function* is given by

$$R(\nu_1, \nu_2) = \overline{P(\nu_1)P(\nu_2)} \tag{4.7}$$

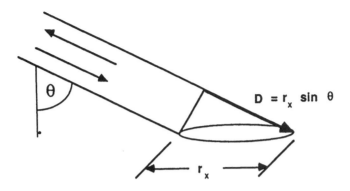

Figure 4.3 Backscattering geometry for an illuminated cell with ground-range dimension r_x.

Birkemeier and Wallace [20] argued that for a target with randomly distributed scatterers, the process is stationary; i.e., $R(\nu_1, \nu_2) = R(\Delta\nu)$, and showed that the *autocovariance function,* defined as

$$R_v(\Delta\nu) = R(\Delta\nu) - \bar{P}^2 \tag{4.8}$$

is given by

$$R_v(\Delta\nu) = \bar{P}^2 \left[\frac{\sin\alpha\,\Delta\nu}{\alpha\,\Delta\nu}\right]^2 \tag{4.9}$$

where $\bar{P} = \bar{P}(\nu_1) = \bar{P}(\nu_2)$ is the mean value of the input power (assumed constant over the frequency separation $\Delta\nu$), and

$$\alpha = \frac{2\pi D}{c} = \frac{2\pi}{c} r_x \sin\theta \tag{4.10}$$

The distance D is defined in Figure 4.3 and c is the speed of light. The correlation coefficient is the normalized autocovariance function. Thus,

$$\rho(\Delta\nu) = \frac{R_v(\Delta\nu)}{R_v(0)} = \frac{R_v(\Delta\nu)}{\bar{P}^2} = \left(\frac{\sin\alpha\,\Delta\nu}{\alpha\,\Delta\nu}\right)^2 \tag{4.11}$$

The two signals $P(\nu_1)$ and $P(\nu_2)$ may be regarded as statistically uncorrelated, and therefore independent, if the separation $\Delta\nu$ corresponds to the first zero of $\rho(\Delta\nu)$ which occurs at $\Delta\nu = \pi/\alpha$. This was called the *critical frequency change* by

Birkemeier and Wallace [20], but we shall refer to it as the *decorrelation bandwidth* $\Delta \nu_d$, which is given by

$$\Delta \nu_d = \frac{\pi}{\alpha} = \frac{c}{2D} \cong \frac{150}{D} \text{ MHz} \tag{4.12}$$

with D, the maximum range extent of the observed resolution cell, measured in meters.

Thus, if N observations of a randomly distributed target are made at N different frequencies such that no pair of frequencies is separated by less than $\Delta \nu_d$, then the average of the N measurements is statistically equivalent to the average of N spatially independent observations of the distributed target. In other words, frequency agility can be used to reduce the fading variance just as spatial averaging can, and the decorrelation bandwidth in the frequency domain is analogous with the decorrelation distance in the spatial domain (see (4.3)).

Next, we shall consider the case of continuous averaging over a fixed bandwidth B.

4-2.2 Continuous Averaging in Frequency

For continuous integration over a swept-frequency bandwidth B extending from ν_1 to ν_2, the variance of

$$P(B) = \frac{1}{B} \int_{\nu_1}^{\nu_2} P(\nu) \, d\nu \tag{4.13}$$

is given by [102]:

$$s_P^2(B) = \frac{2}{B} \int_0^B \left(1 - \frac{\xi}{B} \right) R_v(\xi) \, d\xi \tag{4.14}$$

where ξ is $\Delta \nu$, the variable bandwidth. Use of (4.9) in (4.14) leads to

$$s_P^2(B) = \frac{2\overline{P}^2}{B} \int_0^B \left(1 - \frac{\xi}{B} \right) \left(\frac{\sin \alpha \xi}{\alpha \xi} \right)^2 \, d\xi \tag{4.15}$$

The *effective number of independent samples* realized as a result of frequency averaging may be obtained by relating the variance of P to its mean value as in (3.26):

$$N = \left(\frac{\overline{P}}{s_P}\right)^2$$

$$= \frac{B}{2}\left[\int_0^B \left(1 - \frac{\xi}{B}\right)\left(\frac{\sin \alpha\xi}{\alpha\xi}\right)^2 d\xi\right]^{-1} \tag{4.16}$$

Figure 4.4 presents a plot of N *versus* the ratio $(B/\Delta\nu_d)$, where $\Delta\nu_d = \pi/\alpha$ is the decorrelation bandwidth defined previously by (4.12). We observe that for $(B/\Delta\nu_d) \geq 3$, the plot approaches a linear variation with a slope of 1. This observation is consistent with the result that if $\alpha B \gg 1$, the term ξ/B in (4.16) is negligible over the region where the autocovariance function is of significant size in the integrand, which allows us to integrate the function analytically and obtain the approximate solution:

$$N \approx \frac{\alpha B}{\pi} = \frac{B}{\Delta\nu_d}$$

$$\approx \frac{2DB}{c} \tag{4.17}$$

Here D is the slant range resolution of the radar system. We may show the equivalence of the above result for the chirped pulse radar case (as in an RAR or SAR) by noting that B is the chirp bandwidth and $2D/c$ is the dechirped pulse length τ. Hence,

$$N \approx B\tau$$

$$\approx \frac{B}{B_r} \tag{4.18}$$

where $B_r = 1/\tau$ is the receiver bandwidth. If the transmitted pulse is dechirped in the receiver to obtain the narrowest possible pulse length, the receiver bandwidth B_r must be equal to the modulation bandwidth B. Hence, $N = 1$. However, if we desire to have N larger than 1, the pulse may be dechirped only partially, thereby using the excess bandwidth to provide frequency averaging. This is referred to as *coherent frequency averaging* [86], in contrast with incoherent frequency averaging, wherein the averaging operation is performed after the detection and sampling operations. That is, full dechirping is performed to retrieve the best range resolution possible, and after the image is produced, several range pixels are averaged together to increase N.

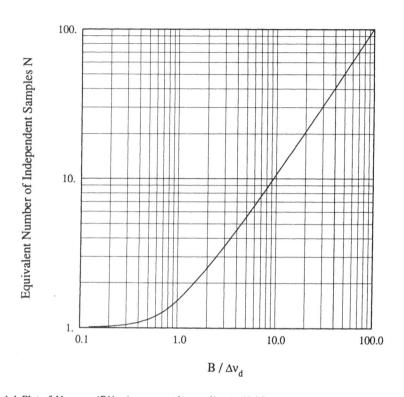

Figure 4.4 Plot of N *versus* $(B/\Delta \nu_d)$, computed according to (4.16).

4-2.3 Examples of Experimental Observations

Figure 4.5 displays a typical example of the measured frequency spectrum of the power backscattered from a given footprint of a snow-covered soil surface. This spectrum is statistically similar to many others that were observed for an asphalt surface, a grassy surface, and other types of distributed targets. The spectra were measured by the University of Michigan's truck-mounted Millimeter Wave Polarimeter [140], which sweeps in frequency from 34 to 36 GHz in 401 discrete steps and, in a similar fashion, from 93 to 95 GHz and from 138 to 140 GHz.

We observe that $P(\nu)$ in Figure 4.5 varies relatively slowly as a function of frequency, implying high correlation between adjacent frequency points, but the overall variation across the 34-36 GHz band is on the order of 23 dB. The improvement (reduction) in spatial variability of the return provided by frequency averaging is demonstrated in Figure 4.6, which shows both single-frequency measurements and the 2-GHz averaged measurements of the return from snow as a function of spatial position (see the sketch shown in Figure 3.2(a)). The associated

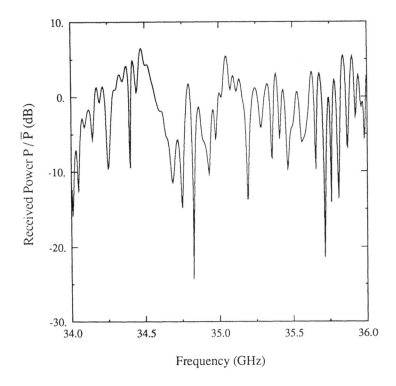

Figure 4.5 Typical trace of the frequency variation from 34 to 36 GHz of the received power for a given footprint of snow. The incidence angle is 40°. (From [143].)

normalized standard deviation is 1.0 for the single-frequency data, compared to 0.27 for the frequency-averaged data.

In the experiment alluded to previously in connection with the spectrum shown in Figures 4.5, 50 such spectra were recorded corresponding to 50 independent footprints. For each spectrum, the autocovariance function $R_v(\Delta \nu)$ was computed using the standard definitions given by (4.7) and (4.8), and then $R_v(\Delta \nu)$ was normalized with respect to its value at $\Delta \nu = 0$ in order to obtain the correlation coefficient $\rho(\Delta \nu)$. This process was repeated for all 50 spectra, and then an average correlation coefficient:

$$\rho_m(\Delta \nu) = \frac{1}{50} \sum_{i=1}^{50} \rho_i(\Delta \nu) \tag{4.19}$$

was obtained by averaging over the spatial index i. The averaging process was used to obtain a more representative estimate of the spectrum of $\rho(\Delta \nu)$ than that

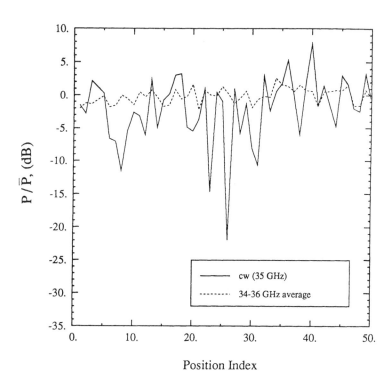

Position Index

Figure 4.6 Reduction of signal variability through frequency averaging.

obtained on the basis of only one or a few samples. A plot of $\rho_m(\Delta\nu)$ is shown in Figure 4.7 for snow. The figure also includes a plot of the theoretical expression given by (4.11). We observe that the measured correlation coefficient decreases with increasing frequency shift $\Delta\nu$ in an exponential-like manner and at a rate slightly faster than the theoretical function. Similar results were observed for asphalt and other surfaces.

The normalized standard deviation associated with the received power P, when averaged over a bandwidth B, is given by

$$\frac{s_P(B)}{\overline{P}} = \left[\frac{2}{B} \int_0^B \left(1 - \frac{\xi}{B} \right) \rho(\xi)\, d\xi \right]^{1/2} \tag{4.20}$$

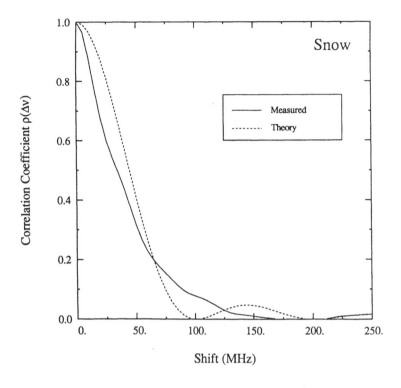

Figure 4.7 Comparison of the theoretical autocorrelation function given by (4.11) with that computed on the basis of spectral measurements of the radar backscatter. (From [143].)

where $\xi = \Delta\nu$. This expression was obtained from (4.14) after replacing $R_v(\xi)$ with $\overline{P}^2 \rho(\xi)$. Figure 4.8 shows plots of the normalized standard deviation as given by (4.20) with $\rho(\Delta\nu)$ as given by (4.11). The figure also includes points representing the normalized standard deviation as computed directly from the measured data; for a given bandwidth B, s_P is based on the values of P measured at all frequencies between $B/2$ below and $B/2$ above 35 GHz. We observe that for both snow and asphalt the data and the curves shown in Figure 4.8 are in close agreement with one another, indicating that the theoretical expressions given by (4.11)—(4.17) indeed provide a reasonable description of the improvement realized by frequency averaging with respect to reducing the fading variance of the backscattered signal.

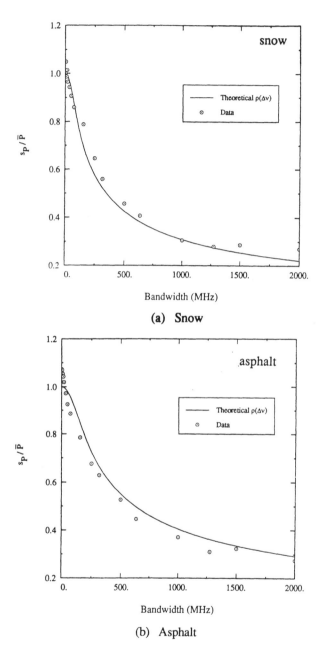

(a) Snow

(b) Asphalt

Figure 4.8 Normalized standard deviation *versus* bandwidth B. "Theoretical" refers to (4.20) with $\rho(\Delta\nu)$ given by (4.11). (From [133].)

PART II

Chapter 5

Data Quality and Presentation Format

5-1 DATA-QUALITY EVALUATION

The literature contains hundreds of articles and reports documenting microwave measurements of the backscatter from terrain. How then do we decide whether to include the data reported in a certain paper in a cumulative data base? To deal with this question, we developed the following criteria.

(1) *Calibration:* If the σ° data is presented using a scale relative to some arbitrary reference and cannot be corrected to normalized radar cross section units of m^2/m^2, then the data are excluded from the data base. Conversely, if the measurement system has been calibrated against a target of known radar cross section so as to convert the receiver output to units of σ°, then the data are considered "calibrated" and warrant further evaluation.

(2) *Precision:* Whereas calibration addresses the absolute magnitude, or accuracy, of a measurement, precision defines the range of uncertainty, or error bars, associated with the measurement. We define the precision as the 95% to 5% confidence interval, and adopt the Rayleigh fading model to establish these limits, which are shown graphically in Figure 3.8 as a function of N, the number of independent samples incorporated in a measurement. More precisely, we consider a measurement of σ° to be *precise* if the 5/95 percent levels are within ± 2 dB relative to the estimated mean. According to Figure 3.8, this condition is satisfied if $N \geq 16$.

Hence, if it can be established from the information provided in support of the data that sufficient spatial or frequency averaging was performed, as a result of which $N \geq 16$, then the data are regarded as *precise* and the evaluation process moves on to the next criterion.

(3) *Target Description:* The terrain (target) description given in association with some of the reported backscattering data, particularly data based on measurements from airborne platforms, is so general that it is very difficult to use the data in

any meaningful terrain classification scheme, unless we simplify them to an all-inclusive single category: terrain. An example of such target descriptors is "rolling hills," which may or may not include trees, other types of vegetation, bare-soil surfaces, ponds, *et cetera*. Consequently, data associated with such poorly described targets were excluded from the data base.

The classification scheme used in this handbook is discussed in Section 5-2, together with a description of the source code and other data identifiers.

(4) *Dynamic-Range Test:* In 1976, Bush and Ulaby [32] reported the results of a study in which they evaluated the variability in radar backscatter data of terrain among eight different measurement programs and systems. [Their report is reprinted in the form of an appendix at the end of this chapter.] They selected data for vegetation under "similar" phenological conditions measured by radars having approximately the same sensor parameters (frequency, polarization configuration, and angle of incidence range). Actually, the systems under consideration operated at frequencies ranging between 8.8 and 15.5 GHz. One of their major observations was that the Ohio State University (OSU) data reported in the 1960 Terrain Handbook [37] were about 5-7 dB lower in level than the data reported by the University of Kansas (KU) and other measurement programs, including the data reported by OSU in later reports (Figure 5.1). A brief account of this comparative study was published in the *IEEE Transactions on Antennas and Propagation* [32], together with comments provided by W. P. Peake, one of the principal authors of the OSU data reports. Peake stated that he agrees with the observations advanced by Bush and Ulaby, although his estimate of the systematic error associated with the data reported in the 1960 report [37] is closer to 4 or 5 dB. He attributed the error to the calibration technique used in the measurement process. He also stated that all the curves in [37], which were later reproduced unchanged in the books by Barton [9] and Long [87], should be raised uniformly by this amount.

Based on this comparative analysis, we decided to raise all the data reported in the 1960 OSU report [37] by 6 dB, and we then used the combined OSU/KU data bases as a reference data base for evaluating the calibration accuracy of other data sets wherever possible. Although the initial comparison of the OSU and KU data sets was based on vegetation targets, the combined data base included a variety of other types of distributed targets including asphalt and concrete surfaces, bare-soil surfaces, and surfaces covered with thin layers of snow (Figure 5.2).

5-2 TERRAIN CATEGORIES

For the purpose of categorizing the radar backscatter from terrain, an ideal classification scheme would incorporate all aspects of the phenomenology. The organizational structure of such a scheme would impart an understanding of the

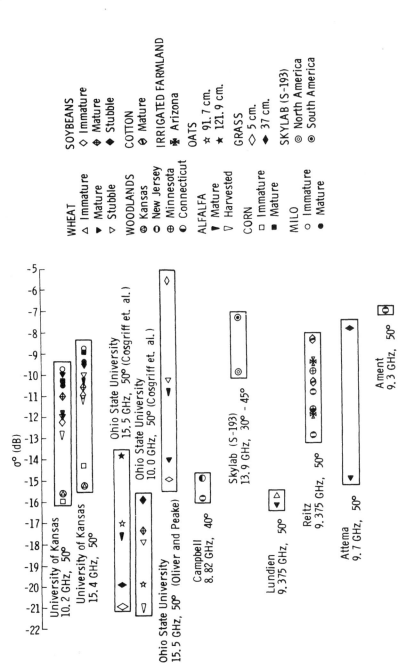

Figure 5.1 Comparison of data acquired by eight programs that examined radar backscatter from vegetation.

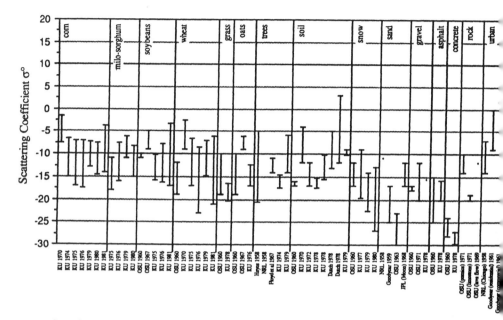

Figure 5.2 Dynamic range of σ° data reported in the literature for various types of distributed targets at X-band, vv polarization, and incidence angle of 50°. The 1960 OSU data [37] have been corrected by raising them 6 dB (see Section 5-1).

scattering mechanisms and permit the identification and aggregation of terrain categories that exhibit similar scattering behaviors. From a strictly phenomenological perspective, a logical structure might utilize a *top down* description of terrain from the sensor viewpoint wherein terrain would be classified by the presence and nature of intervening media (such as snow and vegetation) and subclassified by the nature of the underlying substrate with respect to roughness and dielectric properties.

Although such an approach is valid from the standpoint of the scattering mechanisms, it is not consistent with the traditional terrain classification schemes

Although such an approach is valid from the standpoint of the scattering mechanisms, it is not consistent with the traditional terrain classification schemes used in the earth sciences nor would it be wholly consistent with ancillary descriptions of target conditions as generally reported for backscatter data. A large number of classification systems have evolved within various disciplines, each dealing with a given component of the total landscape in a specific fashion and/or a specific application. Thus, there are distinctive classification systems for the geomorphology of the landscape, the lithology of the substrate, the properties of the soils, the type of natural vegetation, and the economic use of the land.

In recognition of this problem, Anderson [5] acknowledged the need for a generalized classification system for terrain and proposed a land-use and land-cover classification for use with remotely sensed data. Land cover refers to the vegetation or artificial "man-made" constructions covering the land surface, whereas land use refers to the application of a land unit or its intended economic purpose. The generalized classification scheme of Anderson [5] is based upon discrimination of land cover from optical sensing but is largely applicable to radar backscatter as well. A variant of the Anderson [5] classification is applied to the radar backscatter data presented in Part II as shown in Figure 5.3.

Terrain is initially subdivided into four major categories: barren and sparsely vegetated land, vegetated land, urban land, and snow-covered land. Open water surfaces (such as lakes and rivers) and ice are excluded from treatment. However, because the state of water (liquid or frozen) as contained in snow, soils, and vegetation is a prime determinant of dielectric properties, all data are annotated as to the thermal regime at the time of observation. Each of the major categories is further subdivided on the basis of surface roughness, vegetation class, land use, and near-surface dielectric properties.

The data base is organized according to the categories and subcategories shown in the figure; however, because of space limitations the statistical distributions in this handbook are presented for only the following categories: (1) Soil and rock surfaces, (2) trees, (3) grasses, (4) shrubs, (5) short vegetation, (6) road surfaces, (7) urban areas, (8) dry snow, and (9) wet snow.

5-2.1 Barren and Sparsely Vegetated Land

Barren land has a limited ability to support life and is characterized by the absence or scarcity of vegetation. The category includes exposed bedrock (such as volcanic material), gravel, desert pavement, and sparsely vegetated soil surfaces such as those found in arid environments. This category also includes nonvegetated soil surfaces of agricultural lands resulting from tillage practices such as plowing. Subcategories have been defined on the basis of surface roughness whenever possible. These include exposed rock and stony surfaces, smooth soil (rms roughness < 1 cm), medium rough soil (disked or with 1 cm < rms roughness < 3 cm),

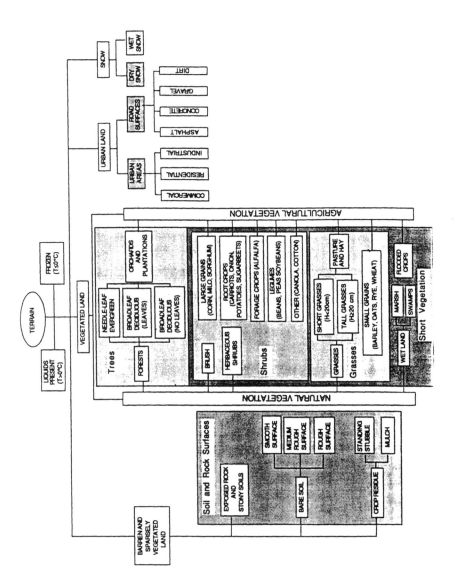

Figure 5.3 Backscatter data classification scheme.

rough soil (plowed or with rms roughness > 3 cm), and soil surfaces with crop residues such as cut stubble or mulches. The data are not subclassified further with respect to near-surface moisture profile or soil composition.

5-2.2 Vegetated Land

Vegetated land is defined as consisting of both naturally occurring vegetation and agricultural crops. Although this distinction is largely one of land use, it can also have important implications with respect to the spatial uniformity and homogeneity of the vegetation cover. The spatial variability of naturally occurring vegetation is controlled by a complex series of soil, plant and atmosphere interrelationships that may vary at the meter scale. Hence, the local variance in vegetation type and physiognomy may be very large when compared to that exhibited by cultivated plant communities. Cultivated agricultural crops generally produce vegetation covers that are even-aged and relatively homogeneous with a well ordered structure and spatial distribution. As a consequence, the preponderance of the available backscatter data represent cultivated agricultural vegetation because these characteristics offer the experimentalist both controlled conditions for comparative studies and relative ease in statistically defining the target conditions. All of the backscatter data from vegetated terrain are tagged with respect to natural occurrence or as a result of cultivation.

The general vegetation category is subdivided into four units on the basis of canopy structure and substrate conditions: trees, shrubs, grasses, and wetlands. Wetlands are differentiated from the other categories on the basis of a water or saturated-soil substrate. Trees, shrubs and grasses are differentiated by gross physiognomy and the relative apportionment of aboveground biomass between woody stems and foliage. The aboveground biomass of trees is dominated by woody material in the boles, branches, and stems. It is not uncommon for mature trees to have 90% or more of their total aboveground biomass contained in woody boles and stems. Shrubs may contain woody stems, but aboveground biomass is not dominated by the bole fraction (if present). Shrubs are bushy in form. Grasses are defined as being dominated by foliar biomass, and the leaves and stems assume a generally vertical orientation.

Each of these subcategories may contain both naturally occurring and agricultural components. For example, the tree category includes both naturally occurring forests and orchards; shrubs include woody bush and agricultural bush crops and large grains; and grasses include natural grasslands, pasture, and small cereal grains.

Trees

The subcategory of trees is differentiated on the basis of foliar material into: needle-leaf evergreens, broadleaf trees with leaves, and deciduous broadleaf trees without leaves. Each subcategory of trees is treated in Part II.

Shrubs and Bushy Plants

This category includes natural herbaceous shrubs, large grains (corn, milo and sorghum), legumes (beans, peas and soybeans), root crops (carrots, onions, potatoes and sugar beets), forage crops, such as alfalfa, and canola and cotton. This category is generally intermediate with respect to trees and grasses, in terms of height and mass fraction of woody material. They are typically bushy in growth habit, with the exception of onions. With respect to phenology, backscatter data in this category range from emergence of cultivated crops until harvest. Leaf geometry, orientation and number density of plants per unit area are highly variable within this category as are the moisture and roughness conditions of the soil substrate. Post-harvest conditions are included within the barren and sparsely vegetated land category.

Grasses

Grasses include natural rangeland, pasture, hay, and small cereal grains. Wherever possible rangeland, pasture, and hay grasses have been differentiated by height into tall grass (height >20 cm) and short grass (height <20 cm). The small cereal grains are subdivided by crop: barley, oats, rye, and wheat. The cereal grain subcategories cover the period from emergence to harvest and thus cover heights up to about 1 meter. In addition, the soil substrate beneath grasses is generally categorized as smooth.

Wetlands

The wetland subcategory consists of marshes, swamps, and flooded agricultural land. Hence, the vegetation cover itself can be highly variable in terms of the amount and nature of biomass and plant geometry. Wetlands are considered to include trees, shrubs and grasses as distinguished by the presence of a water substrate as opposed to soil.

5-2.3 Urban Land

Urban land is defined as either extended surfaces constructed by man or areas of land dominated by man-made structures. Extended surfaces are typified by paving materials used for transportation networks. The backscatter data are primarily from road surfaces as observed by truck-mounted scatterometers. The nature of the surface material is used to subcategorize the data (asphalt, concrete, and gravel).

Man-made structures can behave as point targets. Size, shape, orientation and the nature of the building material are important determinants of the radar cross section of a structure at a given set of wave parameters and viewing geometries. Because the radar backscatter from areas containing such structures is neither uniform nor random, and the sizes of such structures (buildings and bridges) can be quite large, truck-mounted scatterometers are poorly suited for determination of average backscatter. Consequently, source data for urban areas are derived from airborne scatterometry whereby it is practical to average the response over extended areas. This category is subdivided on the basis of land use into residential, commercial and industrial areas.

The definition of these areas is largely arbitrary and, for a given data source, is presumably defined on the basis of the types of structures and their relative spatial density with respect to those of paved surfaces and vegetation covers. As a result, the values reported in the literature, and used herein for each category, are not sufficient for adequate statistical treatment, and are therefore regarded as examples.

5-2.4 Snow

The snow category is defined as any vegetation cover or surface (soil, rock or paving material) that is covered by a continuous layer of snow. No snow-covered man-made structures are included in this category. No minimum or maximum snow depth is established. Although snow can be highly variable with respect to surface roughness and vertical profile development in terms of density and crystal size and shape distributions, the reported snow data are not subclassified on this basis. Snow wetness is the only physical property of the snowpack that has been recorded on a sufficiently common basis to warrant its use in subclassification of snow. Snow wetness is defined as the volume fraction of a given snow layer in the liquid phase. The surface layer of the snow pack is used to differentiate wet and dry snow on the basis of its reported liquid water content (LWC). Snow is classified as wet for all instances where LWC $>1\%$ by volume.

5-2.5 Classification Summary

The terrain classification scheme depicted in Figure 5.3 subclassifies terrain into a large number of discrete categories based upon the nature of the overlying media (vegetation or snow) and the nature of the substrate (soil, water or paving material). Although each subcategory may present a unique clutter environment at a given set of wave parameters, there are a number of similarities in the radar backscatter behavior of generalized groups of terrain conditions. The purpose of

Part II is to present these behaviors for generalized terrain categories. The source data are concatenated into nine general categories: barren and sparsely vegetated land (*Soil and Rock Surfaces*), forests and orchards (*Trees*), grasses (*Grasses*), shrubs, bushy plants and other crops (*Shrubs*), grass, shrubs, and wetlands (*Short Vegetation*), man-made surfaces (*Roads*), residential, commercial, and industrial areas (*Urban Areas*), dry snow (*Snow-dry*), and wet snow (*Snow-wet*). The source data are statistically summarized and plotted for each general category. Similar treatment of specific category and intermediate classification levels is available on a computer diskette.

5-3 FREQUENCY BANDS

For each terrain category, the reported data were grouped into distinct frequency bands, defined as follows:

- L-Band: 1–2 GHz;
- S-Band: 2–4 GHz;
- C-Band: 4–8 GHz;
- X-Band: 8–12 GHz;
- K_u-Band: 12–18 GHz;
- K_a-Band: 30–40 GHz;
- W-Band: 90–100 GHz.

Although some measurements have been reported at frequencies above 100 GHz, the volume of available data is not large enough to warrant arranging it in the form of statistical distributions.

Next, for each category and frequency band, the data are arranged by polarization configuration and incidence angle (relative to nadir). Only linear polarization configurations are considered, namely hh, vv, and hv (which also represents vh because $\sigma_{hv}^{\circ} = \sigma_{vh}^{\circ}$).

5-4 STATISTICAL DISTRIBUTIONS

The data in this handbook are organized by terrain category in Appendices A to I. For each category, the appendix contains: statistical distribution tables (SDTs), angular plots of the mean value of σ°, $\overline{\sigma}^{\circ}$, and of the 90-percent confidence interval, and histograms of the data distributions. As stated earlier, the purpose of the handbook is twofold: to provide ready access to statistical summaries of available radar backscatter data of terrain for applications in system design and signal processing and to elucidate the strengths and weaknesses inherent in the available source data. The presence of gaps in the data indicates the need for additional measurements for certain terrain categories and frequency bands. Tables 5.1 to 5.3 illustrate the numerical strengths of the source data used in the compilation of this handbook.

Table 5.1   Summary of data entries compiled for HH polarized backscatter.

							Frequency Band						
	L		S		C		X		Ku		Ka		W
Category	N	$\Delta\theta$	N	$\Delta\theta$	N	$\Delta\theta$	N	$\Delta\theta$	N	$\Delta\theta$	N	$\Delta\theta$	N	$\Delta\theta$
Soil & Rock Surfaces	3,380	0-50	2,057	0-50	6,979	0-60	912	0-80	254	0-70	77	0-80		
Trees	338	0-60	14	0-80	138	0-55	658	0-80	593	0-80				
Grasses	1,114	0-80	901	0-80	2,401	0-80	5,341	0-80	7,282	0-80	279	0-80		
Shrubs	3,078	0-80	2,633	0-80	7,042	0-80	14,237	0-80	16,256	0-80	398	10-80		
Short Vegetation	4,213	0-80	3,558	0-80	9,518	0-80	19,607	0-80	23,557	0-80	678	0-80		
Road Surfaces	7	10-70					168	0-80	158	0-80	216	0-80		
Urban Areas	23	0-80	35	0-80	8	15-80	142	0-80						
Dry Snow	544	0-80	654	0-80	1,068	0-80	983	0-80	1,695	0-80	551	0-80	84	0-75
Wet Snow	497	0-80	561	0-80	842	0-80	857	0-80	1,514	0-80	350	0-80	26	0-75

N = number of observations of σ° (over the angular range $\Delta\theta$) available in the source data.
$\Delta\theta$ = minimum and maximum angle of incidence observed.

Table 5.2   Summary of data entries compiled for HV polarized backscatter.

							Frequency Band						
	L		S		C		X		Ku		Ka		W
Category	N	$\Delta\theta$	N	$\Delta\theta$	N	$\Delta\theta$	N	$\Delta\theta$	N	$\Delta\theta$	N	$\Delta\theta$	N	$\Delta\theta$
Soil & Rock Surfaces	2,722	0-50	1,776	0-50	5,738	0-60	354	0-70	121	0-50	10	5-80		
Trees	608	0-60	14	0-80	156	0-80	780	0-80	558	0-80	6	10-70		
Grasses	785	0-80	840	0-80	2,219	0-80	3,746	0-70	2,465	0-70	19	0-70		
Shrubs	2,486	0-80	2,483	0-80	6,414	0-80	7,529	0-70	6,707	0-70				
Short Vegetation	3,301	0-80	3,345	0-80	8,700	0-80	11,187	0-70	9,205	0-70	20	0-70		
Road Surfaces							62	10-80	67	0-80	91	0-80		
Urban Areas	26	0-80	26	0-80			26	0-80						
Dry Snow	322	0-80	500	0-80	769	0-75	406	0-80	1,184	0-80	226	0-80	149	0-75
Wet Snow	327	0-80	463	0-80	654	0-80	490	0-80	1,159	0-80	149	0-80	156	0-75

N = number of observations of σ° (over the angular range $\Delta\theta$) available in the source data.
$\Delta\theta$ = minimum and maximum angle of incidence observed.

Table 5.3   Summary of data entries compiled for VV polarized backscatter.

							Frequency Band						
	L		S		C		X		Ku		Ka		W
Category	N	$\Delta\theta$	N	$\Delta\theta$	N	$\Delta\theta$	N	$\Delta\theta$	N	$\Delta\theta$	N	$\Delta\theta$	N	$\Delta\theta$
Soil & Rock Surfaces	1,953	0-50	2,056	0-50	5,074	0-60	436	0-70	1,662	0-70	114	0-80		
Trees	338	0-60	14	0-80	163	0-80	595	0-80	626	0-80	31	10-70	17	10-70
Grasses	1,029	0-80	901	0-80	2,114	0-80	5,834	0-80	7,306	0-80	284	0-80	35	0-70
Shrubs	2,654	0-80	2,629	0-80	5,530	0-80	15,389	0-80	16,777	0-80	418	10-80		
Short Vegetation	3,659	0-80	3,554	0-80	7,667	0-80	21,284	0-80	24,129	0-80	703	0-80	35	0-70
Road Surfaces	6	20-70					168	0-80	158	0-80	221	0-80	27	0-70
Urban Areas	13	0-80	13	0-80			37	0-80						
Dry Snow	524	0-80	568	0-80	855	0-80	463	0-80	982	0-80	492	0-80	378	0-80
Wet Snow	475	0-80	523	0-80	731	0-80	487	0-80	981	0-80	332	0-80	211	0-80

N = number of observations of σ° (over the angular range $\Delta\theta$) available in the source data.
$\Delta\theta$ = minimum and maximum angle of incidence observed.

5-4.1 Statistical Distribution Tables

Each terrain-category appendix of this handbook contains statistical distribution tables (SDTs) organized by frequency band, polarization configuration and incidence angle. At each incidence angle for which data are available, the following information is provided in the SDT (Table 5.4 is included here as an example):

1. N, the number of source $\sigma°$ data points available in the data base. This should not be confused with the number of independent samples used in the measurement process that led to each $\sigma°$ data point.

2. The maximum, median, and minimum values of $\sigma°$ in the distribution, denoted $\sigma°_{max}$, Median, $\sigma°_{min}$, respectively, all expressed in dB.

3. The percentile values of the $\sigma°$ distribution; $\sigma°_x$ represents the value of $\sigma°$ exceeded by $x\%$ of the points in the distribution.

4. The mean value of $\sigma°$ in dB.

5. The standard deviation of the $\sigma°$ (dB) distribution, s, expressed in dB.

If data are available at all frequency bands and polarization configurations, each category appendix should contain 21 SDTs. If a certain entry is blank, it means that the number of available data points is less than 16, and therefore is too small to justify the calculation of the various first-order statistics (such as the 5% level) with a reasonable degree of confidence.

For certain categories and specific frequency bands, we note that the mean value of $\sigma°$ does not always follow a monotonic behavior as a function of incidence angle θ. This is generally the result of having a nonuniform distribution of data points over the angular range and is attributable to data sources associated with experiments that focused on specific angles of incidence or specific target conditions. Consequently, the data from certain experiments may dominate the calculated mean $\sigma°$ at specific angles of incidence.

5-4.2 Angular Plots

Each terrain category appendix in this handbook contains angular plots for each frequency band and polarization (with sufficient data). The plots display mean $\sigma°$ and the 90% occurrence interval around the mean as a function of incidence angle. These plots are not simply graphic displays of the information in the SDTs; they represent the product of smoothing functions applied to the source data as a function of angle.

In the example shown in Figure 5.4, the middle curve represents the angular variation of $\bar{\sigma}°$ and the top and bottom curves represent the 5% and 95% occurrence levels, respectively. The process by which these curves were generated is explained next.

Table 5.4   Sample statistical distribution table.

Statistical Distribution Table for Dry Snow

X Band, HH Polarization

Angle	N	σ^o_{max}	σ^o_{5}	σ^o_{25}	Median	σ^o_{75}	σ^o_{95}	σ^o_{min}	Mean	Std. Dev.
0°	117	14.5	12.7	11.4	9.5	5.7	3.0	1.8	8.6	3.3
5°	52	5.1	2.3	-2.4	-3.4	-4.5	-8.0	-9.0	-3.2	2.9
10°	36	3.2	-2.5	-3.6	-4.8	-5.7	-9.8	-10.1	-4.5	2.2
20°	114	-2.5	-4.3	-6.3	-7.8	-8.7	-10.8	-15.2	-7.6	2.0
25°	48	-5.4	-8.2	-9.7	-10.5	-11.6	-16.2	-17.1	-10.9	2.3
30°	30	-6.9	-7.3	-8.4	-9.6	-11.8	-16.3	-16.3	-10.2	2.4
40°	55	-6.4	-6.5	-6.9	-9.1	-10.7	-15.3	-16.1	-9.4	2.7
50°	316	-4.4	-7.6	-10.2	-11.9	-14.0	-17.3	-20.5	-12.1	2.9
55°	53	-10.6	-12.5	-13.7	-14.9	-16.6	-19.3	-21.0	-15.2	2.1
60°	17	-5.5	-9.9	-14.6	-16.8	-18.7	-21.1	-21.1	-15.9	4.2
70°	82	-4.1	-14.3	-15.3	-16.8	-18.4	-21.6	-24.7	-16.9	3.0
75°	52	-13.6	-14.4	-14.7	-15.3	-19.2	-20.8	-25.7	-16.8	2.6
80°	11	-16.1						-30.6	-23.7	3.9

X Band, HV Polarization

Angle	N	σ^o_{max}	σ^o_{5}	σ^o_{25}	Median	σ^o_{75}	σ^o_{95}	σ^o_{min}	Mean	Std. Dev.
0°	9	3.6						-5.5	-0.8	3.1
20°	69	-6.6	-9.5	-10.3	-11.6	-12.8	-16.4	-21.5	-11.9	2.3
30°	20	-9.3	-10.4	-12.1	-14.0	-16.7	-21.2	-21.2	-14.7	3.4
40°	45	-11.4	-11.8	-12.2	-13.2	-14.1	-22.0	-22.0	-13.8	2.7
50°	120	-12.2	-14.5	-15.6	-17.3	-18.5	-21.9	-24.4	-17.4	2.3
55°	12	-15.1						-24.3	-19.5	2.8
60°	7	-19.5						-27.0	-23.8	3.3
70°	71	-14.6	-18.1	-21.1	-22.0	-23.5	-28.4	-30.1	-22.2	2.8
75°	49	-15.9	-19.9	-21.3	-21.7	-25.7	-27.3	-33.1	-22.9	3.0
80°	4	-25.1						-32.4	-30.0	3.4

X Band, VV Polarization

Angle	N	σ^o_{max}	σ^o_{5}	σ^o_{25}	Median	σ^o_{75}	σ^o_{95}	σ^o_{min}	Mean	Std. Dev.
0°	73	15.9	14.6	12.4	9.3	6.2	3.4	2.0	9.3	3.7
5°	17	6.4	6.3	0.8	-2.8	-6.7	-9.4	-10.2	-2.0	5.4
10°	29	3.4	-0.8	-2.8	-4.1	-5.7	-8.0	-10.1	-3.8	2.6
20°	74	-1.0	-1.4	-3.4	-4.8	-7.3	-10.7	-12.7	-5.3	2.6
30°	28	-4.8	-6.1	-8.3	-9.6	-11.7	-14.0	-14.0	-9.8	2.6
35°	11	-7.9						-19.3	-12.5	4.1
40°	13	-6.9						-16.7	-11.8	2.9
50°	92	-6.4	-8.9	-11.1	-12.4	-13.6	-17.5	-18.5	-12.6	2.4
55°	12	-9.0						-21.4	-15.1	3.7
60°	15	-5.5	-9.8	-13.0	-14.9	-19.9	-20.5	-20.5	-15.0	4.4
70°	78	-4.1	-12.9	-15.1	-15.9	-17.4	-21.5	-24.7	-16.3	3.0
75°	12	-15.0						-26.9	-17.8	3.2
80°	9	-18.6						-30.6	-23.6	4.2

For a given category, frequency band, and polarization, the angular behavior of mean σ^o is determined by an iterative least-squares and non-linear curve-fitting routine. The assumed function consists of exponential and cosine terms as given by:

$$\sigma^o_{mean}(dB) = f(\theta) = P_1 + P_2 \exp(-P_3\theta) + P_4 \cos(P_5\theta + P_6) \tag{5.1}$$

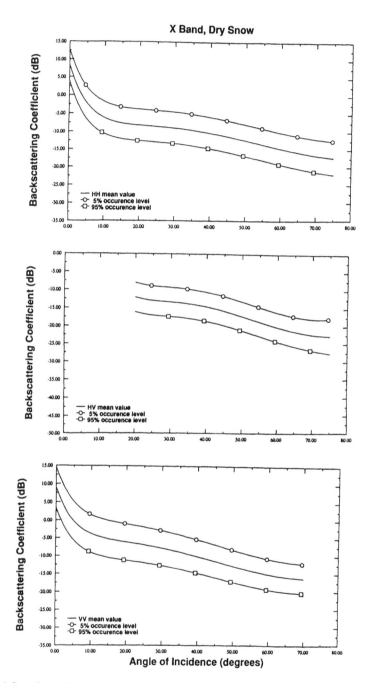

Figure 5.4 Sample angular plots showing mean σ° (from (5.1)) and the 90% occurrence interval (from (5.3) to (5.4)) for dry snow at X-band.

where incidence angle θ is given in radians. The curve-fitting approach effectively weights each angle θ by its constituent sample size and hence reduces the non-monotonic angular behavior cited earlier as caused by target condition biases. Satisfactory convergence of the iterative solution is defined by attainment of residual errors $\ll 1$ dB at most angles of incidence. The data at a given angle of incidence θ are only included in the curve-fitting routine (5.1) if the number of observations N is greater than or equal to 16. Hence, a small sample population ($N < 16$) is judged to be insufficient for adequate determination of mean σ° at a given θ. Furthermore, (5.1) is solved only for those cases wherein there is sufficient data ($N \geq 16$) at four or more angles of incidence. The resultant parameter values (P_1 to P_6) for each frequency band and polarization are given in tabular form for each terrain category in its respective appendix. These tables, which will be referred to as *parameter loadings* tables, also list the angular range ($\Delta\theta$) over which the parameter loadings are valid.

Examples of the curve fitting solutions for mean σ° are shown in Figures 5.5 and 5.6. These figures show the smoothed curves of mean σ° as derived using (5.1), with the parameter values given in tables A.2 and H.2 in Appendices A and H, respectively. The data points shown in Figures 5.5 and 5.6 are the average values of observed σ° as given at each angle of incidence in the SDTs. For the HH-polarized C-band response of bare soil shown in Figure 5.5, the fitted solution is shown to be in very good agreement with the data, with residual errors <1.12 dB. The σ° data analyzed to construct this figure consisted of 6,979 σ° observations derived from several extensive experimental programs. The observed soil conditions cover a wide variety of soil types, surface roughness, and soil moisture with an approximately uniform sampling distribution of target conditions at each angle θ. Hence, both the observed and calculated (from (5.1)) values of mean σ° behave in a monotonic fashion.

Figure 5.6, on the other hand, shows a "worst case" example of the curve fitting procedure, which corresponds to using HH polarized X-band data for dry snow conditions. The source data consists of 983 observations of σ°, but these observations are not uniformly distributed with respect to angle of incidence θ or with respect to target conditions. The majority of the source data are shown by the SDT to be at 0°, 20°, and 50° (and comes from a wide variety of experiments). The more limited sample populations at other angles of incidence come from more limited experimental target conditions and are shown to be the angles with the highest residuals between the observed mean σ° and the smoothed angular response calculated using (5.1). In general, the magnitudes of the residuals are inversely related to sample size.

The 90% occurrence interval is determined in a stochastic manner and assumes that the true target populations yield Gaussian distributions of σ° (dB) at any given sensor configuration when considered over the range of naturally occurring target conditions. The observed standard deviations s reported in the SDTs

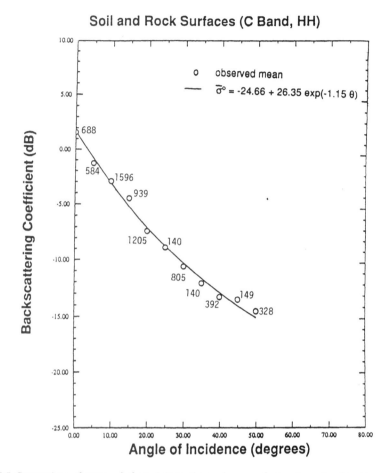

Soil and Rock Surfaces (C Band, HH)

Figure 5.5 Comparison of mean $\sigma°$ (from 5.1) to observed mean $\sigma°$ at each angle of incidence for soil and rock surfaces at C-band with HH polarization. Each mean is annotated with the size of the sample population.

are treated as a sample population with each angular sample weighted by its constituent sample size. A least-squares nonlinear regression approach is used to define s as a function of θ for each terrain category, frequency band, and polarization. The modeled function is:

$$s(\theta) = M_1 + M_2 \exp(-M_3\theta) \quad (dB) \tag{5.2}$$

where θ is in radians. The resultant coefficients (M_1 to M_3) for the "smoothed" standard deviation functions are also given in the parameter loadings tables. In instances where the sample size of the source data is small (i.e., $N < 6$), the M

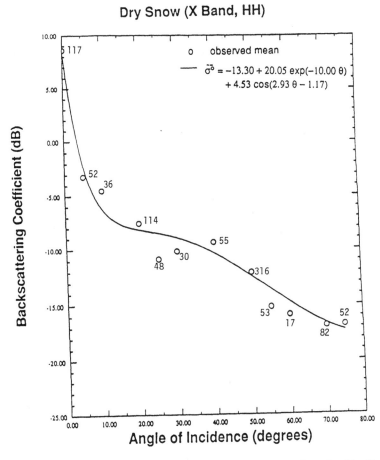

Figure 5.6 Comparison of mean $\sigma°$ (from 5.1) to observed mean $\sigma°$ at each angle of incidence for dry snow at X-band with HH polarization. Each mean $\sigma°$ is annotated with the size of the sample population (from Table 5.4).

parameter values obtained by this method are not statistically significant and are therefore listed as null values.

The 90% occurrence intervals plotted in each terrain category section are based upon the assumption of Gaussian distributions. This assumption leads to

$$\sigma°_{5\%}(dB) = f(\theta) + 1.645\,s(\theta) \tag{5.3}$$

and

$$\sigma°_{95\%}(dB) = f(\theta) - 1.645\,s(\theta) \tag{5.4}$$

The solutions to (5.1), (5.3), and (5.4) are plotted in the angular plots similar to those shown in Figure 5.4.

While the preceding analysis assumed Gaussian distributions of $\sigma°$ dB for computational convenience, it should not be tacitly assumed that all terrain categories truly possess Gaussian distributions of physical and biophysical characteristics.

5-4.3 Histograms

Individual histograms of the $\sigma°$ distributions contained in the data base are given in each terrain category section for each frequency band, angle, and polarization combination at specific angles of incidence. Typically, these are given at $\theta = 20°$, $40°$, and $60°$. The example shown in Figure 5.7 contains histograms for dry snow at X-band for HH, HV, and VV polarizations, all at $\theta = 50°$. Histograms are not presented for cases where the sample population is insufficient in size ($N < 16$). This occurs most commonly for HV polarized data.

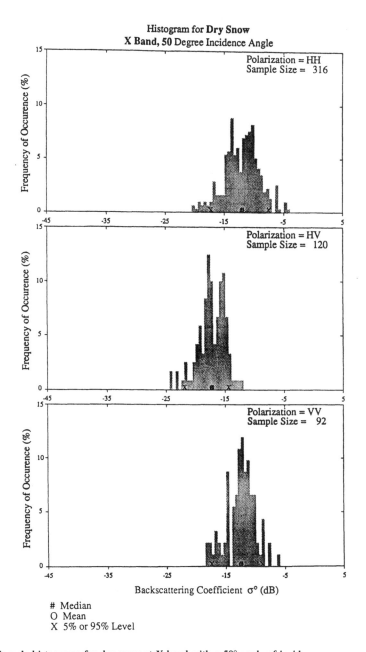

Table 5.7 Sample histograms for dry snow at X-band with a 50° angle of incidence.

Appendix 5.A

Variability in the Measurement of Radar Backscatter

THOMAS F. BUSH, STUDENT MEMBER, IEEE, AND
FAWWAZ T. ULABY, SENIOR MEMBER, IEEE

Abstract—A variety of systems and platforms have been used over the past three decades to acquire radar backscatter data of terrain. The variability in the reported data was evaluated for agricultural crops under "similar" phenological conditions and for approximately the same sensor parameters (frequency, polarization, and angle of incidence). The evaluation reveals wide variations in the magnitude of the scattering coefficient among different measurement programs. While it is difficult to determine the exact causes of these variations it is quite evident that extreme care must be employed in 1) monitoring the measurement system transfer function, 2) calibrating the system on an absolute basis, and 3) acquiring and reporting detailed target parameter information.

Since its conception, radar has been a useful tool, helpful in providing solutions to numerous problems. While one of radar's

Reprinted from IEEE Transactions on Antennas and Propagation, Vol. AP-24, 1976, pp. 896-899.

earliest applications was that of a simple target detector as used by the military, radar has since evolved into a sophisticated tool often used for probing and interrogating remote objects and scenes. Presently radar is being seriously considered as a potential tool for remotely sensing croplands and woodlands from satellite altitudes in hopes of providing a cost effective method for monitoring these renewable resources.

In describing the radar backscatter from area extended targets, the radar backscattering coefficient $\sigma°$ is normally employed. $\sigma°$, in contrast to σ, the radar cross section, does not have units of area, but rather is an average value of the scattering cross section per unit area and therefore is a unitless quantity. The "dimensionless" feature of $\sigma°$ is desirable in that $\sigma°$ should be independent of the radar system used in acquiring backscattering data from an area extended target. Thus the relative size of the area illuminated by the radar should be of no consequence, although to make an accurate estimate of $\sigma°$, the absolute size of the illuminated area must be determined to as high a degree of accuracy as is practical.

In recent decades numerous investigators [1]–[12] have been active in experimental campaigns to quantify the scattering coefficient $\sigma°$ of various types of vegetation. Because of the potential of radar as a remote monitor of croplands, much of the data collected has been from agricultural fields. In all cases either $\sigma°$ or $\gamma°$, where $\gamma° = \sigma°/\cos\theta$ (θ is the angle of incidence as measured from nadir), have been used as backscatter quantifiers. However, in spite of the consistent use of $\sigma°$ and $\gamma°$, there appear to be discrepancies among the data reported by independent researchers.

Fig. 1 presents a cursory view of the ranges of $\sigma°$ of vegetated scenes as reported by nine researchers. All data presented were acquired in the range of incidence angles between 30° and 50°. Furthermore, all data were acquired using X-band radar excepting the Ohio State University and the University of Kansas data in which case both X- and Ku-band data are presented. Attempts were made to present only data collected with a horizontally polarized antenna configuration $\sigma_H°$. When this was not possible,

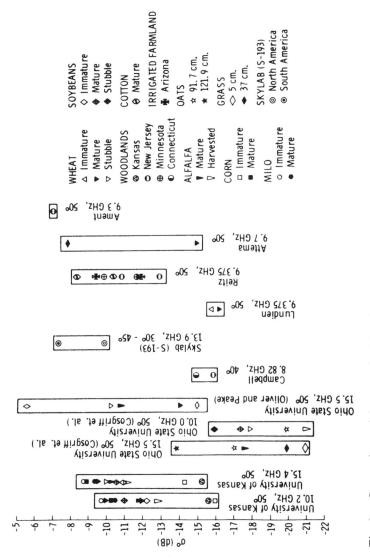

Fig. 1. Comparison of data acquired by nine programs studying radar backscatter from vegetation.

$\sigma_V{}^\circ$ is presented. Table 1 presents a summary of the characteristics of the various systems. In most cases the data presented in Fig. 1 have been acquired from croplands. The notable exceptions are data collected from woodlands and from satellite altitudes (S193) in which case the average radar scattering coefficients of North and South America are plotted. While not all studies involved the same vegetation type, it is possible to find and compare data collected under "similar" target conditions. However, the degree of similarity cannot, in most cases, be determined for lack of complete and quantative ground truth data. In any case, it is felt that the data presented in Fig. 1 are reasonably representative of the vegetation types which have been studied with radar within the last two decades. Moreover no discrimination has been made in terms of the type of system used to acquire data (i.e., ground based, airborne, doppler, pulse, etc.).

Viewing Fig. 1 as a whole, it can be seen that the apparent dynamic range of σ° (as measured within the frequency and angular limits noted) is approximately 17 dB; the highest being data acquired over South America by Skylab and the lowest being harvested alfalfa and grass, as measured by the Ohio State program (Cosgriff *et al.*). When the data generated by each program are viewed on a comparative basis, however, it appears that a certain amount of bias is present in all the data. Consider, for example, the data of the University of Kansas (KU) and Ohio State University (OSU) (Cosgriff *et al.*). It is apparent that the OSU (Cosgriff *et al.*) data are consistently lower than that of KU at both frequencies. Note that there exists very little overlap between the two sets of data, even when the highest OSU (Cosgriff *et al.*) data presented in Fig. 1 are compared to the lowest KU data. Furthermore, a gross comparison of the data collected from similar targets (mature soybeans, wheat stubble, mature and harvested alfalfa) shows that the OSU (Cosgriff *et al.*) data averaged about 7 dB lower than the KU data. This is not true, however, when a comparison is made between the KU data and OSU data reported by Oliver and Peake [6] in which case a general agreement is noted. Similar comparisons

TABLE I
Summary of System Characteristics

Reference	Affiliation	Platform	Frequencies	Ranging Method
1-4	University of Kansas	Truck	1-18 GHz **	FM-CW
5,6	Ohio State University	Truck	10.0 GHz* 15.5 GHz* 33.0 GHz	Doppler
7	General Precision Laboratory	Airborne	8.82 GHz*	Pulse
8	NASA	Spaceborne (Skylab S193)	13.9 GHz*	Pulse
9	U.S. Army Engineers	Arch Mounted	0.297 GHz 5.870 GHz 9.375 GHz* 34.543 GHz	Pulse
10	Goodyear	Airborne	9.375 GHz*	Pulse
11	NRL	Airborne	0.425 GHz 1.250 GHz 3.300 GHz 9.300 GHz*	Pulse
12	Delft University of Technology and NIWARS	Ground Based	9.7 GHz*	FM-CW

**10.2 and 15.4 GHz data are incorporated herein.   *Frequencies of data incorporated herein.

can be made among the other data presented. In terms of absolute level, for example, the KU data and Goodyear data (Reitz) agree rather well. Furthermore the data acquired by Skylab S193 over North America agrees with the Goodyear and KU values. Certainly the S193 data represents a gross ensemble average over the North American continent but it does represent a means of comparison with other data. Overall, the OSU (Cosgriff *et al.*) data appear to be significantly lower in magnitude than all the other data shown.

The sources of the apparent bias noted between the various data sets can probably be divided into three categories: a) a lack of adequate ground truth data, b) a lack of accurate system calibration, and c) a lack of accurate data processing. The lack of adequate ground truth is perhaps the easiest source of bias to remove although this certainly does not diminish its importance. Only very gross comparisons can be made between data collected from two independent fields of "mature corn" for example. The term "mature" should be quantified as much as possible as should the remaining target descriptors such as soil surface roughness, soil moisture, plant moisture, plant height, the presence of crop stress, etc.

In terms of system calibration a number of factors must be included. Perhaps the most important is the measurement of targets of known radar cross section to provide the absolute system calibration necessary for the comparison of data acquired using independent systems. Another factor to be considered is the implementation of a method for monitoring the inevitable drifting in the overall transfer function of the radar. If such system monitoring is employed then the effects of transfer function drift can be markedly reduced. To be able to accurately estimate the illuminated target area it is necessary to measure the antenna patterns and to calibrate the ranging capabilities (if employed) of the radar system. Finally the algorithm for the processing of the data must be carefully designed and implemented, particularly that portion of the algorithm which estimates the area of illumination which can be the source of considerable error.

REFERENCES

[1] F. T. Ulaby and T. F. Bush, "Corn growth as monitored by radar," University of Kansas Center for Research, Inc., RSL Technical Report 177-57, Lawrence, November 1975.

[2] T. F. Bush and F. T. Ulaby, "Radar return from a continuous vegetation canopy," *IEEE Trans. Antennas and Propagation*, 1976.

[3] F. T. Ulaby and T. F. Bush, "Monitoring wheat growth with radar," *Photogrammetric Engineering and Remote Sensing*, 1976.

[4] T. F. Bush, F. T. Ulaby, and T. Metzler, "Radar backscatter properties of milo and soybeans," University of Kansas Center for Research, Inc., RSL Technical Memorandum 177-52, Lawrence, October 1975.

[5] R. L. Cosgriff, W. H. Peake, and R. C. Taylor, "Terrain scattering properties of sensor system design," *Terrain Handbook III*, Engr. Expt. Sta., Ohio State University Bull. 181, May 1960.

[6] T. L. Oliver and W. H. Peake, "Radar backscattering data for agricultural surfaces," Ohio State University, ElectroScience Lab., Tech. Rept. 1903-9, 1969.

[7] J. P. Campbell, "Backscattering characteristics of land and sea at *X*-band," *Proc. Natl. Conf. Aeron. Electron.*, May 1958.

[8] R. K. Moore, F. T. Ulaby, A. Sobti, S. T. Ulaby, E. C. Davison, and S. Siriburi, "Design data collection with Skylab microwave radiometer-scatterometer S-193, Final Report," University of Kansas Center for Research, Inc., RSL Technical Report 243-12, Lawrence, September 1975.

[9] J. R. Lundien, "Terrain analysis by electromagnetic means," Technical Report No. 3-693, Report 2, U.S. Army Engineer Waterways Experiment Station, Vicksburg, Mississippi, 55 pp., 1966.

[10] E. A. Reitz, "Radar terrain return study, final report: Measurements of terrain backscattering coefficients with an airborne *X*-band radar," *Goodyear Aerospace Corp. Rept. GERA-463*, 1959.

[11] W. Ament, F. MacDonald, and R. Shewbridge, "Radar terrain reflections for several polarization and frequencies," *Trans. 1959 Symposium on Radar Return*, Pt. 2, May 11-12, 1959, University of New Mexico N.O.T.S. TP 2339, U.S. Naval Ordnance Test Station, China Lake, California.

[12] E. P. W. Attema and J. van Kuilenburg, "Short range vegetation scatterometry," *Proc. URSI Specialist Meeting on Microwave Scattering and Emission from the Earth*, Berne, Switzerland, September 1974.

Additional Comments by W. H. Peake

The points made by Bush and Ulaby are well taken, and should be borne in mind by all those who must use, or make terrain measurements.

As Bush and Ulaby have pointed out, there are a number of factors which contribute to the variability in the reported data, the most important being the variability in the surface itself, and in the inability of the experimenter to adequately specify or describe the surface. The kind of "ground truth" given in [1], [4], or [6] indicates the currently acceptable level of detail for measurements on vegetated surfaces. The simple surface description used in the older literature can no longer be considered adequate, if the measurements are to have any value at all.

A second difficulty lies in the fact that the parameter σ_0 is not measurable in itself, but must be derived from an observed power or power ratio. The conversion always involves at least a convolution over range and antenna pattern, and may involve another convolution over pulse shape or doppler filter response. For ground based radars mounted on towers or booms monitoring fairly tall vegetation, a further problem is that the range to the "surface" may be ill defined and the deconvolution process ("estimating the area of illumination") can produce systematic errors.[1] Furthermore, if the diameter of the illuminated area is not several times the height of the vegetation, multiply scattered radiation may not be properly collected. The consequence, in operational terms, is that the derived parameter σ_0, which should be independent of range and antenna pattern, may fail to be so if the experimental geometry is inappropriate.

Of course, the problem of data intercomparison would be greatly simplified (especially for imaging radars with absolute gray scale calibration) if there existed a stable, preferably calculable, standard surface as a reference. At one time [15], [16] an attempt was made to provide at least a relative standard calibration surface, but that particular test range does not seem to have been used for the purpose in recent years, and no other suitable surface appears to be available.

These general difficulties in data comparison led the author of a recent monograph [13], in discussing sea clutter, to suggest that "the wide variability of the data makes analysis difficult—it is virtually hopeless to obtain meaningful results by comparing data at one wavelength from one observer with those at another wavelength from another observer." In our own report [14] we commented along the same lines, that "in general, it may be assumed that the calibration procedures of any one investigator will be consistent, but there may be systematic differences between the absolute levels measured with different experimental configurations." The examples exhibited by Bush and Ulaby amply justify these cautionary statements as they apply to data on vegetated surfaces.

As to the specific problem of the absolute levels of the data reported in [5], the following comments may be of interest, particularly since the entire monograph has recently been reprinted by Barton [17], and much of the data has been utilized by Long [13]. There have been three major series of CW terrain return measurements made at the ElectroScience Laboratory (formerly the Antenna Laboratory). The first [5], [18] used a corner reflector on a rotating arm as a calibration standard. The second [19], [20], a set of bistatic measurements, used a stationary sphere. The third set [6], [21] used an oscillating sphere for direct polarized calibration, and a dipole array rotating in its own plane for the cross-polarized calibration. All were made at short range (~ 20 ft). It has been clear for some time that "the absolute levels of our earlier data [5], [18] are consistently lower than our more recent data [6], [21] for similar surfaces. This difference may be due to a systematic error in the earlier calibration technique" [14]. It is hardly possible, after 15 years, to reconstruct the earlier system, (though a finger of suspicion has been pointed [22] at the dynamic response of the pen recorders used to monitor the transient calibration signal). Thus it can never really be determined whether the actual systematic error is 7 dB, as Bush and Ulaby suggest. My own estimate, (based on a comparison of data from ostensibly similar asphalt surfaces) is that the systematic error is 4 or 5 dB, i.e., all the curves in [5] and [18], (and these are reproduced unchanged by Barton [17], Long [13], and Peake and Oliver [14]) should be raised uniformly by this amount.

On the other hand, a study of Figs. 5A-1 to 5A-20, 5B-1 to 5B-6 of [5], and of Figs. 2–4 of [18] shows that for nonvegetated surfaces, from smooth concrete and sand to large limestone rubble, the dependence of σ_0 on surface roughness, frequency, polarization, and look angle is entirely consistent with theoretical expectations. These measurements, which were actually made by R. C. Taylor, were carried out with great care and attention to detail, and still justify, in my opinion, his estimate that the internal self consistency, or precision of the measurements is about ± 1 dB, even though the absolute levels are a few decibels too low. Thus the many parametric dependencies and theoretical interpretations based on Taylor's measurements [5], [18]should remain unaffected by the uniform systematic error.

ADDITIONAL REFERENCES

[13] M. W. Long, *Radar Reflectivity of Land and Sea.* Lexington Books, D. C. Heath and Co., 1975.

[14] W. H. Peake and T. L. Oliver, "The response of terrestrial surfaces at microwave frequencies," Report 2440-7, The Ohio State University ElectroScience Laboratory, May 1971.

[15] G. J. Shepard, "Radar geology test area, Wilcox Playa, Arizona," U.S. Army Corps of Engineers Report, January 1966.

[16] W. D. Meyer, "Analysis of radar calibration data. Supplement," Goodyear Aerospace Corp., May 1968 (AD 836 934).

[17] D. K. Barton (Ed.), *Radars, Vol. 5 Radar Clutter.* Dedham, Mass.: Aertech House Inc., 1975.

[18] W. H. Peake and R. C. Taylor, "Radar back-scattering from moonlike surfaces," Report 1388-9, Antenna Laboratory, The Ohio State University, May 1963.

[19] S. T. Cost, "Measurements of the bistatic echo area of terrain at X-Band," Report 1822-2, Antenna Laboratory, The Ohio State University, May 1965.

[20] W. H. Peake and S. T. Cost, "The bistatic echo area of terrain at 10 GHz," IEEE 1968 Wescon Technical Paper—Session 22/2 Terrain Radar Scatter-Experimental and Theoretical.

[21] C. H. Schultz, T. L. Oliver, and W. H. Peake, "Radar backscattering data for surfaces of geological interest," Report 1903-7, ElectroScience Laboratory, The Ohio State University, December 1969.

[22] R. C. Taylor, private communication.

Manuscript received April 8, 1976.

The author is with the ElectroScience Laboratory, The Ohio State University, Columbus, OH 43212.

[1] The best known example of a systematic error caused by an incorrect deconvolution is the "land-sea" bias effect in doppler navigation systems.

REFERENCES

1. Aitchison, J. and J. A. C. Brown, *The Lognormal Distribution,* New York: Cambridge University Press, 1973.

2. Allen, C. T. and F. T. Ulaby, "Characterization of the Microwave Extinction Properties of Vegetation Canopies," Radiation Laboratory Report 022132-1-T, The University of Michigan, Ann Arbor, November 1984.

3. Allen, C. T., B. Brisco, and F. T. Ulaby, "Modeling the Temporal Behavior of the Microwave Backscattering Coefficient of Agricultural Crops," RSL Technical Report 360-21, Remote Sensing Laboratory, Univ. of Kansas Center for Research, Inc., Lawrence, KS, 1984.

4. Ament, W., F. MacDonald, and R. Shewbridge, "Radar Terrain Reflections for Several Polarizations and Frequencies," *Trans. Symp. on Radar Return,* Part 2, University of New Mexico, NOTS TP 2339, U. S. Naval Ordnance Test Station, China Lake, CA, May 11–12, 1959.

5. Anderson, J. R., E. E. Hardy, J. T. Roach, and R. E. Witmer, "A Land Use and Land Cover Classification System for Use with Remote Sensor Data," U.S. Geological Survey Professional Paper 964, 1976.

6. Aslam, A., F. T. Ulaby, B. Jung, M. Hemmat, M. C. Dobson, D. R. Brunfeldt, and J. Paris, "Multi-Channel MARS C/X-Band Scatterometer System, Volume 2: System Specification and Experimental Data," Remote Sensing Laboratory Technical Report 580-1, University of Kansas Center for Research, Inc., Lawrence, KS, April 1983.

7. Attema, E. P. W., and F. T. Ulaby, "Vegetation Modeled as a Water Cloud," *Radio Science,* Vol. 13, 1978, pp. 357–364.

8. Baars, E. P., and H. Essen, "Millimeter-Wave Backscatter Measurements on Snow-Covered Terrain," *IEEE Trans. Geosci. Remote Sensing,* Vol. 26, No. 3, May 1988, pp. 282–299.

9. Barton, D. K., ed., "Radar Clutter," *Radars,* Vol. 5, Dedham, MA: Artech House, 1975.

10. Batlivala, P. P., and J. Cihlar, "Joint Soil Moisture Experiment (TEXAS): Documentation of Radar Backscatter and Ground Truth Data," Remote Sensing Laboratory Technical Report 264-1, University of Kansas Center for Research, Inc., Lawrence KS, April 1975.

11. Batlivala, P. P., and M. C. Dobson, "Soil Moisture Experiment (KANSAS): Documentation of Radar Backscatter and Ground Truth Data," Remote Sensing Laboratory Technical Report 264-7, University of Kansas Center for Research, Inc., Lawrence, KS, March 1976.

12. Batlivala, P. P., and F. T. Ulaby, "Radar Look Direction and Row Crops," *Photogram. Eng. Remote Sensing,* Vol. 42, 1976, pp. 233–238.

13. Beckmann, P., *Probability in Communication Engineering,* New York, New York: Harcourt, Brace, and World, 1967, p. 122.

14. Berger, R., R. Layman, T. Van Zandt, J. Walsh, and J. Knox, "Observations of the Backscatter from Snow at Millimeter Wavelengths," *Snow Symposium V,* Hanover, NH, August 1985, Vol. 1, U.S. Army Cold Regions Research and Engineering Laboratory, Hanover, NH, CREL Special Report 86-15, pp. 311–316.

15. Bernard, R. and D. Vidal-Madjar, "ERSME: Diffusiometre Heliportable en Bande c. Application a la Mesure de l'Humidite des Sols," Presente au *Symposium EARSEL/ESA,* Remote Sensing Applications for Environmental Studies, 26–29 April 1983, Brussels.

16. Bernard, R., O. Taconet, D. Vidal-Madjar, J. L. Thony, M. Vauclin, A. Chapoton, F. Wattrelot, and A. Lebrun, "Comparison of Three In Site Surface Soil Moisture Measurements and Applications to C-Band Scatterometer Calibration," *IEEE Trans. Geosci. Remote Sensing,* Vol. GE-22, No. 4, July 1984, pp. 388–391.

17. Bernard, R., J. V. Soares, and D. Vidal-Madjar, "Differential Bare Field Drainage Properties from Airborne Microwave Observations," *Water Resources Research,* Vol. 22, No. 6, 1986, pp. 869–875.

18. Bernard, R., P. Lancelin, G. Laurent, "Radar Observation of the Guyana Rain Forest," Centre de Recherches en Physique de Environment Terrestre et Planetaire Campaign Report, CRPE/156, 1987.

19. Bernard, R., Ph. Martin, J. L. Thony, M. Vauclin, and D. Vidal-Madjar, "C-Band Radar for Determining Surface Soil Moisture," *Remote Sensing Environ.,* Vol. 12, 1982, pp. 189–200.

20. Birkemeier, W. P. and N. D. Wallace, "Radar Tracking Accuracy Improvement by Means of Pulse-to-Pulse Frequency Modulation," *IEEE Trans. on Communication and Electronics,* January 1963, pp. 571–575.

21. Birrer, I. J., E. M. Bracalente, G. J. Dome, J. Sweet, and G. Berhold, "$\sigma°$ Signature of the Amazon Rain Forest Obtained from the Seasat Scatterometer," *IEEE Trans. Geosci. Remote Sensing,* Vol. GE-20, 1982, pp. 11–17.

22. Blanchard, A. J., and A. T. C. Chang, "Estimation of Soil Moisture from Seasat SAR Data," *Water Resources Bull.,* No. 19, 1983, pp. 803–810.

23. Bohren, C. F., and D. R. Huffman, *Absorption and Scattering of Light by Small Particles,* New York, NY: John Wiley & Sons, 1983.

24. Boothe, R. R., "The Weibull Distribution Applied to the Ground Clutter Backscatter Coefficient," in *Automatic Detection and Radar Data Processing,* D. C. Schleher, Ed. Dedham, MA: Artech House, 1980.

25. Bradley, G. A. and P. F. Gabel, "Calibration of the NASA/JSC C-130 Aircraft 13.3 GHz and 1.6 GHz Scatterometers with the MAS 1-18 GHz," RSL Technical Report 264-10, University of Kansas Center for Research, Inc., Lawrence, KS, June 1978.

26. Bradley, G. A., "1978 Agriculture Soil Moisture Experiment (ASME) Data Documentation," Remote Sensing Laboratory Technical Report 460-3, University of Kansas Center for Research, Inc., Lawrence, KS, October 1980.

27. Bradley, G. A. and F. T. Ulaby, "A Comparison of Ground and Airborne Remote Sensing Radar Measurements," Remote Sensing Laboratory Technical Report 460-1, University of Kansas Center for Research, Inc., Lawrence, KS, October 1980.

28. Bradley, G. A., and F. T. Ulaby, "Aircraft Radar Response to Soil Moisture," *Remote Sensing Environ.,* No. 11, 1981, pp. 419–438.

29. Brunfeldt, D. R., and F. T. Ulaby, "Measured Microwave Emission and Scattering in Vegetation Canopies," *IEEE Trans. Geosci. Remote Sensing,* Vol. GE-22, 1984, pp. 520–524.

30. Bush, T. F. and F. T. Ulaby, "Remotely Sensed Wheat Maturation with Radar," Remote Sensing Laboratory Technical Report 177-55, University of Kansas Center for Research, Inc., Lawrence, KS, May 1975.

31. Bush, T. F., and F. T. Ulaby, "Fading Characteristics of Panchromatic Radar Backscatter from Selected Agricultural Targets," *IEEE Trans. Geosci. Electron.,* Vol. GE-13, No. 4, October 1975, pp. 149–157.

32. Bush, T. F. and F. T. Ulaby, "Variability in the Measurement of Radar Backscatter," *IEEE Trans. Antennas Propag.,* Vol. AP-24, 1976, pp. 896–899.

33. Bush, T. F., F. T. Ulaby, T. Metzler, and H. Stiles, "Seasonal Variations of the Microwave Scattering Properties of Deciduous Trees as Measured in the 1-18 GHz Spectral Range," Remote Sensing Laboratory Technical Report 177-60, University of Kansas Center for Research, Inc., Lawrence, KS, June, 1976.

34. Bush, T. F. and F. T. Ulaby, "Radar Backscatter Properties of Milo and Soybeans," Remote Sensing Laboratory Technical Report 177-59, University of Kansas Center for Research, Inc., Lawrence, KS, May 1975.

35. Chi, C. Y., D. G. Long, and F. K. Li, "Radar Backscatter Measurement Accuracies Using Digital Doppler Processors in Spaceborne Scatterometers," *IEEE Trans. on Geosci. Remote Sensing,* Vol. GE-24, 1986, pp. 426–437.

36. Cimino, J., M. C. Dobson, D. Gates, E. Kasischke, R. Lang, J. Norman, J. Paris, F. T. Ulaby, S. Ustin, V. Vanderbilt, and J. Weber, "Eos Synergism Study Data Report," JPL Technical Report, May 1988.

37. Cosgriff, R. L., W. H. Peake, and R. C. Taylor, "Terrain Scattering Properties for Sensor System Design" (Terrain Handbook II), Engineering Experiment Station Bulletin 181, Ohio State University, 1960.

38. Currie, N. C., F. B. Dyer, and G. W. Ewell, "Characteristics of Snow at Millimeter Wavelengths," *Proceedings of the IEEE APS 1976 International Symposium,* 1976, pp. 579–582.

39. Currie, N. C., J. R. Teal, Jr., F. B. Dyer, and G. W. Ewell, "Radar Millimeter Backscatter Measurements from Snow," Georgia Institute of Technology Final Report AFATL-TR-77-4, Air Force Armament Laboratory (DLMT), Eglin Air Force Base, FL, March–April 1976.

40. Currie, N. C., D. J. Lewinski, M. S. Applegate, R. D. Hays, "Radar Millimeter Backscatter Measurements—Volume 1: Snow and Vegetation," Georgia Institute of Technology Final Report AFATL-TR-77-92, Air Force Armament Laboratory (DLMT), Eglin Air Force Base, FL, July 1977.

41. Currie, N. C., J. D. Echard, M. J. Gary, A. H. Green, T. L. Lane, and J. M. Trostel, "Millimeter-Wave Measurements and Analysis of Snow-Covered Ground," *IEEE Trans. on Geosci. Remote Sensing,* Vol. 26, No. 3, May 1988, pp. 307–318.

42. Cutrona, L. J., E. N. Leith, C. J. Palermo, and L. J. Porcello, "Optical Data Processing and Filtering Systems," *IEEE Trans. Inform. Theory,* Vol. II-6, pp. 386–400, 1980.

43. Daley, J. C., W. T. Davis, J. R. Duncan, and M. B. Laing, "NRL Terrain Clutter Study, Phase II," Naval Research Laboratory Report 6749, October 21, 1968.

44. de Loor, G. P., A. A. Jurriens, and H. Gravesteijn, "The Radar Backscatter from Selected Agricultural Crops," *IEEE Trans. Geosci. Electron.,* Vol. GE-25, 1974, pp. 70–77.

45. de Loor, G. P., P. Hoogeboom, and E. P. W. Attema, "The Dutch ROVE Program," *IEEE Trans. Geosci. Remote Sensing,* Vol. GE-20, 1982, pp. 3–11.

46. Dobson, M. C., H. Stiles, D. Brunfeldt, T. Metzler, and S. McMeekin, "Data Documentation: 1975 MAS 1-8 and MAS 8-18 Vegetation Experiments," Remote Sensing Laboratory Technical Report 264-15, University of Kansas Center for Research, Inc., Lawrence, KS, December 1977.

47. Dobson, M. C., "1977 Bare Soil Moisture Experiment Data Documentation," Remote Sensing Laboratory Technical Report 264-26, University of Kansas Center for Research, Inc., Lawrence, KS, May 1979.

48. Dobson, M. C., D. Brunfeldt, G. Burns, S. Ulaby, and S. McMeekin, "Data Documentation: MAS 1-8 and MAS 8-18 1976 Vegetation Experiments," RSL Technical Report 264-16, University of Kansas Center for Research, Inc., Lawrence, Kansas, May 1978.

49. Dobson, M. C., and F. T. Ulaby, "Microwave Backscatter Dependence on Surface Roughness, Soil Moisture, and Soil Texture: Part III—Soil Tension," *IEEE Trans. Geosci. Remote Sensing,* Vol. GE-19, 1981, pp. 51–61.

50. Eger, G. W., F. T. Ulaby, and E. T. Kanemasu, "A Three-Part Geometric Model to Predict the Radar Backscattering from Wheat, Corn and Sorghum," Remote Sensing Laboratory Technical Report 360-18, University of Kansas Center for Research, Inc., Lawrence, KS, April 1982.

51. Ericson, L. O., "Terrain Return Measurements with an Airborne X-Band Radar Station," 6th Conference of the Swedish National Committee on Scientific Radio, March 13, 1963.

52. Everett, M. E., *Airborne Millimetre Snow Measurements,* British Aerospace, Preston, Lancashire, Report No. ATG 98, March 1983.

53. Floyd, W. C., "Scatterometer Data Analysis Program, Final Report," Report No. 57667-2, September 21, 1967, Ryan Aeronautical Co., San Diego, CA, N68-23721.

54. Fung, A. K., W. H. Stiles, F. T. Ulaby, "Surface Effects on the Microwave Backscatter and Emission of Snow," *Proceedings of the International Communications Conference,* Seattle, Washington, June 1980, pp. 49.6.1–49.6.7.

55. Gallagher, J. G., T. H. Saunders, A. E. Dick, and L. H. Bartlett, "Millimetric Radar Backscatter Trials: II. Results at 93 GHz," presented at *Snow Symposium VI,* Hanover, NH, August 1986.

56. Goodyear Aircraft Corporation, "Final Report: Measurements of Terrain Backscattering Coefficients with an Airborne X-Band Radar," GERA-463, September 30, 1959, AD-229104.

57. Green, A. H., *Millimeter Radar Reflectivity Measurements of Snow Covered Terrains,* U.S. Army Missile Command, Redstone Arsenal, Alabama, Technical Report RE-84-9, 1984.

58. Guinard, N. W. and J. C. Daley, "An Experimental Study of a Sea Clutter Model," *Proceedings of the IEEE,* vol. 58, pp. 543–50, April 1970.

59. Hayes, R. D., J. R. Walsh, D. F. Eagle, H. A. Ecker, M. W. Long, J. G. B. Rivers, and C. W. Stuckey, "Study of Polarization Characteristics of Radar Targets," Engineering Experiment Station, Georgia Institute of Technology, Final Report, Contract DA-36-039-sc-64713, October 1958.

60. Hayes, D. T., U. H. W. Lammers, and R. A. Marr, "Scattering from Snow Backgrounds at 35, 98, and 140 GHz," Rome Air Development Center Technical Report 84-69, Hanscom Air Force Base, MA, April 1984.

61. Hirosawa, H., S. Komiyama, and Y. Matsuzaka, "Cross-Polarized Radar Backscatter from Moist Soil," *Remote Sensing Environ.,* Vol. 7, 1978, pp. 211–217.

62. Hirosawa, H. and Y. Matsuzaka, "Measurement of Scattering Coefficient $\sigma°$ of Trees," *International Journal of Remote Sensing,* Vol. 8, No. 4, April 1987, pp. 609–620.

63. Hirosawa, H., Y. Matsuzaka, M. Daito, and N. Nakamura, "Measurement of Backscatter from Conifers in the C and X Bands," *International Journal of Remote Sensing,* Vol. 8, No. 11, November 1987, pp. 1687–1694.

64. Hoekman, D. H., E. Attema, and L. Krul, "A Multilayer Model for Radar Backscattering by Vegetation Canopies," Vol II, *1982 IEEE Int. Geosci. Remote Sensing Symp.* (IGARSS'82) Digest, Munich, FRG, June 1–4, 1982. (IEEE Cat. No. 82CH14723-6).

65. Hoekman, D. H., "Radar Backscattering of Forest Parcels," *Int. Symp. Microwave Signatures in Remote Sensing,* Toulouse, France, 16–20 January 1984.

66. Hoekman, D. H., "Experiments of Modelling Radar Backscatter of Forest Stands and Research on Classification," *3rd International Colloquium Spectral Signatures of Objects in Remote Sensing,* Les Arcs, France, December 16–20, 1985.

67. Hoekman, D. H., "Multiband Scatterometer Data Analysis of Forests," *International Journal of Remote Sensing,* Vol. 8, No. 11, November 1987, pp. 1695–1708.

68. Huppi, R. A., RASAM: A Radiometer-Scatterometer To Measure Microwave Signatures of Soil, Vegetation and Snow," Ph.D. Dissertation, University of Bern (Bern, Switzerland), December 1987.

69. Ivey, H. D., M. W. Long, and V. R. Widerquist, "Some Polarization Properties of K_a- and X_b-Band Echoes from Vehicles and Trees," *Record of the First Annual Radar Symposium,* University of Michigan, 1955.

70. Jackson, T. J., and T. J. Schmugge, "Aircraft Active Microwave Measurements for Estimating Soil Moisture," *Photogram. Eng. Remote Sensing,* Vol. 47, 1981, pp. 801–805.

71. Jackson, T. J., P. E. O'Neil, T. C. Coleman, and T. J. Schmugge, "Aircraft Remote Sensing of Soil Moisture and Hydrologic Parameters, Chickasha, Oklahoma Data Report," 1982, USDA Agricultural Research Results No. 14, Beltsville, MD, 16 pages.

72. Jao, J. K., "Amplitude Distribution of Composite Terrain Radar Clutter and the K-Distribution, *IEEE Trans. Antennas and Prop.*, Vol. AP-32, 1984, pp. 1049–1062.

73. Johnson, J. W., L. A. Williams, Jr., E. M. Bracalante, F. B. Beck, and W. L. Grantham, "Seasat-A Satellite Scatterometer Instrument Evaluation," *IEEE J. Oceanic Engineering,* Vol. OE-5, pp. 138–144, 1980.

74. Kashihara, H., K. Nakada, M. Murata, M. Hiroguchi, and H. Aiba, "A Study of Amplitude Distribution of Space-Borne Synthetic Aperture Radar (SAR) Data," *Proceedings of ISNCR-84 Symposium,* 22–24 October, 1984, Tokyo.

75. Kerr, D. E., *Propagation of Short Radar Waves,* New York: McGraw-Hill, 1951, p. 554.

76. Khamsi, H. R., A. K. Fung, and F. T. Ulaby, "Rough Surface Scattering Based on Facet Model," Remote Sensing Laboratory Technical Report 177-52, University of Kansas Center for Research, Inc. Lawrence, KS, November 1974.

77. King, C., "Agricultural Terrain Scatterometer Observations with Emphasis on Soil Moisture Variations," CRES Technical Report 177-44, University of Kansas Center for Research, Inc., Lawrence, KS, August 1973.

78. Kobayashi, T. and H. Hirosawa, "Measurement of Radar Backscatter from Rough Soil Surfaces Using Linear and Circular Polarization," *International Journal of Remote Sensing,* Vol. 6, No. 2, 1985, pp. 345–352.

79. Koolen, A. J., F. F. R. Koenigs, and W. Bouten, "Remote Sensing of Surface Roughness and Top Soil Moisture of Bare Tilled Soil with an X-Band Radar," *Neth. Agric. Sci.,* Vol. 27, 1979.

80. Kouyate, F., C. Allen, M. C. Dobson, and F. T. Ulaby, "Radar Backscatter Measurements of Periodic Soil Surfaces: Data Documentation," Remote Sensing Laboratory Technical Report 460-15, University of Kansas Center for Research, Inc., Lawrence, KS, August 1982.

81. Le Toan, T. and M. Pausader, "Active Microwave Signatures of Soil and Vegetation-Covered Surfaces: Results of Measurement Programs," *Proc. ISP Int. Colloq. Spectral Signatures of Objects in Remote Sensing,* Avignon, France, September 8–11, 1981, pp. 303–314.

82. Le Toan, T., "Active Microwave Signatures of Soil and Crops: Significant Results of Three Years of Experiments," *1982 IEEE Int. Geosci. Remote Sensing Symp.* (IGARSS'82) Digest, Vol. 1, Munich, FRG, 1–4 June 1982. (IEEE Cat. No. 82CH14723-6).

83. LeToan, T., A. Lopes, and A. Malavaud, "Relationships Between Radar Backscattering and the Characteristics of a Crop Canopy: Considerations on the Effect of Structure," *Proc. ISP Int. Colloq. Spectral Signatures of Objects in Remote Sensing,* Bordeaux, France, September 1983.

84. Le Toan, T., G. Flouzat, M. Pausader, A. Fluhr, and A. Lopes, "Soil Moisture Content and Microwave Backscattering in the 1-9 GHz Region," *International Geoscience and Remote Sensing Symposium,* Washington, DC, 1981.

85. Lewinski, D. J., "Nonstationary Probabilistic Target and Clutter Scattering Models," *IEEE Trans. on Antennas and Prop.,* Vol. AP-31, No. 3, May 1983, pp. 490–498.

86. Li, F. K., C. Croft, and D. N. Held, "Comparison of Several Techniques to Obtain Multiple-Look SAR Imagery," *IEEE Trans. Geosci. and Remote Sensing,* Vol. GE-21, pp. 370–375, 1973.

87. Long, M. W., *Radar Reflectivity of Land and Sea,* Second Edition, Dedham, MA: Artech House, 1983.

88. Lopes, A., M. Huet, and T. Le Toan, "Experience Micro-Ondes Sur Un Couvert de Ble: Premiers Resultats," *Spectral Signature of Objects in Remote Sensing International Colloquium,* Avignon, France, September 8–11, 1981.

89.   Luebbert, S., M. Hemmat, and M. C. Dobson, "Monitoring Crop Development with Radar," Remote Sensing Laboratory Technical Report 605-2, University of Kansas Center for Research, Inc., Lawrence, KS, March 1984.

90.   Luebbert, S., M. Hemmat, and M. C. Dobson, "Absolute Calibration of the Canadian SAR-580," Remote Sensing Laboratory Technical Report 605-3, University of Kansas Center for Research, Inc., Lawrence, KS, March 1984.

91.   Lundien, J. R., "Terrain Analysis by Electromagnetic Means: Radar Responses to Laboratory Prepared Soil Samples," Technical Report No. 3-693, U.S. Army Engineer Waterways Experiment Station, Corps. of Engineers, Vicksburg, Mississippi.

92.   MacDonald, F. C., "The Correlation of Radar Sea Clutter on Vertical and Horizontal Polarizations with Wave Height and Slope," *IRE Convention Record*, Part 1, pp. 29–32, 1956.

93.   Maetzler, C. and E. Schanda, "Snow Mapping with Active Microwave Sensors," *Int. J. Remote Sensing*, Vol. 5, No. 2, 1984, pp. 409–422.

94.   Mougin, E. T. Le Toan, A. Lopes, P. Borderies, and A. Sarramejan, "Backscattering Measurements at X-band on Young Coniferous Trees," *IEEE International Geoscience and Remote Sensing Symposium (IGARSS '87) Digest*, Vol. I, The University of Michigan, Ann Arbor, Michigan, May 18-21, 1987, pp. 287–292.

95.   Narayanan, R. M., C. C. Borel, and R. E. McIntosh, "Radar Backscatter Characteristics of Trees at 215 GHz," *IEEE Trans. Geosci. Remote Sensing*, Vol. 26, No. 3, May 1988, pp. 217–228.

96.   Oliver, T. L., and W. H. Peake, "Radar Backscattering Data for Agricultural Surfaces at Microwave Frequencies," The Ohio State University, ElectroScience Laboratory Technical Report 1903-9, February 13, 1969.

97.   Paris, J. F., "Radar Backscattering Properties of Corn and Soybeans at Frequencies of 1.6, 4.75, and 13.3 GHz," *IEEE Trans. Geosci. Remote Sensing*, Vol. GE-21, 1983, pp. 392–400.

98.   Peake, W. H., and R. C. Taylor, "Radar Back-Scattering Measurements from Moon-Like Surfaces," Report 1388-9, ElectroScience Laboratory, Department of Electrical Engineering, Ohio State University, Columbus, OH, May 1, 1963.

99.   Peake, W. H. and T. L. Oliver, "The Response of Terrestrial Surfaces at Microwave Frequencies," Air Force Systems Command, Air Force Avionics Laboratory Technical Report AFAL-TR-70-301 (AD 884 106), May 1971.

100.   Porcello, L. J., N. G. Massey, R. B. Innes, and J. M. Marks, "Speckle Reduction in Synthetic Aperture Radar," *J. Opt. Soc. Amer.*, Vol. 66, 1976, pp. 1305–1311.

101.   Prevot, L., R. Bernard, O. Taconet, and D. Vidal-Madjar, "Evaporation from a Bare Soil Evaluated from a Soil Water Transfer Model Using Remotely Sensed Soil Moisture Data," *Water Resources Research*, Vol. 20 No. 2, February 1984, pp. 311–316.

102.   Rice, S. O., "Mathematical Analysis of Random Noise," *Bell System Technical Journal*, Vol. 23, 1944, p. 282.

103.   Ridenour, L. N., *Radar System Engineering*, MIT Radiation Laboratory Series, Vol. I, New York, New York: McGraw-Hill, 1947, Fig. 3.8.

104.   Schleher, D. C., "Radar Detection in Log Normal Clutter," presented at the *IEEE 1975 Int. Radar Conf.*, Washington, D.C., 1975.

105.   Schleher, D. C., "Radar Detection in Weibull Clutter," *IEEE Trans. on Aerospace Electron. Syst.*, Vol. AES-12, No. 6, November 1976, pp. 736–743.

106.   Shultz, C. H., T. L. Oliver, and W. H. Peake, "Radar Backscattering Data for Surfaces of Geological Interest," Report 1903-7, ElectroScience Laboratory, Department of Electrical Engineering, Ohio State University, Columbus, OH, December 2, 1969.

107.   Soares, J. V., R. Bernard, and D. Vidal-Madjar, "Spatial and Temporal Behaviour of a Large Agricultural Area as Observed from Airborne C-band Scatterometer and Thermal Infrared Radiometer," *Int. J. Remote Sensing*, Vol. 8, No. 7, 1987, pp. 981–996.

08.   Stiles, W. H., F. T. Ulaby, B. Hanson, and L. Dellwig, "Snow Backscatter in the 1-8 GHz Region," Remote Sensing Laboratory Technical Report 177-61, University of Kansas Center for Research, Inc., Lawrence, KS, June 1976.

09.   Stiles, W. H., and F. T. Ulaby, "Backscatter Response of Roads and Roadside Surfaces," Remote Sensing Laboratory Technical Report 377-1, University of Kansas Center for Research, Inc., Lawrence, KS, July 1978.

10.   Stiles, W. H., and F. T. Ulaby, "The Active and Passive Microwave Response to Snow Parameters: Part I—Wetness," *J. Geophys. Res.*, No. 85, 1980, pp. 1037–1044.

11.   Stiles, W. H. and F. T. Ulaby, "Microwave Remote Sensing of Snowpacks," Remote Sensing Laboratory Technical Report 340-3, University of Kansas Center for Research, Inc., Lawrence, KS, June 1980.

12.   Stiles, W. H., F. T. Ulaby, A. K. Fung, and A. Aslam, "Progress in Radar Snow Research," Remote Sensing Laboratory Technical Report 410-1, University of Kansas Center for Research, Inc., Lawrence, KS, February 1981.

113.   Stiles, W. H., F. T. Ulaby, A. Aslam, and M. Abdelrazik, "Active Microwave Investigation of Snowpacks: Experimental Data Documentation," Colorado 1979–1980, Remote Sensing Laboratory Technical Report 410-3, University of Kansas Center for Research, Inc., Lawrence, KS, July 1981.

114.   Swanson, R. R. Zoughi. R. K. Moore, L. F. Dellwig, J. Jiang, and A. K. Soofi, "Backscatter and Dielectric Measurements from Rocks of Southeastern Utah at C-, X-, and Ku-bands," *International Journal of Remote Sensing*, Vol. 9, No. 4, 1988, pp. 625–639.

115.   Trebits, R. N., R. D. Hayes, and L. C. Bomar, "MM-wave Reflectivity of Land and Sea," *Microwave Journal*, Vol. 21, No. 8, August 1978, pp. 49–54.

116   Trostel, J. M., T. L. Lane, N. C. Currie, D. L. Odom, A. H. Green, J. A. Saffold, and M. R. Christian, "MM Wave Tower Snow Data Collection Program." In: *Snow Symposium V*, Hanover, NH, August 1985, Vol. 1, U.S. Army Cold Regions Research and Engineering Laboratory, Hanover, NH, CRREL Special Report 86-15, pp. 291–300.

117.   Trunk, G. V. and S. G. George, "Detection of Targets in NonGaussian Sea Clutter," *IEEE Trans. Aerospace Electron Syst.*, Vol. AES-6, September 1970, pp. 620–628.

118.   Ulaby, F. T., R. K. Moore, R. Moe, and J. Holtzman, On Microwave Remote Sensing of Vegetation, *Proc. 8th Intl. Symp. Rem. Sens. Env.*, University of Michigan, Ann Arbor, MI, 1972.

119.   Ulaby, F. T., "Radar Measurements of Soil Moisture Content," *IEEE Trans. Antennas Prop.*, Vol. AP-22, 1974, pp. 257–265.

120.   Ulaby, F. T., J. Cihlar, and R. K. Moore, "Active Microwave Measurement of Soil Water Content," *Remote Sensing Environ.*, No. 3, 1974, pp. 185–203.

121.   Ulaby, F. T. and P. P. Batlivala, "Measurements of Radar Backscatter from a Hybrid of Sorgum," Remote Sensing Laboratory Technical Report 264-2, University of Kansas Center for Research, Inc., Lawrence, KS, May 1975.

122.   Ulaby, F. T., T. F. Bush, and P. P. Batlivala, "Radar Response to Vegetation II: 8-18 GHz Band," *IEEE Trans. Antennas Prop.*, Vol. AP-23, 1975, pp. 608–618.

123.   Ulaby, F. T., and P. P. Batlivala, "Optimum Radar Parameters for Mapping Soil Moisture," *IEEE Trans. Geosci. Electron.*, Vol. GF-14, 1976, pp. 81–93.

124.   Ulaby, F. T., and T. F. Bush, "Monitoring Wheat Growth with Radar," *Photogram. Eng. Remote Sensing*, Vol. 42, 1976, pp. 557–568.

125.   Ulaby, F. T., and T. F. Bush, "Corn Growth as Monitored by Radar," *IEEE Trans. Antennas Prop.*, Vol. AP-24, 1976, pp. 819–828.

126.   Ulaby, F.T., P. P. Batlivala, and M. C. Dobson, "Microwave Backscatter Dependence on Surface Roughness, Soil Moisture, and Soil Texture: Part I—Bare Soil," *IEEE Trans. Geosci. Electron.*, Vol. GE-16, 1978, pp. 286–295.

127. Ulaby, F. T., and J. E. Bare, "Look-Direction Modulation Function of the Radar Backscattering Coefficient of Agricultural Fields," *Photogram. Eng. Remote Sensing,* Vol. 45, 1979, pp. 1495–1506.

128. Ulaby, F. T., W. H. Stiles, D. Brunfeldt, and E. Wilson (1979), "1-35 GHz Microwave Scatterometer," *Proc. 1979 IEEE/MTT-S Intl. Microwave Symp.,* Orlando, FL, 1979.

129. Ulaby, F. T., G. A. Bradley, and M. C. Dobson, "Microwave Backscatter Dependence on Surface Roughness, Soil Moisture, and Soil Texture: Part II—Vegetation-Covered Soil," *IEEE Trans. Geosci. Electron.,* Vol. GE-17, 1979, pp. 33–40.

130. Ulaby, S., "1977 Vegetation-Covered Soil Moisture Experiment Data Documentation," Remote Sensing Laboratory Technical Report 264-27, University of Kansas Center for Research, Inc., Lawrence, KS, April 1979.

131. Ulaby, F. T. and W. H. Stiles, "Backscatter Response of Snow Covered Roads and Roadside Surfaces," Remote Sensing Laboratory Technical Report 377-2, University of Kansas Center for Research, Inc., Lawrence, KS, May 1979.

132. Ulaby, F. T., and W. H. Stiles, "Microwave Response of Snow," *Adv. Space Res.,* No. 1, 1981, pp. 131–149.

133. Ulaby, F. T., R. K. Moore, and A. K. Fung, *Microwave Remote Sensing: Active and Passive—Microwave Remote Sensing Fundamentals and Radiometry,* Vol. 1, Reading, MA: Addison-Wesley, 1981.

134. Ulaby, F. T., A. Aslam, and M. C. Dobson, "Effects of Vegetation Cover on the Radar Sensitivity to Soil Moisture," *IEEE Trans. Geosci. Remote Sensing,* Vol. GE-20, 1982, pp. 476–481.

135. Ulaby, F. T., R. K. Moore, and A. K. Fung, *Microwave Remote Sensing: Active and Passive—Radar Remote Sensing and Surface Scattering and Emission Theory,* Vol. II, Reading, MA: Addison-Wesley, 1982.

136. Ulaby, F. T., W. H. Stiles, A. K. Fung, H. J. Eom, and M. Abdelrazik, "Observations and Modeling of the Radar Backscatter from Snowpacks," Remote Sensing Laboratory Technical Report 527-4, University of Kansas Center for Research, Inc., Lawrence, KS, July 1982.

137. Ulaby, F. T., C. T. Allen, G. Eger III, and E. Kanemasu, "Relating the Microwave Backscattering Coefficient to Leaf Area Index," *Remote Sensing Environ.,* Vol. 14, 1984, pp. 113–133.

138. Ulaby, F. T., R. K. Moore, and A. K. Fung, *Microwave Remote Sensing: Active and Passive—From Theory to Applications,* Vol. III, Dedham, MA: Artech House, Inc., 1986.

139. Ulaby, F. T., F. Kouyate, B. Brisco and T. H. L. Williams, "Textural Information in SAR Images," *IEEE Trans. Geoscience and Remote Sensing,* Vol. GE-24, No. 2, 1986, pp. 235–245.

140. Ulaby, F. T., T. F. Haddock, J. East, and M. W. Whitt, "A Millimeter-Wave Network Analyzer Based Scatterometer," *IEEE Trans. on Geosci. and Remote Sensing,* Vol. 26, No. 1, January 1988, pp. 75–81.

141. Ulaby, F. T., M. W. Whitt, and M. C. Dobson, "Polarimetric Radar Detection of Point Targets in a Forest Environment," Radiation Laboratory Final Report 024825-1-F, The University of Michigan, Ann Arbor, MI, April 1988.

142. Ulaby, F. T., T. E. van Deventer, J. R. East, T. F. Haddock, and M. E. Coluzzi, "Millimeter-Wave Bistatic Scattering from Ground and Vegetation Targets," *IEEE Trans. Geosci. Remote Sensing,* Vol. 26, No. 3, May 1988, pp. 229–243.

143. Ulaby, F. T., T. F. Haddock, and R. T. Austin, "Fluctuation Statistics of Millimeter-Wave Scattering from Distributed Targets," *IEEE Trans. Geosci. Remote Sensing,* Vol. 26, No. 3, May 1988, pp. 268–281.

144. University of Michigan Millimeter Wave Data Base, Radiation Laboratory, University of Michigan, Ann Arbor, Michigan.

145. Valenzuela, G. R., and M. B. Laing, "Point-Scatterer Formulation of Terrain Clutter Statistics," Naval Research Laboratory Report 7459, September 27, 1972, Washington, D.C.

146. Vallese F., and J. A. Kong, "Correlation Function Studies for Snow and Ice," *J. Appl. Phys.*, Vol. 52, 1981, pp. 4921–4925.

147. Van Kasteren, H. W. J., "Radar Signature of Crops, the Effect of Weather Conditions and the Possibilities of Crop Discrimination with Radar," *Spectral Signature of Objects in Remote Sensing International Colloquium*, Avignon, France, September 8–11, 1981.

148. van Zyl, J. J., H. A. Zebker, and C. Elachi, "Imaging Radar Polarization Signatures: Theory and Observation," *Radio Sci.*, Vol. 22, No. 4, July–August 1987, pp. 529–543.

149. Warden, M. P., "An Experimental Study of Some Clutter Characteristics," *AGARD Conf. Proc. No. 66 on Advanced Radar Systems*, November, 1970.

150. Weinstock, W. W., "Radar Cross-Section Target Models," and "Illustrative Problems in Radar Detection Analysis," In *Modern Radar Analysis, Evaluation, and System Design*, R. S. Berkowitz, Ed., New York: Wiley, 1965.

151. Westinghouse Defense and Space Center, "Backscatter Flight Program," Report No. 1878A, October 1964, AD 608274.

152. Whitt, M. W. and F. T. Ulaby, "Millimeter-wave Polarimetric Measurements of Artificial and Natural Targets," *IEEE Trans. on Geosci. and Remote Sensing*, Vol. 26, No. 5, September 1988.

153. Williams, L. D., D. E. Sugden, and R. V. Birnie, "Millimetric Radar Backscatter from Snowcover," Final Report to Royal Signals and Radar Establishment, Department of Geology, University of Edinburgh, Edinburgh, UK, March 1987.

154. Williams, L. D., and J. G. Gallagher, "The Relation of Millimeter-Wavelength Backscatter to Surface Snow Properties," *IEEE Trans. Geosci. Remote Sensing*, Vol. GE-25, No. 2, March 1987, pp. 188–193.

155. Williams, L. D., J. G. Gallagher, D.E. Sugden, and R. V. Birnie, "Surface Snow Properties Effects on Millimeter-Wave Backscatter," *IEEE Trans. Geosci. Remote Sensing*, Vol. 26, No. 3, May 1988, pp. 300–306.

156. Wilson, E. A., D. R. Brunfeldt, F. T. Ulaby, and J. C. Holtzman, "Circularly Polarized Measurements of Radar Backscatter from Terrain," Remote Sensing Laboratory Technical Report 393-1 ETL-0201, University of Kansas Center for Research, Inc., Lawrence, KS, February 1980.

157. Zebker, H. A., J. J. van Zyl, and D. N. Held, "Imaging Radar Polarimetry from Wave Synthesis," *J. Geophys. Res.*, Vol. 92, No. 81, January 1987, pp. 683–701.

158. Zelenka, J. S., "Comparison of Continuous and Discrete Mixed-Integrator Processors," *J. Opt. Soc. Amer.*, 66, 1976, pp. 1295–1304.

APPENDIX A
BACKSCATTERING DATA FOR SOIL AND ROCK SURFACES

A-1 Data Sources and Parameter Loadings

Table A.1   Data sources for soil and rock surfaces.

Band	References
L	11, 25, 26, 28, 47, 80, 99
S	10, 11, 26, 47, 68
C	6, 10, 11, 17, 19, 26, 28, 47, 80, 89, 90, 107, 114
X	6, 26, 37, 43, 45, 51, 56, 89, 90, 96, 98, 99, 106, 109, 114, 119
Ku	25, 26, 28, 37, 54, 96, 98, 106, 109, 114, 119
Ka	37, 45, 96, 98, 99, 109
W	

Table A.2   Mean and standard deviation parameter loadings for soil and rock surfaces.

Band	Pol.	Angular Range θ_{min}	θ_{max}	$\sigma°$ Parameters P_1	P_2	P_3	P_4	P_5	P_6	SD(θ) Parameters M_1	M_2	M_3
L	HH	0	50	-85.984	99.0	0.628	8.189	3.414	-3.142	5.600	-5.0×10^{-4}	-9.0
	HV	0	50	-30.200	15.261	3.560	-0.424	0.0	0.0	4.675	-0.521	3.187
	VV	0	50	-94.360	99.0	0.365	-3.398	5.0	-1.739	4.618	0.517	-0.846
S	HH	0	50	-91.20	99.0	0.433	5.063	2.941	-3.142	4.644	2.883	15.0
	HV	0	40	-46.467	31.788	2.189	-17.990	1.340	1.583	4.569	0.022	-6.708
	VV	0	50	-97.016	99.0	0.270	-2.056	5.0	-1.754	14.914	-9.0	-0.285
C	HH	0	50	-24.855	26.351	1.146	0.204	0.0	0.0	14.831	-9.0	-0.305
	HV	0	50	-26.700	15.055	1.816	-0.499	0.0	0.0	4.981	1.422	15.0
	VV	0	50	-24.951	28.742	1.045	-1.681	0.0	0.0	4.361	4.080	15.0
X	HH	0	80	4.337	6.666	-0.107	-29.709	0.863	-1.365	1.404	2.015	-0.727
	HV	0	70	-99.0	96.734	0.304	6.780	-2.506	3.142	3.944	0.064	-2.764
	VV	10	70	-42.553	48.823	0.722	5.808	3.000	-3.142	3.263	11.794	8.977
Ku	HH	0	60	-95.843	94.457	0.144	-2.351	-3.556	2.080	14.099	-9.0	-0.087
	HV	10	50	-99.0	46.475	-0.904	-30.0	2.986	-3.142	5.812	2.0×10^{-4}	-9.0
	VV	0	60	-98.320	99.0	0.129	-0.791	5.0	-3.142	13.901	-9.0	-0.273
Ka	HH											
	HV	Insufficient Data										
	VV											
W	HH											
	HV	Insufficient Data										
	VV											

$$\sigma° = P_1 + P_2 \exp(-P_3\theta) + P_4 \cos(P_5\theta + P_6)$$
$$SD(\theta) = M_1 + M_2 \exp(-M_3\theta)$$
where θ is the angle of incidence in radians.

A-2   L-Band Data:

Statistical Distribution Table for Soil and Rock Surfaces

L Band, HH Polarization

Angle	N	σ^o_{max}	σ^o_5	σ^o_{25}	Median	σ^o_{75}	σ^o_{95}	σ^o_{min}	Mean	Std. Dev.
0°	231	20.5	16.5	9.5	3.5	-1.1	-3.9	-19.0	4.5	6.8
5°	373	12.9	8.0	3.9	0.4	-3.:	-7.2	-11.1	0.4	4.7
10°	548	12.3	3.2	-0.9	-3.7	-7.1	-12.5	-19.4	-4.2	4.8
15°	549	12.0	0.6	-3.6	-6.9	-10.8	-17.2	-23.3	-7.3	5.8
20°	551	8.7	1.0	-5.8	-9.2	-13.1	-20.0	-26.8	-9.3	6.2
25°	142	8.0	-3.0	-6.2	-10.4	-14.3	-20.8	-24.4	-10.4	5.5
30°	349	7.5	-4.6	-10.2	-13.9	-17.5	-25.1	-28.9	-13.9	5.9
35°	141	1.6	-5.3	-9.6	-13.3	-17.1	-23.9	-27.1	-13.6	5.7
40°	148	2.0	-7.8	-12.3	-15.3	-18.9	-24.0	-28.8	-15.4	5.1
45°	141	-5.0	-12.5	-15.8	-18.6	-21.0	-27.2	-30.1	-18.6	4.3
50°	207	-6.7	-13.5	-18.0	-21.0	-23.4	-29.4	-32.1	-20.7	4.6

L Band, HV Polarization

Angle	N	σ^o_{max}	σ^o_5	σ^o_{25}	Median	σ^o_{75}	σ^o_{95}	σ^o_{min}	Mean	Std. Dev.
0°	188	4.0	-6.6	-12.6	-17.1	-19.8	-23.3	-37.4	-16.0	5.5
5°	334	-2.8	-13.5	-16.3	-18.3	-20.4	-23.6	-27.9	-18.3	3.2
10°	355	-8.9	-17.5	-20.6	-23.1	-25.5	-28.6	-32.8	-23.0	3.6
15°	367	-12.3	-18.0	-21.9	-24.7	-28.7	-34.2	-38.6	-25.3	5.0
20°	397	-10.5	-18.7	-22.4	-25.2	-29.6	-35.5	-42.2	-26.0	5.2
25°	144	-18.0	-20.6	-23.2	-25.8	-29.1	-33.7	-36.2	-26.4	4.0
30°	322	-16.8	-21.5	-25.4	-28.2	-32.6	-38.3	-42.3	-29.0	5.2
35°	143	-17.9	-21.6	-24.4	-26.8	-30.8	-36.2	-37.4	-27.5	4.6
40°	149	-18.9	-22.6	-25.0	-27.4	-30.4	-36.1	-37.6	-28.0	4.2
45°	144	-21.3	-24.4	-26.9	-29.2	-31.7	-35.8	-39.4	-29.5	3.6
50°	179	-21.9	-24.9	-27.8	-30.4	-33.9	-40.4	-44.2	-31.2	4.6

L Band, VV Polarization

Angle	N	σ^o_{max}	σ^o_5	σ^o_{25}	Median	σ^o_5	σ^o_{95}	σ^o_{min}	Mean	Std. Dev.
0°	230	19.9	16.3	9.1	4.1	-0.4	-3.9	-19.5	4.9	6.5
5°	230	13.2	7.9	4.3	1.1	-1.9	-5.7	-10.6	1.2	4.4
10°	409	10.8	2.9	-0.6	-3.2	-6.4	-11.7	-16.3	-3.6	4.5
15°	406	10.0	-1.1	-3.9	-7.2	-11.2	-17.7	-23.8	-7.9	5.2
20°	408	7.5	-2.6	-6.7	-9.9	-13.2	-20.3	-25.7	-10.4	5.4
30°	205	7.7	-4.5	-10.9	-14.5	-17.9	-26.9	-30.0	-14.7	6.4
50°	65	-7.1	-12.2	-16.9	-20.2	-22.5	-26.7	-29.6	-19.5	4.9

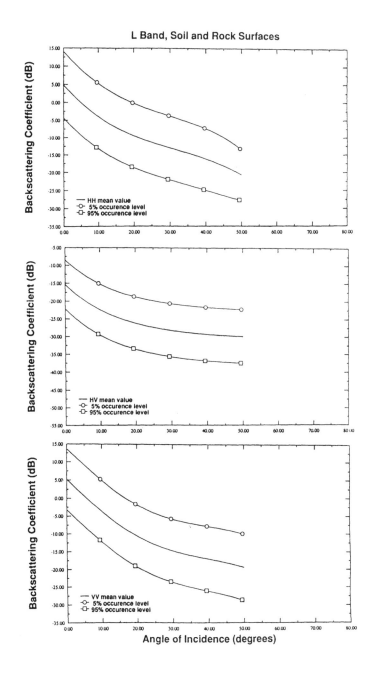

L Band, Soil and Rock Surfaces

Angle of Incidence (degrees)

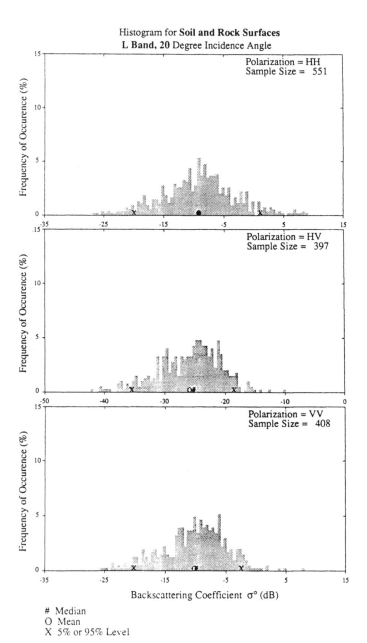

Histogram for **Soil and Rock Surfaces**
L Band, 20 Degree Incidence Angle

Median
O Mean
X 5% or 95% Level

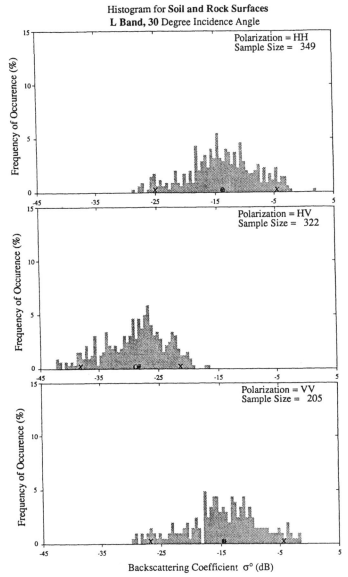

Histogram for **Soil and Rock Surfaces**
L Band, 30 Degree Incidence Angle

Median
O Mean
X 5% or 95% Level

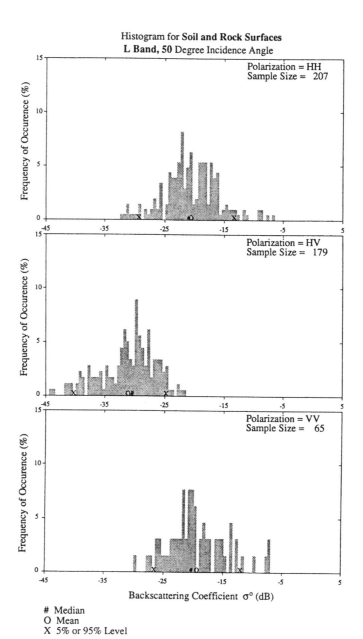

Histogram for **Soil and Rock Surfaces**
L Band, 50 Degree Incidence Angle

Median
O Mean
X 5% or 95% Level

A-3   S-Band Data:

Statistical Distribution Table for Soil and Rock Surfaces

S Band, HH Polarization

Angle	N	σ^o_{max}	σ^o_5	σ^o_{25}	Median	σ^o_{75}	σ^o_{95}	σ^o_{min}	Mean	Std. Dev.
0°	264	20.1	16.7	7.2	1.7	-3.0	-8.1	-11.3	2.6	7.5
5°	186	12.4	7.5	4.0	0.3	-4.9	-10.1	-14.7	-0.4	5.7
10°	442	6.0	3.3	-0.5	-3.6	-7.2	-12.5	-16.0	-3.9	4.7
15°	361	5.3	-0.1	-2.6	-5.4	-9.5	-14.1	-19.4	-6.2	4.5
20°	442	0.9	-2.0	-5.8	-8.8	-12.0	-16.2	-21.5	-9.0	4.5
30°	264	0.8	-2.9	-8.8	-12.7	-15.4	-20.6	-25.5	-12.2	5.1
40°	82	-5.3	-9.0	-11.6	-14.3	-20.2	-26.0	-29.2	-15.8	5.7
50°	16	-15.0	-16.4	-18.0	-18.9	-21.3	-22.1	-22.4	-19.2	2.2

S Band, HV Polarization

Angle	N	σ^o_{max}	σ^o_5	σ^o_{25}	Median	σ^o_{75}	σ^o_{95}	σ^o_{min}	Mean	Std. Dev.
0°	258	-1.7	-5.2	-10.6	-14.8	-18.8	-22.9	-25.5	-14.5	5.5
5°	181	-6.7	-11.9	-14.0	-16.9	-22.1	-25.3	-28.8	-17.8	4.6
10°	311	-8.7	-14.1	-17.1	-19.8	-23.8	-27.3	-31.1	-20.3	4.2
15°	275	-14.5	-16.0	-19.2	-22.6	-25.5	-30.2	-36.0	-22.6	4.3
20°	393	-8.9	-15.6	-20.0	-23.2	-26.2	-30.2	-36.4	-23.1	4.6
30°	264	-8.7	-16.0	-19.7	-25.4	-28.6	-34.0	-38.6	-24.7	5.7
40°	80	-9.9	-15.6	-19.9	-23.8	-29.8	-36.1	-40.3	-24.9	6.7
50°	14	-25.9						-35.3	-31.7	2.7

S Band, VV Polarization

Angle	N	σ^o_{max}	σ^o_5	σ^o_{25}	Median	σ^o_{75}	σ^o_{95}	σ^o_{min}	Mean	Std. Dev.
0°	264	19.7	16.2	7.2	1.0	-3.1	-8.1	-11.7	2.4	7.5
5°	186	11.7	7.4	3.2	0.0	-5.6	-10.0	-14.2	-0.9	5.7
10°	441	6.2	3.1	-0.4	-3.8	-7.3	-12.5	-16.9	-4.0	4.7
15°	362	4.9	-0.1	-2.7	-5.4	-9.8	-14.3	-20.1	-6.4	4.7
20°	442	1.5	-1.9	-5.7	-8.6	-12.2	-16.6	-23.3	-9.0	4.6
30°	263	0.8	-3.1	-8.8	-13.5	-16.3	-20.3	-24.1	-12.6	5.2
40°	82	-3.4	-8.0	-10.4	-14.3	-17.6	-23.4	-26.0	-14.2	4.9
50°	16	-10.0	-13.5	-16.9	-18.5	-19.7	-20.8	-21.8	-17.8	3.0

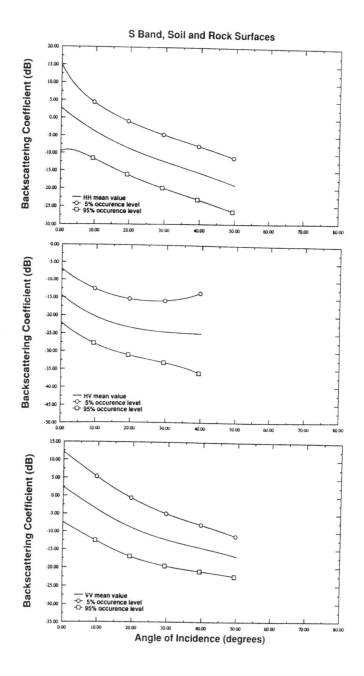

S Band, Soil and Rock Surfaces

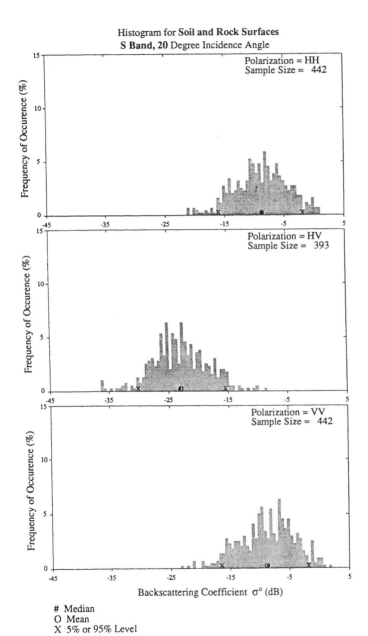

Histogram for **Soil and Rock Surfaces**
S Band, 20 Degree Incidence Angle

Median
O Mean
X 5% or 95% Level

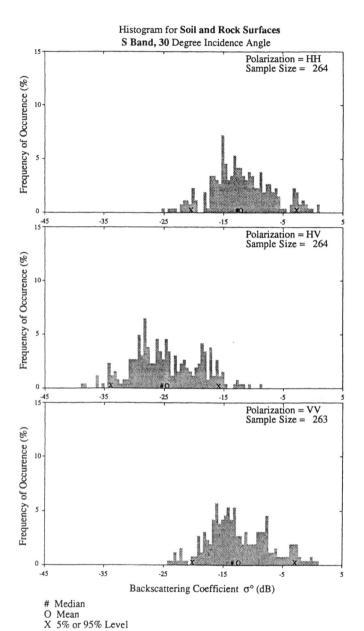

Histogram for **Soil and Rock Surfaces**
S Band, 30 Degree Incidence Angle

Median
O Mean
X 5% or 95% Level

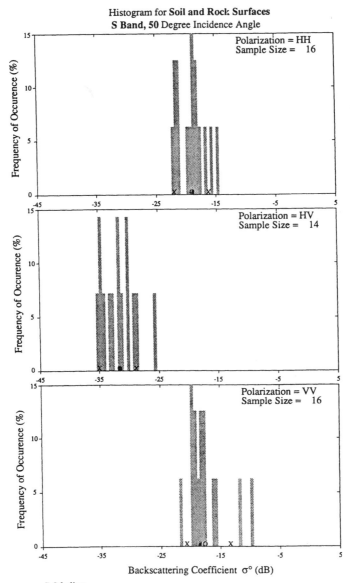

Histogram for **Soil and Rock Surfaces**
S Band, 50 Degree Incidence Angle

Median
O Mean
X 5% or 95% Level

A-4 C-Band Data:

Statistical Distribution Table for Soil and Rock Surfaces

C Band, HH Polarization

Angle	N	σ^o_{max}	σ^o_5	σ^o_{25}	Median	σ^o_{75}	σ^o_{95}	σ^o_{min}	Mean	Std. Dev.
0°	688	26.1	18.1	6.8	0.0	-4.9	-8.7	-16.1	1.6	8.1
5°	584	14.8	6.8	2.4	-1.1	-5.7	-8.7	-12.6	-1.3	5.1
10°	1596	14.1	4.3	0.9	-2.8	-7.0	-10.0	-16.1	-2.9	4.7
15°	939	13.5	2.3	-0.8	-4.2	-8.0	-11.7	-19.8	-4.4	4.6
20°	1205	13.2	0.0	-4.4	-7.3	-10.4	-14.1	-22.0	-7.3	4.5
25°	140	2.4	0.5	-6.1	-9.2	-12.2	-17.2	-18.6	-8.8	4.9
30°	805	8.6	-2.5	-7.7	-10.7	-13.6	-17.6	-25.6	-10.5	4.5
35°	140	-4.0	-6.8	-9.2	-11.7	-14.3	-18.7	-21.1	-12.0	3.6
40°	392	0.6	-7.3	-10.0	-12.5	-15.8	-21.3	-27.0	-13.2	4.4
45°	149	-8.3	-9.8	-10.8	-13.1	-15.5	-19.5	-21.6	-13.4	2.9
50°	328	-5.8	-9.3	-11.5	-13.8	-16.8	-21.5	-24.6	-14.4	3.7
60°	13	-6.0						-20.5	-14.6	4.8

C Band, HV Polarization

Angle	N	σ^o_{max}	σ^o_5	σ^o_{25}	Median	σ^o_{75}	σ^o_{95}	σ^o_{min}	Mean	Std. Dev.
0°	623	9.2	0.3	-8.0	-13.5	-17.2	-20.8	-24.7	-12.2	6.6
5°	530	0.4	-7.1	-11.0	-14.0	-18.0	-21.5	-25.9	-14.3	4.6
10°	947	-0.1	-8.3	-12.8	-16.8	-19.9	-24.0	-28.8	-16.4	4.9
15°	723	-1.0	-9.2	-13.9	-17.5	-20.8	-25.3	-30.9	-17.3	4.9
20°	997	5.7	-11.1	-15.8	-19.3	-22.4	-28.4	-43.0	-19.3	5.2
25°	143	-12.5	-15.7	-18.4	-20.9	-23.5	-28.8	-31.3	-21.3	4.0
30°	789	-8.9	-12.7	-17.8	-21.0	-24.4	-31.1	-45.6	-21.3	5.6
35°	142	-15.1	-17.4	-19.9	-22.6	-24.9	-31.4	-32.7	-23.1	4.2
40°	386	-10.9	-15.4	-19.0	-22.2	-26.1	-33.2	-45.3	-23.0	5.7
45°	151	-17.4	-18.8	-21.2	-23.6	-26.0	-32.6	-34.7	-24.1	3.9
50°	293	-12.6	-14.1	-20.7	-24.0	-27.2	-31.3	-34.7	-23.6	4.9
55°	7	-24.0						-28.8	-26.4	1.5
60°	7	-24.2						-28.0	-25.5	1.5

C Band, VV Polarization

Angle	N	σ^o_{max}	σ^o_5	σ^o_{25}	Median	σ^o_{75}	σ^o_{95}	σ^o_{min}	Mean	Std. Dev.
0°	682	26.3	19.0	7.7	0.4	-4.4	-8.6	-15.1	2.1	8.4
5°	446	15.4	7.9	3.3	-0.6	-6.2	-9.4	-13.8	-0.9	5.6
10°	1052	13.6	5.4	1.0	-2.5	-6.1	-10.4	-16.8	-2.6	4.8
15°	801	13.5	2.6	-0.5	-3.9	-7.4	-11.2	-21.5	-4.0	4.5
20°	1047	11.9	-0.1	-4.1	-7.0	-10.0	-13.7	-22.6	-7.0	4.2
30°	657	8.8	-2.0	-7.5	-10.6	-13.0	-17.5	-25.6	-10.3	4.5
40°	247	1.1	-6.8	-10.2	-12.6	-15.3	-21.2	-27.5	-12.9	4.2
45°	10	-9.7						-16.4	-13.3	1.7
50°	119	-5.5	-8.2	-11.7	-14.3	-16.7	-20.1	-22.9	-14.2	3.6
55°	6	-12.8						-22.0	-16.2	3.1
60°	7	-13.3						-25.0	-17.5	4.0

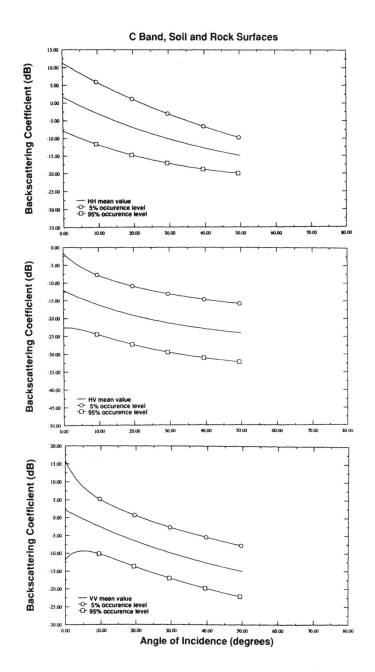

C Band, Soil and Rock Surfaces

Angle of Incidence (degrees)

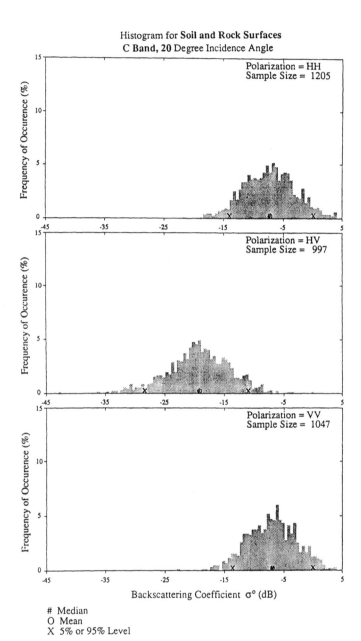

Histogram for **Soil and Rock Surfaces**
C Band, 20 Degree Incidence Angle

Median
O Mean
X 5% or 95% Level

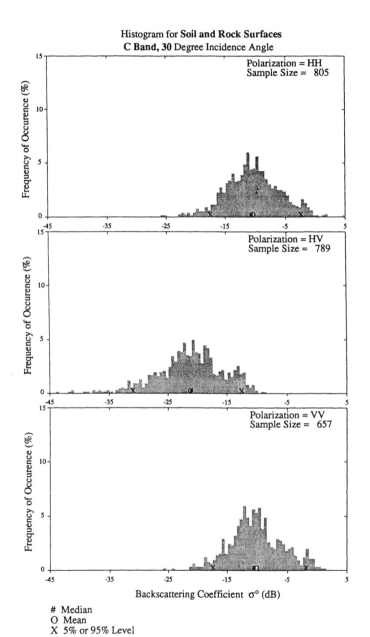

Histogram for **Soil and Rock Surfaces**
C Band, 30 Degree Incidence Angle

Backscattering Coefficient σ° (dB)

Median
O Mean
X 5% or 95% Level

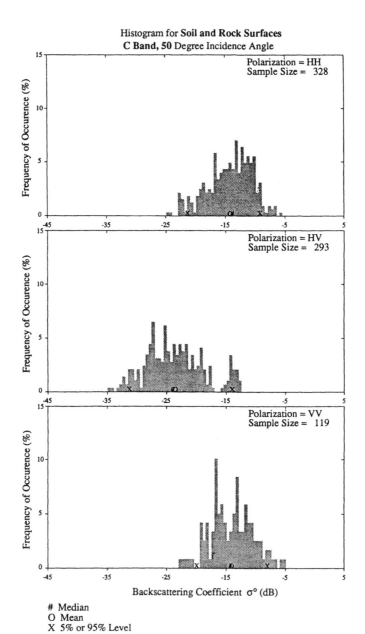

Histogram for **Soil and Rock Surfaces**
C Band, 50 Degree Incidence Angle

Median
O Mean
X 5% or 95% Level

A-5 X-Band Data:

Statistical Distribution Table for Soil and Rock Surfaces

X Band, HH Polarization

Angle	N	σ^o_{max}	σ^o_5	σ^o_{25}	Median	σ^o_{75}	σ^o_{95}	σ^o_{min}	Mean	Std. Dev.
0°	25	15.5	14.3	11.4	5.1	-3.2	-5.7	-6.3	5.0	7.4
5°	8	6.0						-7.5	1.5	4.5
10°	43	10.0	4.3	1.3	-1.6	-4.7	-12.9	-15.1	-1.6	5.1
15°	309	6.5	4.5	2.0	-1.0	-3.0	-5.0	-9.1	-0.5	3.1
20°	55	3.0	1.3	-2.0	-4.5	-6.9	-14.4	-17.0	-4.8	4.2
30°	94	3.5	-1.7	-5.6	-8.0	-10.4	-15.9	-19.1	-8.1	4.3
40°	58	3.0	0.0	-6.1	-9.2	-12.2	-20.4	-21.2	-9.1	5.5
45°	16	-7.3	-9.4	-10.3	-11.1	-12.3	-18.7	-19.2	-12.0	3.2
50°	94	3.0	-2.0	-8.4	-11.5	-13.9	-19.6	-24.9	-10.8	5.2
60°	126	1.0	-6.0	-12.2	-15.0	-21.0	-27.0	-31.0	-16.3	6.4
70°	57	-3.5	-6.4	-12.2	-14.5	-18.8	-29.1	-37.0	-15.2	6.3
80°	27	-7.5	-11.5	-13.2	-16.7	-20.5	-28.3	-29.5	-17.4	5.6

X Band, HV Polarization

Angle	N	σ^o_{max}	σ^o_5	σ^o_{25}	Median	σ^o_{75}	σ^o_{95}	σ^o_{min}	Mean	Std. Dev.
0°	19	-4.0	-5.4	-6.7	-9.2	-11.1	-17.7	-18.0	-9.3	3.8
5°	3	-6.8						-16.5	-10.4	5.3
10°	6	-6.0						-17.9	-10.3	4.0
15°	18	-10.0	-10.7	-12.0	-12.9	-17.6	-19.6	-21.0	-14.2	3.2
20°	30	-8.0	-12.0	-14.2	-15.5	-18.6	-24.4	-30.3	-16.2	4.3
30°	59	-11.6	-13.3	-15.2	-18.5	-21.9	-26.6	-36.6	-18.6	4.6
40°	23	-14.5	-15.8	-16.3	-18.2	-22.0	-26.3	-37.0	-19.5	4.8
45°	17	-17.5	-18.7	-20.8	-23.1	-25.3	-26.9	-27.0	-22.7	2.9
50°	131	-7.7	-12.5	-16.6	-20.7	-23.8	-26.6	-38.2	-20.3	4.8
60°	29	-17.1	-19.1	-20.5	-24.7	-26.2	-30.8	-41.1	-23.9	4.6
70°	19	-19.6	-21.1	-22.5	-23.0	-27.6	-31.7	-47.7	-25.2	6.1

X Band, VV Polarization

Angle	N	σ^o_{max}	σ^o_5	σ^o_{25}	Median	σ^o_{75}	σ^o_{95}	σ^o_{min}	Mean	Std. Dev.
0°	14	11.3						-6.4	2.0	5.7
5°	12	7.0						-6.3	-0.2	5.3
10°	24	5.7	3.0	-0.2	-5.3	-9.3	-13.6	-14.1	-4.9	5.7
15°	13	1.0						-8.5	-4.6	3.3
20°	48	-0.9	-2.5	-3.7	-6.1	-9.9	-13.6	-17.6	-6.9	3.9
30°	61	-3.6	-5.8	-7.6	-8.7	-11.6	-14.5	-18.5	-9.5	3.1
40°	32	-5.1	-7.1	-8.0	-10.2	-14.1	-18.4	-19.0	-11.0	3.6
45°	15	-7.7	-9.5	-10.2	-11.5	-13.8	-15.0	-16.3	-11.6	2.2
50°	144	-2.6	-7.1	-8.7	-11.0	-13.0	-16.7	-20.0	-11.1	3.2
60°	42	-9.6	-11.4	-12.3	-13.9	-17.3	-20.6	-22.6	-14.8	3.2
70°	31	-11.7	-13.1	-14.3	-16.0	-20.5	-25.3	-27.1	-17.0	3.7

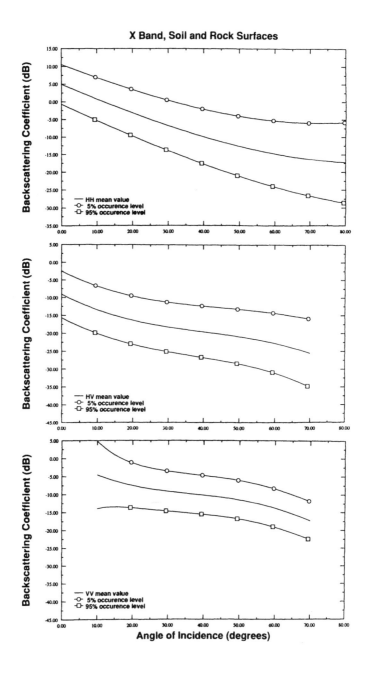

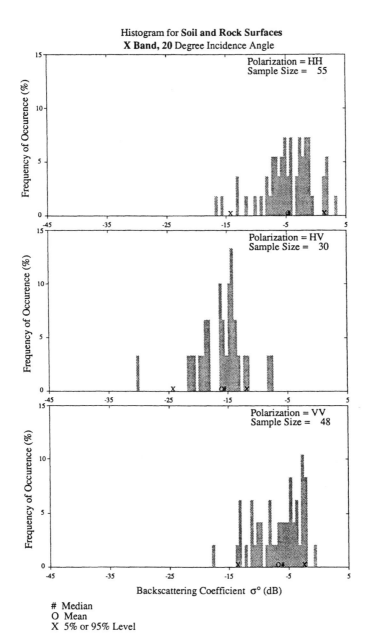

Histogram for **Soil and Rock Surfaces**
X Band, 20 Degree Incidence Angle

Median
O Mean
X 5% or 95% Level

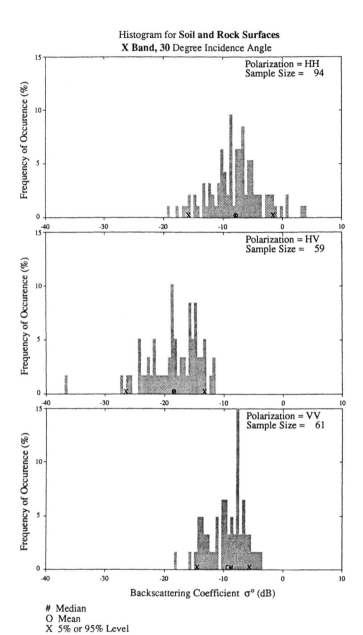

Histogram for **Soil and Rock Surfaces**
X Band, 30 Degree Incidence Angle

Median
O Mean
X 5% or 95% Level

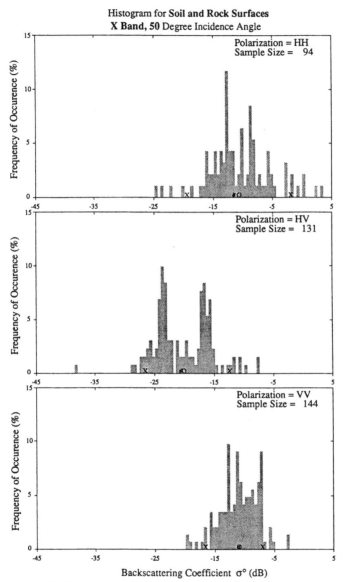

Histogram for **Soil and Rock Surfaces**
X Band, 50 Degree Incidence Angle

Median
O Mean
X 5% or 95% Level

A-6   Ku-Band Data:

Statistical Distribution Table for **Soil and Rock Surfaces**

Ku Band, **HH** Polarization

Angle	N	σ^o_{max}	σ^o_5	σ^o_{25}	Median	σ^o_{75}	σ^o_{95}	σ^o_{min}	Mean	Std. Dev.
0°	24	11.4	6.5	3.3	-0.8	-3.6	-9.3	-9.6	-0.3	5.7
5°	21	6.1	3.4	1.6	-2.8	-6.4	-7.9	-8.4	-2.0	4.5
10°	27	3.8	1.8	-0.1	-4.9	-8.1	-12.1	-13.0	-4.3	4.7
15°	22	6.8	1.3	-1.5	-5.1	-10.2	-12.2	-13.5	-4.9	5.5
20°	34	6.0	-0.8	-5.0	-8.1	-11.1	-16.4	-18.4	-7.8	5.1
30°	32	-1.3	-5.0	-8.4	-12.0	-14.1	-16.6	-16.7	-11.2	4.0
40°	20	-5.3	-7.2	-7.6	-9.9	-13.1	-18.1	-18.2	-11.0	4.1
50°	40	-6.6	-7.2	-10.9	-14.8	-17.9	-20.5	-21.1	-14.2	4.3
60°	22	-7.0	-8.7	-10.2	-13.7	-17.7	-23.0	-23.0	-14.2	4.9
70°	12	-8.5						-28.2	-13.9	5.7

Ku Band, **HV** Polarization

Angle	N	σ^o_{max}	σ^o_5	σ^o_{25}	Median	σ^o_{75}	σ^o_{95}	σ^o_{min}	Mean	Std. Dev.
0°	11	-9.1						-23.2	-16.8	4.7
5°	12	-9.3						-22.5	-18.1	4.4
10°	16	-9.5	-11.8	-13.9	-19.3	-23.3	-24.9	-25.8	-18.6	5.1
15°	14	-12.4						-25.9	-21.1	4.4
20°	21	-4.0	-9.6	-16.9	-22.5	-25.9	-28.3	-30.8	-20.1	7.3
30°	18	-12.9	-17.9	-21.9	-24.9	-27.6	-30.8	-31.9	-24.2	4.7
50°	29	-12.7	-13.6	-18.5	-22.3	-28.1	-31.8	-32.2	-22.5	6.3

Ku Band, **VV** Polarization

Angle	N	σ^o_{max}	σ^o_5	σ^o_{25}	Median	σ^o_{75}	σ^o_{95}	σ^o_{min}	Mean	Std. Dev.
0°	30	11.2	6.1	2.9	-1.7	-4.6	-10.1	-10.3	-1.0	5.3
5°	163	10.0	6.4	3.1	0.4	-3.6	-7.4	-9.5	-0.1	4.5
10°	169	8.9	6.0	2.7	-0.3	-4.3	-8.2	-13.9	-0.8	4.7
15°	164	10.2	5.7	2.2	-0.8	-4.1	-8.3	-13.5	-1.0	4.7
20°	175	5.8	1.9	-1.4	-4.2	-7.6	-11.9	-18.4	-4.2	4.5
25°	136	2.2	1.2	-2.5	-4.7	-6.8	-9.4	-12.7	-4.6	3.2
30°	174	0.6	-2.0	-5.0	-7.1	-8.9	-14.0	-17.1	-7.2	3.5
35°	137	-1.6	-4.0	-6.3	-8.2	-9.6	-11.9	-14.1	-7.9	2.4
40°	158	-2.1	-5.1	-6.7	-8.5	-9.8	-13.1	-17.6	-8.5	2.5
45°	139	-3.9	-5.5	-7.4	-8.8	-9.8	-12.6	-13.8	-8.8	2.0
50°	182	-3.2	-5.5	-8.1	-9.7	-11.6	-17.7	-22.9	-10.2	3.3
60°	22	-5.6	-7.2	-9.3	-13.5	-15.9	-19.3	-21.2	-12.7	4.3
70°	13	-9.2						-20.0	-12.9	3.5

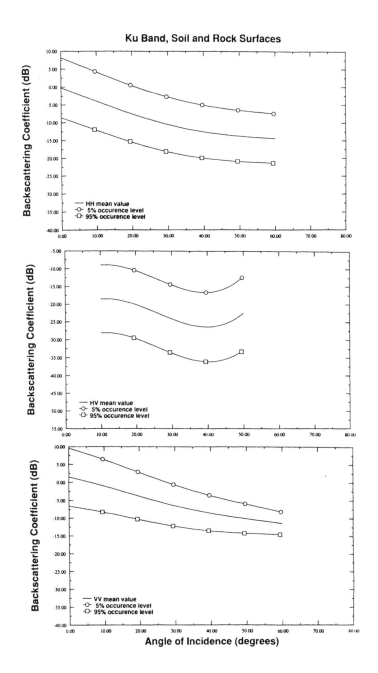

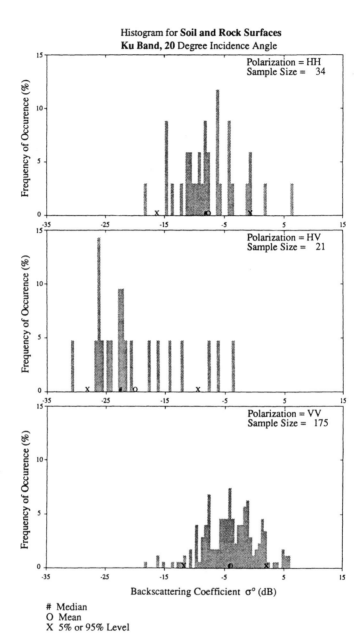

Histogram for **Soil and Rock Surfaces**
Ku Band, 20 Degree Incidence Angle

Median
O Mean
X 5% or 95% Level

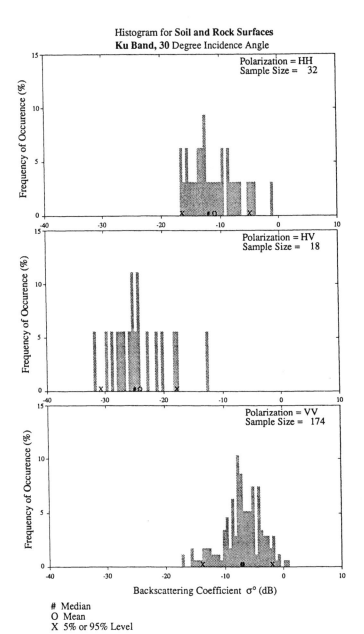

Histogram for **Soil and Rock Surfaces**
Ku Band, 30 Degree Incidence Angle

Polarization = HH
Sample Size =   32

Polarization = HV
Sample Size =   18

Polarization = VV
Sample Size =   174

Backscattering Coefficient σ° (dB)

\# Median
O Mean
X 5% or 95% Level

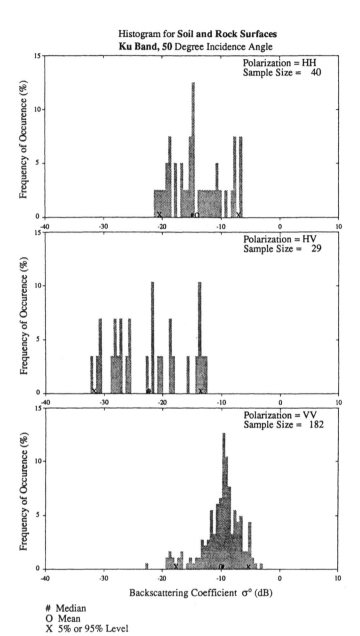

Histogram for **Soil and Rock Surfaces**
Ku Band, 50 Degree Incidence Angle

Polarization = HH
Sample Size =   40

Polarization = HV
Sample Size =   29

Polarization = VV
Sample Size =   182

Backscattering Coefficient σ° (dB)

Median
O Mean
X 5% or 95% Level

A-7 Ka-Band Data:

Statistical Distribution Table for **Soil and Rock Surfaces**

Ka Band, HH Polarization

Angle	N	σ^o_{max}	σ^o_5	σ^o_{25}	Median	σ^o_{75}	σ^o_{95}	σ^o_{min}	Mean	Std. Dev.
0°	1	-7.3						-7.3	-7.3	0.0
5°	1	-8.9						-8.9	-8.9	0.0
10°	7	-2.0						-10.0	-6.0	3.3
15°	2	-3.2						-8.0	-5.6	3.4
20°	10	-3.3						-11.5	-6.3	3.0
30°	10	-3.6						-12.2	-6.8	2.7
40°	10	-4.2						-14.0	-8.3	2.9
50°	10	-5.0						-16.0	-10.0	3.1
60°	10	-6.0						-18.0	-11.7	3.6
65°	1	-10.8						-10.8	-10.8	0.0
70°	9	-9.5						-21.0	-14.7	3.5
80°	6	-12.0						-19.6	-15.5	2.8

Ka Band, HV Polarization

Angle	N	σ^o_{max}	σ^o_5	σ^o_{25}	Median	σ^o_{75}	σ^o_{95}	σ^o_{min}	Mean	Std. Dev.
5°	1	-16.4						-16.4	-16.4	0.0
10°	1	-13.0						-13.0	-13.0	0.0
15°	1	-15.5						-15.5	-15.5	0.0
20°	1	-15.2						-15.2	-15.2	0.0
30°	1	-14.4						-14.4	-14.4	0.0
40°	1	-15.7						-15.7	-15.7	0.0
50°	1	-17.0						-17.0	-17.0	0.0
60°	1	-21.1						-21.1	-21.1	0.0
70°	1	-21.4						-21.4	-21.4	0.0
80°	1	-24.3						-24.3	-24.3	0.0

Ka Band, VV Polarization

Angle	N	σ^o_{max}	σ^o_5	σ^o_{25}	Median	σ^o_{75}	σ^o_{95}	σ^o_{min}	Mean	Std. Dev.
0°	1	-7.7						-7.7	-7.7	0.0
5°	1	-9.3						-9.3	-9.3	0.0
10°	12	-2.1						-19.6	-9.6	5.6
15°	2	-5.2						-6.9	-6.1	1.2
20°	15	-3.3	-5.8	-7.0	-8.4	-13.6	-19.3	-19.8	-9.9	5.0
30°	15	-4.6	-6.1	-7.5	-8.1	-14.5	-19.8	-20.6	-10.0	5.1
40°	15	-6.2	-7.9	-8.6	-9.5	-16.5	-20.9	-21.7	-11.7	4.9
50°	15	-7.9	-8.6	-9.4	-11.6	-17.4	-22.1	-23.0	-12.9	4.9
60°	15	-9.6	-10.0	-10.5	-13.7	-19.1	-23.5	-24.0	-14.4	4.9
65°	1	-10.0						-10.0	-10.0	0.0
70°	14	-11.3						-25.6	-16.9	4.9
80°	8	-14.8						-29.8	-19.7	5.5

APPENDIX B
BACKSCATTERING DATA FOR TREES

B-1   Data Sources and Parameter Loadings

Table B.1   Data sources for trees.

Band	References
L	33, 36, 141
S	33
C	18, 33, 36, 63, 67
X	33, 36, 43, 54, 62, 63, 66, 156
Ku	21, 33, 156
Ka	144
W	144

Table B.2   Mean and standard deviation parameter loadings for trees.

Band	Pol.	Angular Range θ_{min}	θ_{max}	σ° Parameters P_1	P_2	P_3	P_4	P_5	P_6	SD(θ) Parameters M_1	M_2	M_3
L	HH HV VV	Insufficient Data										
S	HH HV VV	Insufficient Data										
C	HH HV VV	Insufficient Data										
X	HH	0	80	-12.078	1.0×10^{-6}	-10.0	4.574	1.171	0.583	13.144	-9.0	-0.073
	HV	0	80	88.003	-99.0	-0.050	1.388	6.204	-2.003	12.471	-9.0	-0.125
	VV	0	80	-11.751	2.0×10^{-6}	-10.0	3.596	2.033	0.122	0.816	3.349	0.347
Ku	HH	0	80	-39.042	1.0×10^{-6}	-10.0	30.0	0.412	0.207	13.486	-9.0	-0.154
	HV	0	80	-40.926	1.0×10^{-6}	-10.0	30.0	0.424	0.138	12.614	-9.0	-0.124
	VV	0	80	-39.612	1.0×10^{-6}	-10.0	30.0	0.528	0.023	13.475	-9.0	-0.154
Ka	HH HV VV	Insufficient Data										
W	HH HV VV	Insufficient Data										

$$\sigma^\circ = P_1 + P_2 \exp(-P_3\theta) + P_4 \cos(P_5\theta + P_6)$$
$$SD(\theta) = M_1 + M_2 \exp(-M_3\theta)$$

where θ is the angle of incidence in radians.

B-2   L-Band Data:

Statistical Distribution Table for **Trees**

L Band, **HH** Polarization

Angle	N	σ^o_{max}	σ^o_5	σ^o_{25}	Median	σ^o_{75}	σ^o_{95}	σ^o_{min}	Mean	Std. Dev.
0°	2	-10.5						-11.4	-10.9	0.6
10°	2	-13.2						-14.5	-13.9	0.9
20°	3	-8.2						-16.1	-12.6	4.0
30°	4	-8.5						-18.2	-14.6	4.6
35°	52	-8.9	-14.8	-15.3	-16.7	-18.5	-21.4	-21.8	-16.9	2.7
40°	7	-10.4						-18.1	-14.0	2.7
45°	5	-7.7						-14.0	-11.2	2.8
50°	24	-10.2	-10.8	-11.0	-11.3	-11.5	-13.3	-15.5	-11.4	1.0
55°	235	-7.4	-8.0	-8.8	-9.7	-12.8	-14.4	-15.0	-10.7	2.2
60°	4	-11.4						-20.9	-16.5	5.2

L Band, **HV** Polarization

Angle	N	σ^o_{max}	σ^o_5	σ^o_{25}	Median	σ^o_{75}	σ^o_{95}	σ^o_{min}	Mean	Std. Dev.
0°	2	-17.6						-20.4	-19.0	2.0
10°	2	-19.8						-22.9	-21.3	2.2
20°	4	-13.0						-21.9	-16.7	4.3
30°	6	-13.7						-22.8	-18.7	4.1
35°	56	-18.6	-19.3	-21.6	-22.4	-23.2	-24.7	-24.9	-22.2	1.6
40°	12	-15.3						-24.7	-20.7	3.1
45°	10	-17.1						-23.6	-19.4	2.4
50°	47	-16.2	-17.4	-18.6	-19.2	-19.9	-21.4	-23.3	-19.2	1.3
55°	463	-13.6	-14.3	-14.9	-15.7	-17.6	-18.9	-22.1	-16.3	1.6
60°	6	-15.4						-26.0	-18.8	4.5

L Band, **VV** Polarization

Angle	N	σ^o_{max}	σ^o_5	σ^o_{25}	Median	σ^o_{75}	σ^o_{95}	σ^o_{min}	Mean	Std. Dev.
0°	2	-10.8						-12.6	-11.7	1.3
10°	2	-11.9						-13.3	-12.6	1.0
20°	3	-7.0						-14.4	-11.1	3.8
30°	4	-7.2						-17.0	-13.1	4.3
35°	52	-12.8	-14.9	-16.2	-17.2	-19.2	-20.7	-21.0	-17.5	2.1
40°	7	-10.0						-19.9	-14.5	3.1
45°	5	-10.3						-17.1	-13.0	3.3
50°	24	-8.9	-9.6	-10.0	-10.4	-11.4	-14.7	-18.3	-10.9	1.8
55°	235	-7.7	-8.5	-9.9	-11.0	-13.2	-13.9	-14.3	-11.3	1.8
60°	4	-10.2						-24.7	-16.9	7.5

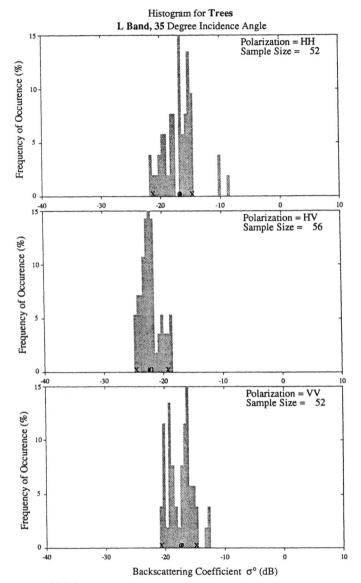

Histogram for **Trees**
L Band, 35 Degree Incidence Angle

Median
O Mean
X 5% or 95% Level

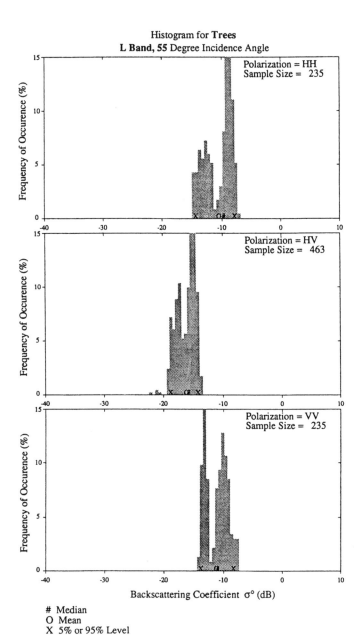

Histogram for **Trees**
L Band, 55 Degree Incidence Angle

Median
O Mean
X 5% or 95% Level

B-3   S-Band Data:

Statistical Distribution Table for Trees

S Band, HH Polarization

Angle	N	σ^o_{max}	σ^o_5	σ^o_{25}	Median	σ^o_{75}	σ^o_{95}	σ^o_{min}	Mean	Std. Dev.
0°	2	-10.0						-11.4	-10.7	1.0
10°	2	-12.0						-16.8	-14.4	3.4
20°	2	-11.0						-16.1	-13.6	3.6
30°	2	-12.5						-16.3	-14.4	2.7
40°	2	-12.3						-15.6	-14.0	2.3
60°	2	-18.0						-22.9	-20.5	3.5
80°	2	-20.1						-21.8	-21.0	1.2

S Band, HV Polarization

Angle	N	σ^o_{max}	σ^o_5	σ^o_{25}	Median	σ^o_{75}	σ^o_{95}	σ^o_{min}	Mean	Std. Dev.
0°	2	-18.0						-19.1	-18.5	0.8
10°	2	-18.5						-23.1	-20.8	3.3
20°	2	-18.0						-20.4	-19.2	1.7
30°	2	-17.9						-21.7	-19.8	2.7
40°	2	-18.1						-21.5	-19.8	2.4
60°	2	-22.7						-25.0	-23.9	1.6
80°	2	-24.2						-28.0	-26.1	2.7

S Band, VV Polarization

Angle	N	σ^o_{max}	σ^o_5	σ^o_{25}	Median	σ^o_{75}	σ^o_{95}	σ^o_{min}	Mean	Std. Dev.
0°	2	-11.0						-12.2	-11.6	0.8
10°	2	-9.9						-15.2	-12.5	3.7
20°	2	-10.9						-15.7	-13.3	3.4
30°	2	-13.0						-15.4	-14.2	1.7
40°	2	-11.8						-14.4	-13.1	1.8
60°	2	-21.6						-22.1	-21.9	0.4
80°	2	-19.0						-21.1	-20.0	1.5

B-4   C-Band Data:

Statistical Distribution Table for Trees

C Band, HH Polarization

Angle	N	σ^o_{max}	σ^o_5	σ^o_{25}	Median	σ^o_{75}	σ^o_{95}	σ^o_{min}	Mean	Std. Dev.
0°	4	-9.0						-12.2	-10.3	1.6
10°	4	-9.1						-14.5	-12.9	2.6
20°	7	-7.0						-14.4	-11.0	2.6
40°	6	-7.5						-13.0	-11.0	2.5
55°	117	-7.7	-7.9	-8.4	-8.8	-9.5	-10.1	-10.4	-8.9	0.7

C Band, HV Polarization

Angle	N	σ^o_{max}	σ^o_5	σ^o_{25}	Median	σ^o_{75}	σ^o_{95}	σ^o_{min}	Mean	Std. Dev.
0°	5	-14.7						-17.5	-16.3	1.1
10°	5	-14.8						-22.8	-19.0	3.3
20°	7	-12.0						-20.7	-17.1	2.9
30°	7	-14.0						-18.8	-16.5	2.0
40°	7	-14.0						-19.8	-17.4	2.4
55°	117	-13.8	-14.0	-14.4	-14.7	-15.0	-15.4	-15.6	-14.7	0.4
60°	4	-20.6						-24.0	-22.0	1.5
80°	4	-23.3						-26.1	-24.8	1.2

C Band, VV Polarization

Angle	N	σ^o_{max}	σ^o_5	σ^o_{25}	Median	σ^o_{75}	σ^o_{95}	σ^o_{min}	Mean	Std. Dev.
0°	5	-8.0						-12.0	-10.2	1.7
10°	5	-9.0						-15.8	-12.3	2.8
20°	7	-7.0						-15.1	-10.8	2.8
40°	7	-8.0						-13.6	-10.6	2.3
45°	6	2.5						-9.6	-2.3	5.5
55°	119	-7.2	-7.5	-8.1	-8.6	-9.1	-9.8	-10.3	-8.6	0.7
60°	8	0.6						-20.0	-9.1	9.4
75°	2	-3.9						-4.2	-4.1	0.2
80°	4	-17.2						-20.5	-19.0	1.5

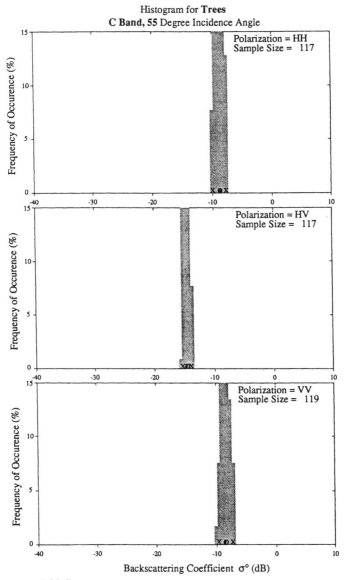

Histogram for **Trees**
C Band, 55 Degree Incidence Angle

Median
O Mean
X 5% or 95% Level

B-5 X-Band Data:

Statistical Distribution Table for Trees

X Band, HH Polarization

Angle	N	σ^0_{max}	σ^0_5	σ^0_{25}	Median	σ^0_{75}	σ^0_{95}	σ^0_{min}	Mean	Std. Dev.
0°	52	-1.9	-2.1	-5.0	-8.6	-12.3	-15.2	-16.5	-8.5	4.2
10°	52	-2.6	-2.9	-4.1	-7.7	-11.1	-17.4	-20.2	-8.3	4.4
20°	54	-2.3	-4.8	-7.3	-9.0	-11.0	-17.8	-19.1	-9.8	4.0
25°	4	-9.4						-10.9	-10.1	0.9
30°	58	-4.4	-4.9	-8.7	-10.2	-12.1	-17.0	-17.4	-10.4·	3.5
40°	65	-4.2	-6.4	-10.1	-11.2	-15.0	-17.5	-18.7	-11.6	3.6
50°	85	-2.4	-6.1	-10.2	-13.3	-14.1	-16.8	-17.5	-12.1	3.4
55°	95	-2.4	-9.5	-16.5	-17.0	-17.4	-18.0	-18.9	-16.2	3.0
60°	74	-2.7	-7.6	-10.8	-13.1	-15.3	-18.9	-19.3	-12.9	3.6
70°	66	-3.9	-8.7	-10.3	-15.0	-17.5	-20.1	-22.9	-14.0	4.2
80°	53	-6.6	-10.1	-11.7	-13.3	-16.3	-18.5	-18.6	-13.5	2.9

X Band, HV Polarization

Angle	N	σ^0_{max}	σ^0_5	σ^0_{25}	Median	σ^0_{75}	σ^0_{95}	σ^0_{min}	Mean	Std. Dev.
0°	59	-4.8	-7.2	-8.8	-11.2	-13.6	-17.5	-20.0	-11.4	3.2
10°	59	-5.1	-6.5	-8.5	-10.2	-12.6	-18.5	-23.1	-10.9	3.5
20°	61	-7.7	-8.9	-10.0	-11.7	-13.3	-16.9	-22.2	-12.0	2.9
30°	62	-8.6	-9.9	-11.0	-12.9	-14.5	-18.0	-19.4	-13.0	2.4
40°	72	-9.5	-10.7	-12.7	-13.9	-16.0	-25.9	-26.9	-14.8	3.9
50°	100	-10.3	-11.6	-13.7	-16.2	-22.6	-24.0	-24.2	-17.8	4.6
55°	183	-21.8	-22.4	-22.9	-23.3	-23.7	-24.2	-26.3	-23.3	0.6
60°	65	-12.3	-13.2	-14.2	-15.0	-16.8	-19.4	-19.8	-15.5	2.0
70°	59	-11.7	-13.2	-14.8	-16.1	-18.0	-23.4	-25.0	-16.7	2.7
80°	60	-11.3	-13.8	-15.2	-16.8	-18.6	-20.5	-21.1	-16.8	2.1

X Band, VV Polarization

Angle	N	σ^0_{max}	σ^0_5	σ^0_{25}	Median	σ^0_{75}	σ^0_{95}	σ^0_{min}	Mean	Std. Dev.
0°	50	-1.9	-2.2	-5.7	-8.3	-11.7	-15.2	-16.5	-8.4	4.0
10°	50	-2.6	-2.9	-4.1	-7.7	-10.3	-17.4	-20.2	-8.0	4.3
20°	53	-2.8	-5.3	-7.3	-9.1	-11.3	-17.8	-19.1	-9.7	3.7
30°	53	-4.4	-4.9	-8.7	-10.0	-12.2	-16.9	-17.4	-10.4	3.5
40°	60	-4.2	-6.1	-9.1	-11.1	-15.0	-17.2	-19.8	-11.6	3.6
50°	76	-5.6	-7.2	-11.0	-13.9	-15.4	-17.2	-18.2	-13.0	3.2
55°	92	-14.7	-16.6	-17.3	-17.9	-18.3	-19.3	-19.5	-17.8	0.8
60°	62	-9.5	-9.8	-11.7	-13.9	-16.4	-17.6	-20.5	-13.7	2.7
70°	55	-8.7	-9.3	-11.4	-15.2	-17.7	-21.8	-22.9	-14.7	3.8
80°	44	-10.1	-10.2	-11.7	-13.2	-15.4	-18.0	-18.5	-13.5	2.5

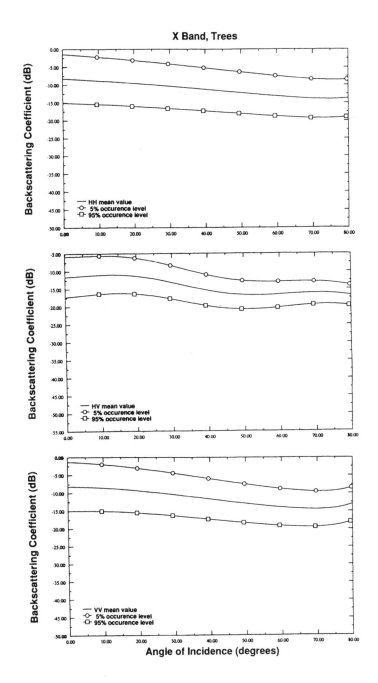

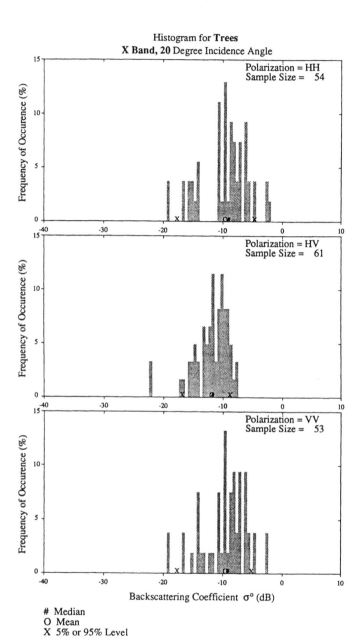

Histogram for **Trees**
X Band, 20 Degree Incidence Angle

Polarization = HH
Sample Size =   54

Polarization = HV
Sample Size =   61

Polarization = VV
Sample Size =   53

Frequency of Occurence (%)

Backscattering Coefficient σ° (dB)

Median
O Mean
X 5% or 95% Level

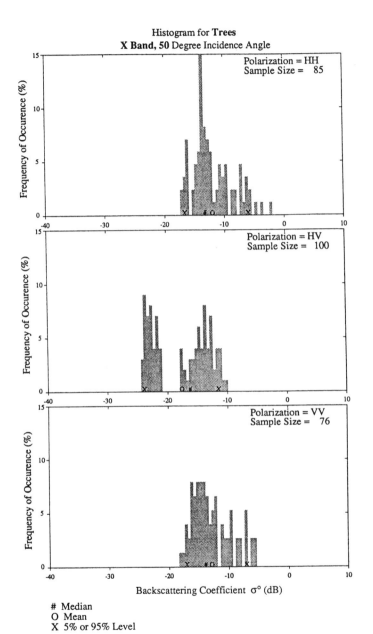

Histogram for **Trees**
X Band, 50 Degree Incidence Angle

Polarization = HH
Sample Size = 85

Polarization = HV
Sample Size = 100

Polarization = VV
Sample Size = 76

Backscattering Coefficient σ° (dB)

\# Median
O Mean
X 5% or 95% Level

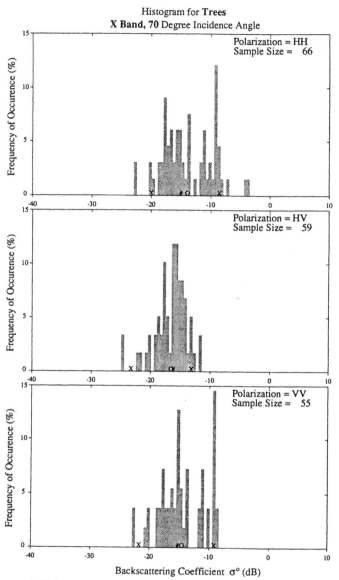

Histogram for **Trees**
X Band, 70 Degree Incidence Angle

Polarization = HH
Sample Size =   66

Polarization = HV
Sample Size =   59

Polarization = VV
Sample Size =   55

Backscattering Coefficient σ° (dB)

\# Median
O Mean
X 5% or 95% Level

B-6 Ku-Band Data:

Statistical Distribution Table for Trees

Ku Band, HH Polarization

Angle	N	σ^{o}_{max}	σ^{o}_{5}	σ^{o}_{25}	Median	σ^{o}_{75}	σ^{o}_{95}	σ^{o}_{min}	Mean	Std. Dev.
0°	66	-1.5	-3.5	-6.9	-10.4	-13.0	-15.9	-19.4	-9.8	4.2
10°	70	-3.7	-4.3	-6.9	-9.1	-12.6	-18.3	-23.9	-9.8	4.5
20°	64	-6.2	-6.7	-7.8	-9.0	-13.6	-18.2	-21.4	-10.9	4.0
30°	65	-6.2	-7.4	-8.7	-11.3	-15.7	-19.8	-21.5	-12.3	4.1
40°	70	-7.2	-8.3	-9.3	-13.0	-15.8	-18.1	-20.1	-12.8	3.5
50°	66	-7.6	-8.3	-11.2	-14.0	-15.3	-17.3	-17.6	-13.3	2.8
60°	70	-8.7	-10.2	-12.4	-14.5	-16.0	-18.8	-19.4	-14.3	2.7
70°	64	-11.6	-12.9	-14.8	-16.6	-18.3	-21.3	-22.4	-16.7	2.6
80°	58	-9.6	-11.6	-14.6	-16.7	-18.2	-20.3	-21.2	-16.2	2.6

Ku Band, HV Polarization

Angle	N	σ^{o}_{max}	σ^{o}_{5}	σ^{o}_{25}	Median	σ^{o}_{75}	σ^{o}_{95}	σ^{o}_{min}	Mean	Std. Dev.
0°	64	-4.3	-6.8	-8.9	-10.6	-12.7	-17.2	-21.6	-11.2	3.4
10°	64	-6.3	-7.7	-9.0	-11.1	-12.4	-20.2	-23.6	-11.4	4.0
20°	64	-8.6	-9.0	-10.2	-11.1	-12.9	-19.4	-23.6	-12.2	3.3
30°	64	-8.3	-9.4	-11.2	-12.8	-14.3	-20.0	-21.8	-13.2	3.0
40°	64	-9.5	-10.6	-12.1	-13.8	-15.6	-18.6	-21.7	-14.0	2.5
50°	58	-9.8	-10.4	-13.1	-14.2	-15.7	-18.1	-19.2	-14.2	2.1
60°	64	-11.0	-12.2	-13.9	-15.1	-16.8	-18.4	-20.7	-15.2	2.0
70°	58	-13.7	-14.0	-15.8	-16.8	-19.1	-24.0	-25.0	-17.7	2.8
80°	58	-12.7	-14.8	-16.4	-17.6	-19.1	-20.7	-21.4	-17.6	2.0

Ku Band, VV Polarization

Angle	N	σ^{o}_{max}	σ^{o}_{5}	σ^{o}_{25}	Median	σ^{o}_{75}	σ^{o}_{95}	σ^{o}_{min}	Mean	Std. Dev.
0°	66	-1.5	-3.5	-6.7	-10.4	-12.3	-15.9	-19.4	-9.6	4.2
10°	70	-3.7	-4.3	-6.8	-9.1	-11.4	-18.4	-23.9	-9.6	4.5
20°	69	-5.2	-6.4	-7.2	-8.9	-12.9	-18.2	-21.4	-10.3	3.9
30°	72	-6.3	-6.9	-8.0	-10.7	-14.8	-19.8	-21.5	-11.5	4.1
40°	80	-6.8	-7.2	-8.6	-12.1	-14.9	-17.2	-20.1	-12.0	3.6
50°	77	-7.6	-8.2	-9.6	-12.9	-14.8	-17.0	-17.4	-12.5	2.8
60°	70	-8.7	-10.2	-12.3	-14.5	-15.9	-17.9	-18.1	-14.0	2.5
70°	64	-11.6	-12.9	-14.1	-16.3	-18.3	-21.3	-22.4	-16.4	2.7
80°	58	-9.6	-11.6	-14.6	-16.7	-18.2	-20.3	-21.1	-16.1	2.6

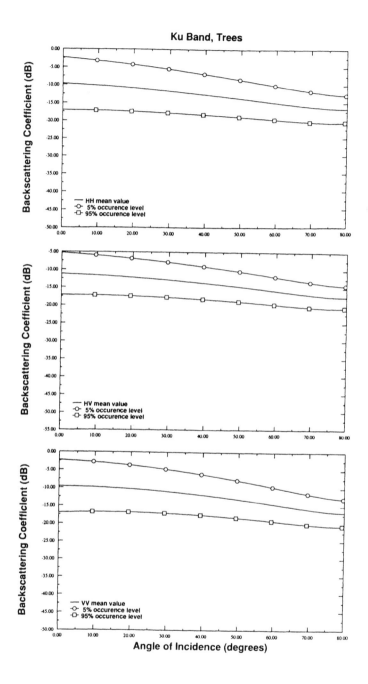

Histogram for **Trees**
Ku Band, 20 Degree Incidence Angle

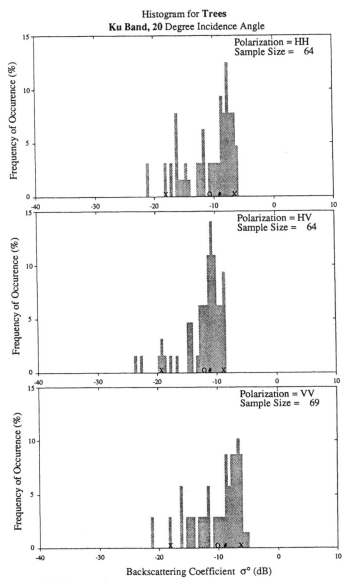

Backscattering Coefficient σ° (dB)

Median
O Mean
X 5% or 95% Level

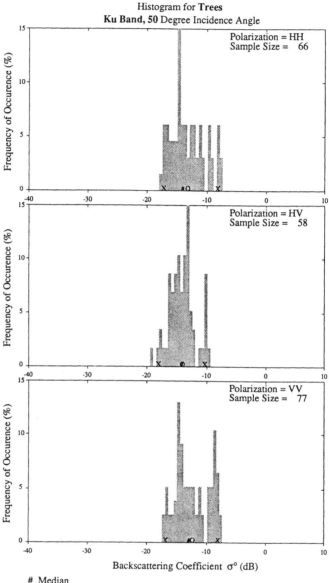

Histogram for **Trees**
Ku Band, 50 Degree Incidence Angle

Median
O Mean
X 5% or 95% Level

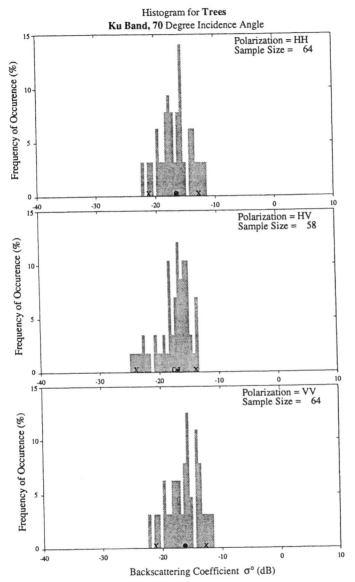

Histogram for **Trees**
Ku Band, 70 Degree Incidence Angle

Median
O Mean
X 5% or 95% Level

B-7   Ka-Band Data:

Statistical Distribution Table for Trees

Ka Band, HV Polarization

Angle	N	σ^o_{max}	σ^o_5	σ^o_{25}	Median	σ^o_{75}	σ^o_{95}	σ^o_{min}	Mean	Std. Dev.
10°	1	-17.3						-17.3	-17.3	0.0
20°	1	-16.1						-16.1	-16.1	0.0
30°	1	-17.1						-17.1	-17.1	0.0
40°	1	-18.7						-18.7	-18.7	0.0
50°	1	-21.4						-21.4	-21.4	0.0
70°	1	-23.3						-23.3	-23.3	0.0

Ka Band, VV Polarization

Angle	N	σ^o_{max}	σ^o_5	σ^o_{25}	Median	σ^o_{75}	σ^o_{95}	σ^o_{min}	Mean	Std. Dev.
10°	4	-7.9						-13.7	-10.6	2.6
20°	5	-7.8						-15.3	-11.2	2.9
30°	5	-9.6						-15.9	-12.1	3.2
40°	5	-10.9						-14.6	-12.7	1.5
50°	5	-8.7						-20.4	-15.2	4.4
60°	4	-13.0						-18.8	-15.5	2.5
70°	3	-13.4						-16.3	-14.8	1.5

B-8 W-Band Data:

Statistical Distribution Table for **Trees**

W Band, **VV** Polarization

Angle	N	σ^o_{max}	σ^o_5	σ^o_{25}	Median	σ^o_{75}	σ^o_{95}	σ^o_{min}	Mean	Std. Dev.
10°	3	-2.1						-6.4	-4.2	2.2
20°	3	-4.7						-6.9	-5.7	1.1
30°	3	-3.4						-6.7	-5.1	1.7
40°	2	-7.8						-8.0	-7.9	0.1
50°	2	-9.0						-9.8	-9.4	0.6
60°	2	-10.2						-11.2	-10.7	0.7
70°	2	-12.2						-13.6	-12.9	1.0

APPENDIX C
BACKSCATTERING DATA FOR GRASSES

C-1 Data Sources and Parameter Loadings

Table C.1   Data sources for grasses.

Band	References
L	26, 28, 46, 71, 96, 99, 108, 130
S	26, 46, 108, 130
C	26, 28, 46, 61, 71, 89, 90, 91, 108, 130
X	6, 34, 37, 45, 46, 48, 50, 89, 90, 91, 96, 99, 109
Ku	26, 28, 34, 37, 45, 46, 48, 50, 96, 109
Ka	37, 45, 50, 96, 99, 109, 144
W	144

Table C.2   Mean and standard deviation parameter loadings for grasses.

Band	Pol.	Angular Range θ_{min}	θ_{max}	σ° Parameters P_1	P_2	P_3	P_4	P_5	P_6	SD(θ) Parameters M_1	M_2	M_3
L	HH	0	80	-29.235	37.550	2.332	-2.615	5.0	-1.616	-9.0	14.268	-0.003
	HV	0	80	-40.166	26.833	2.029	-1.473	3.738	-1.324	-9.0	13.868	0.070
	VV	0	80	-28.022	36.590	2.530	-1.530	5.0	-1.513	-9.0	14.239	-0.001
S	HH	0	80	-20.361	25.727	2.979	-1.130	5.0	-1.916	3.313	3.076	3.759
	HV	0	80	-29.035	18.055	2.80	-1.556	4.534	-0.464	0.779	3.580	0.317
	VV	0	80	-21.198	26.694	2.828	-0.612	5.0	-2.079	3.139	3.413	3.042
C	HH	0	80	-15.750	17.931	2.369	-1.502	4.592	-3.142	1.706	4.009	1.082
	HV	0	80	-23.109	13.591	1.508	-0.757	4.491	-3.142	-9.0	14.478	0.114
	VV	0	80	-93.606	99.0	0.220	-5.509	-2.964	1.287	2.796	3.173	2.107
X	HH	0	80	-33.288	32.980	0.510	-1.343	4.874	-3.142	2.933	1.866	3.876
	HV	20	70	-48.245	47.246	10.0	-30.0	-0.190	3.142	-9.0	12.529	0.008
	VV	0	80	-22.177	21.891	1.054	-1.916	4.555	-2.866	3.559	1.143	5.710
Ku	HH	0	80	-88.494	99.0	0.246	10.297	-1.360	3.142	2.000	1.916	1.068
	HV	40	70	-22.102	68.807	4.131	-4.570	1.952	0.692	3.453	-2.926	3.489
	VV	0	80	-16.263	16.074	1.873	1.296	5.0	-0.695	-9.0	12.773	0.032
Ka	HH	10	70	-99.0	92.382	0.038	1.169	5.0	-1.906	3.451	-1.118	1.593
	HV											
	VV	10	70	-99.0	91.853	0.038	1.100	5.0	-2.050	2.981	-2.604	5.095
W	HH											
	HV	Insufficient Data										
	VV											

$$\sigma^\circ = P_1 + P_2\exp(-P_3\theta) + P_4\cos(P_5\theta + P_6)$$
$$SD(\theta) = M_1 + M_2\exp(-M_3\theta)$$
where θ is the angle of incidence in radians.

C-2 L-Band Data:

Statistical Distribution Table for Grasses

L Band, HH Polarization

Angle	N	σ^o_{max}	σ^o_5	σ^o_{25}	Median	σ^o_{75}	σ^o_{95}	σ^o_{min}	Mean	Std. Dev.
0°	141	17.8	15.5	11.4	9.0	5.4	-1.3	-6.5	8.1	5.2
5°	59	12.8	10.2	4.9	1.0	-1.2	-7.0	-16.0	1.8	5.5
10°	178	3.1	0.1	-3.4	-5.5	-8.8	-14.9	-19.0	-6.1	4.3
15°	59	-2.5	-3.8	-8.9	-12.9	-14.4	-18.2	-21.0	-11.6	4.4
20°	170	-1.7	-5.7	-12.6	-16.7	-19.4	-23.1	-28.7	-15.6	5.5
30°	166	-2.7	-8.9	-16.4	-20.8	-23.5	-26.5	-30.0	-19.3	5.7
40°	113	-4.0	-9.9	-18.6	-22.5	-25.3	-28.9	-35.0	-21.2	6.1
50°	37	-6.1	-7.1	-11.2	-26.5	-28.2	-35.0	-35.0	-21.1	9.8
60°	105	-8.4	-13.3	-22.5	-25.5	-27.7	-31.0	-36.8	-24.0	5.8
80°	86	-21.2	-25.3	-27.0	-29.1	-30.8	-34.3	-38.1	-29.2	3.1

L Band, HV Polarization

Angle	N	σ^o_{max}	σ^o_5	σ^o_{25}	Median	σ^o_{75}	σ^o_{95}	σ^o_{min}	Mean	Std. Dev.
0°	104	1.7	-5.8	-11.1	-15.3	-17.4	-20.3	-25.6	-14.2	4.7
5°	18	-3.3	-4.6	-8.0	-14.5	-19.2	-20.7	-21.2	-13.2	6.3
10°	124	-9.7	-14.7	-20.5	-22.9	-24.3	-28.8	-35.3	-22.4	4.2
15°	31	-16.0	-17.7	-21.4	-24.7	-27.8	-31.3	-31.7	-24.3	4.4
20°	122	-16.4	-20.7	-27.4	-30.9	-33.6	-35.9	-41.0	-29.9	4.8
30°	122	-16.5	-22.8	-28.1	-31.8	-35.6	-38.6	-42.3	-31.5	5.1
40°	92	-20.4	-28.5	-31.1	-33.2	-36.9	-39.8	-40.8	-33.8	3.8
60°	86	-25.7	-31.3	-33.8	-36.1	-39.0	-40.9	-45.2	-36.1	3.5
80°	86	-22.7	-32.4	-35.6	-37.7	-39.4	-42.7	-48.6	-37.4	3.7

L Band, VV Polarization

Angle	N	σ^o_{max}	σ^o_5	σ^o_{25}	Median	σ^o_{75}	σ^o_{95}	σ^o_{min}	Mean	Std. Dev.
0°	141	18.1	15.7	11.4	9.0	5.3	-4.0	-6.6	8.0	5.7
5°	48	12.5	9.2	6.8	1.3	-1.6	-4.3	-5.0	2.6	4.8
10°	141	3.9	0.8	-3.0	-4.9	-7.8	-10.9	-14.8	-5.2	3.7
15°	49	-2.5	-4.4	-7.6	-13.1	-15.4	-19.3	-21.7	-11.8	4.7
20°	162	-1.9	-7.0	-11.2	-16.2	-18.8	-21.6	-26.6	-15.0	5.0
30°	162	-2.1	-9.3	-13.9	-20.6	-23.0	-26.2	-29.2	-18.6	5.8
40°	105	-3.9	-9.5	-17.1	-23.7	-26.0	-30.0	-36.5	-21.6	6.7
50°	29	-4.6	-11.3	-12.6	-14.1	-27.5	-33.3	-33.5	-18.9	9.1
60°	106	-6.2	-12.9	-23.9	-26.4	-28.7	-31.9	-37.3	-25.1	6.1
80°	86	-19.6	-24.2	-25.8	-27.3	-28.9	-33.9	-38.0	-27.6	2.9

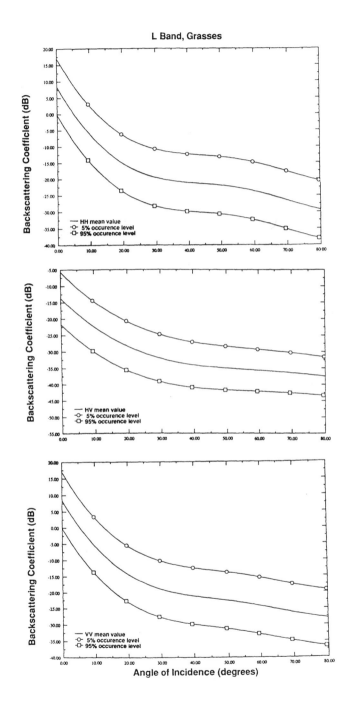

L Band, Grasses

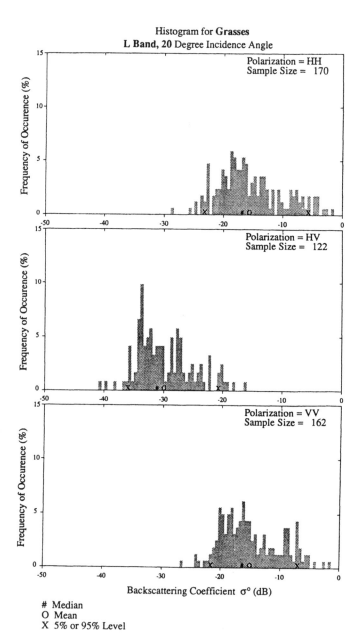

Histogram for **Grasses**
L Band, 20 Degree Incidence Angle

Median
O Mean
X 5% or 95% Level

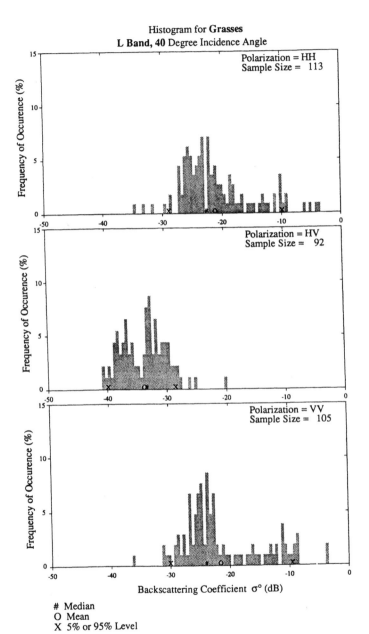

Histogram for **Grasses**
L Band, 40 Degree Incidence Angle

Median
O Mean
X 5% or 95% Level

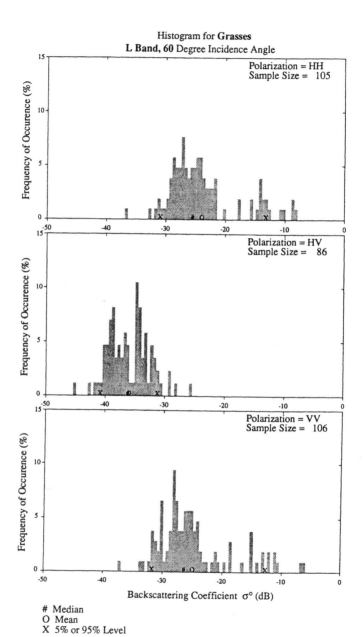

Histogram for **Grasses**
L Band, 60 Degree Incidence Angle

Backscattering Coefficient σ° (dB)

Median
O Mean
X 5% or 95% Level

173

C-3 S-Band Data:

Statistical Distribution Table for Grasses

S Band, HH Polarization

Angle	N	σ^o_{max}	σ^o_5	σ^o_{25}	Median	σ^o_{75}	σ^o_{95}	σ^o_{min}	Mean	Std. Dev.
0°	138	16.9	12.8	10.1	7.2	0.9	-4.1	-25.0	5.7	6.4
5°	46	9.1	3.0	1.7	0.0	-1.8	-10.4	-23.9	-1.0	5.5
10°	137	5.3	1.5	-2.8	-5.1	-7.4	-14.6	-23.5	-5.2	4.7
15°	46	-0.5	-1.7	-5.6	-9.2	-12.5	-20.6	-24.4	-9.3	5.4
20°	138	-2.6	-5.8	-10.5	-12.7	-15.7	-20.0	-24.1	-12.8	4.2
30°	138	-6.7	-10.0	-13.3	-16.0	-18.9	-21.2	-23.9	-15.9	3.6
40°	86	-8.4	-12.3	-14.6	-16.8	-19.2	-22.0	-23.5	-16.7	3.3
60°	86	-10.3	-13.3	-15.9	-18.4	-21.2	-24.2	-26.1	-18.4	3.4
80°	86	-12.1	-14.9	-17.1	-20.4	-22.9	-25.6	-30.0	-20.2	3.5

S Band, HV Polarization

Angle	N	σ^o_{max}	σ^o_5	σ^o_{25}	Median	σ^o_{75}	σ^o_{95}	σ^o_{min}	Mean	Std. Dev.
0°	122	-2.1	-6.3	-10.0	-12.4	-15.3	-18.1	-23.9	-12.4	4.0
5°	27	-8.5	-10.9	-14.8	-15.8	-18.5	-23.9	-24.5	-16.2	3.8
10°	130	-9.1	-13.0	-16.6	-18.9	-22.2	-26.8	-32.9	-19.4	4.4
15°	39	-13.8	-16.5	-19.0	-21.7	-24.3	-35.0	-35.5	-21.7	4.5
20°	132	-10.5	-17.0	-20.0	-22.9	-25.5	-29.6	-38.4	-23.0	4.5
30°	132	-15.3	-18.8	-21.7	-24.0	-26.2	-30.9	-37.0	-24.2	3.8
40°	86	-18.3	-20.7	-23.1	-25.3	-27.0	-30.1	-33.2	-25.2	3.0
60°	86	-21.1	-22.6	-25.2	-27.4	-29.3	-32.0	-35.4	-27.4	3.0
80°	86	-17.4	-25.2	-28.6	-30.4	-31.7	-35.6	-39.1	-30.1	3.4

S Band, VV Polarization

Angle	N	σ^o_{max}	σ^o_5	σ^o_{25}	Median	σ^o_{75}	σ^o_{95}	σ^o_{min}	Mean	Std. Dev.
0°	138	17.7	15.2	10.1	7.5	1.2	-3.9	-24.0	5.8	6.6
5°	46	10.4	3.2	1.6	-0.2	-2.6	-11.0	-24.1	-1.2	5.6
10°	138	7.3	2.3	-1.4	-3.9	-6.6	-14.1	-23.7	-4.4	4.9
15°	46	1.1	-2.7	-6.6	-9.7	-13.3	-20.3	-24.1	-9.9	5.4
20°	137	-1.1	-5.6	-8.6	-12.0	-15.3	-19.7	-25.0	-12.0	4.5
30°	138	-3.1	-9.3	-13.0	-16.1	-18.6	-21.3	-23.2	-15.7	3.8
40°	86	-9.1	-12.7	-14.6	-17.4	-19.7	-22.2	-23.4	-17.3	3.2
60°	86	-13.3	-14.9	-16.9	-19.4	-21.8	-23.9	-27.3	-19.4	3.2
80°	86	-13.8	-15.7	-17.9	-20.9	-23.4	-25.8	-29.4	-20.7	3.4

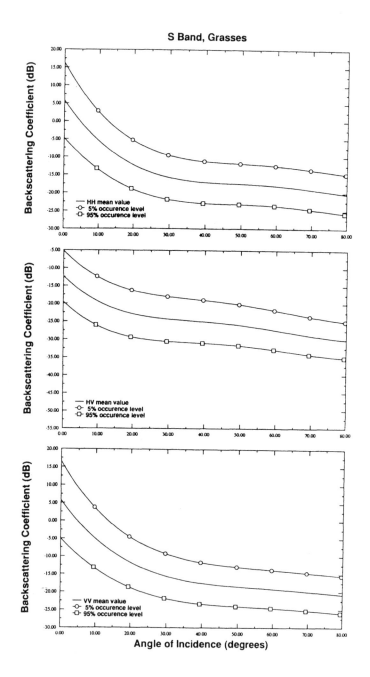

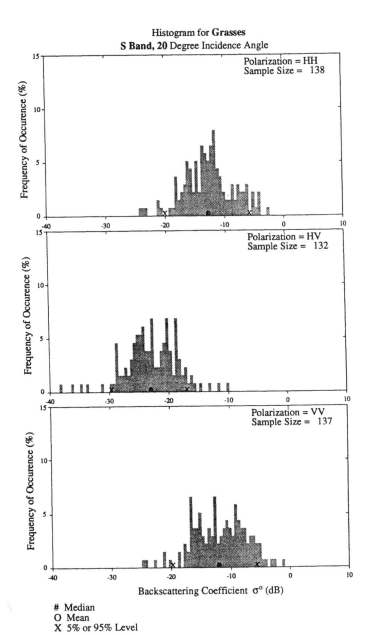

Histogram for **Grasses**
S Band, 20 Degree Incidence Angle

Median
O Mean
X 5% or 95% Level

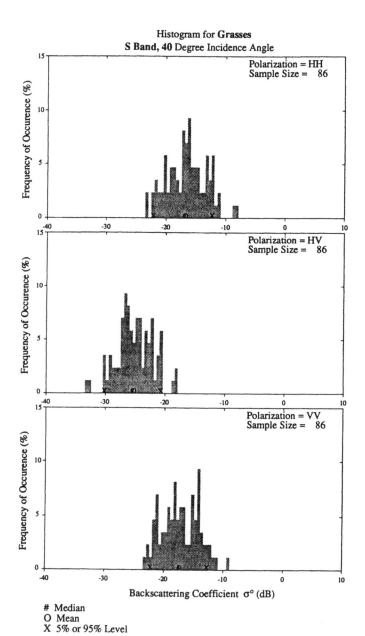

Histogram for **Grasses**
S Band, 40 Degree Incidence Angle

Median
O Mean
X 5% or 95% Level

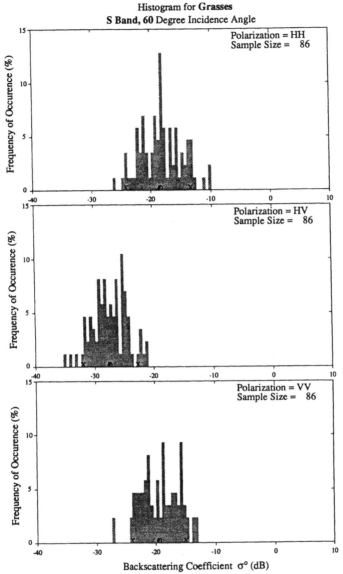

Histogram for **Grasses**
S Band, 60 Degree Incidence Angle

Median
O Mean
X 5% or 95% Level

C-4   C-Band Data:

Statistical Distribution Table for **Grasses**

C Band, **HH** Polarization

Angle	N	σ^o_{max}	σ^o_5	σ^o_{25}	Median	σ^o_{75}	σ^o_{95}	σ^o_{min}	Mean	Std. Dev.
0°	281	15.7	11.6	7.0	5.0	-0.5	-6.7	-12.0	3.6	5.6
5°	105	16.7	7.1	2.6	0.1	-2.1	-4.3	-10.4	0.7	4.3
10°	313	10.7	6.1	1.3	-2.7	-5.9	-12.1	-18.6	-2.7	5.5
15°	106	5.6	1.6	-1.5	-4.9	-7.8	-15.2	-18.5	-5.3	5.0
20°	292	4.6	-1.4	-6.2	-9.3	-11.6	-18.5	-23.6	-9.1	4.7
30°	420	-0.3	-4.9	-8.5	-10.9	-12.9	-18.3	-24.4	-10.9	3.9
40°	184	-4.7	-8.4	-11.1	-12.5	-14.2	-18.7	-22.5	-12.9	3.0
45°	32	-10.1	-11.0	-13.0	-14.8	-18.1	-24.0	-24.0	-15.8	3.9
50°	310	-9.0	-12.3	-13.7	-15.4	-17.5	-22.3	-30.9	-16.0	3.4
60°	186	-5.4	-9.0	-10.7	-12.1	-13.5	-17.6	-22.0	-12.4	2.5
80°	172	-4.8	-10.2	-11.8	-13.2	-15.7	-19.6	-23.2	-13.9	3.0

C Band, **HV** Polarization

Angle	N	σ^o_{max}	σ^o_5	σ^o_{25}	Median	σ^o_{75}	σ^o_{95}	σ^o_{min}	Mean	Std. Dev.
0°	259	1.8	-1.6	-5.9	-8.4	-11.2	-16.1	-23.4	-8.7	4.4
5°	83	1.2	-1.1	-6.9	-9.9	-13.0	-15.7	-17.7	-9.6	4.3
10°	269	0.7	-5.3	-9.3	-12.2	-15.7	-24.8	-28.8	-12.7	5.4
15°	84	-0.2	-3.0	-7.5	-12.3	-17.3	-26.8	-31.7	-13.2	7.5
20°	267	-0.2	-6.5	-12.5	-14.8	-17.9	-28.0	-32.9	-15.4	5.9
30°	395	0.4	-9.1	-14.6	-16.9	-19.6	-26.4	-33.7	-17.1	5.1
40°	178	-9.5	-13.1	-15.3	-17.2	-19.3	-23.0	-30.6	-17.5	3.5
45°	28	-18.5	-19.3	-20.9	-22.8	-25.7	-32.8	-32.8	-24.1	4.4
50°	300	-13.4	-16.5	-19.3	-20.7	-23.2	-29.9	-35.7	-21.6	3.8
60°	184	-12.3	-14.8	-16.5	-18.4	-19.9	-23.7	-27.4	-18.5	2.7
80°	172	-11.2	-14.1	-18.7	-21.1	-23.0	-28.5	-33.8	-21.0	4.0

C Band, **VV** Polarization

Angle	N	σ^o_{max}	σ^o_5	σ^o_{25}	Median	σ^o_{75}	σ^o_{95}	σ^o_{min}	Mean	Std. Dev.
0°	280	15.9	12.9	7.5	4.9	-0.6	-6.9	-12.5	3.8	5.8
5°	95	17.5	7.1	2.2	-0.2	-2.8	-5.3	-12.0	0.3	4.7
10°	280	11.8	6.8	2.0	-1.9	-4.6	-10.8	-18.1	-1.7	5.5
15°	96	8.1	2.6	-1.6	-5.2	-8.1	-14.7	-19.8	-5.4	5.2
20°	279	1.8	-1.1	-5.1	-7.5	-10.1	-17.4	-22.8	-7.8	4.4
30°	396	1.7	-4.8	-8.2	-10.6	-13.1	-17.2	-23.2	-10.7	3.9
40°	172	-5.1	-6.6	-8.8	-10.5	-12.6	-14.9	-17.1	-10.6	2.5
45°	20	-10.5	-10.8	-13.9	-14.3	-15.7	-17.7	-18.3	-14.5	2.0
50°	140	-9.0	-11.7	-13.8	-15.7	-17.3	-24.8	-29.2	-16.2	3.7
60°	184	-5.5	-9.1	-11.0	-12.5	-14.0	-17.6	-20.1	-12.6	2.5
80°	172	-7.9	-11.4	-13.3	-14.7	-17.0	-24.4	-27.0	-15.6	3.8

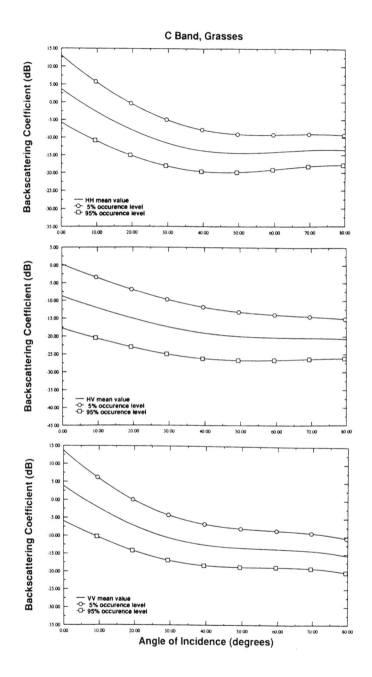

C Band, Grasses

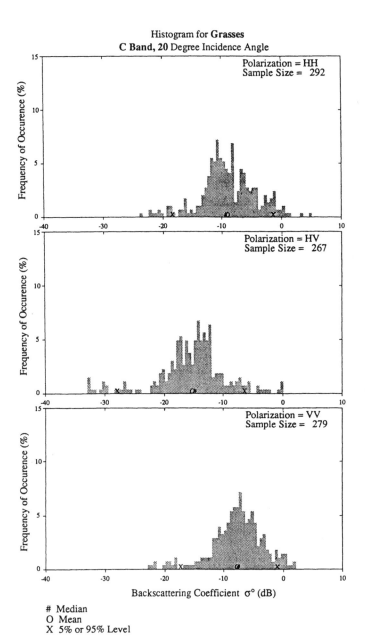

Histogram for **Grasses**
C Band, 20 Degree Incidence Angle

Median
O Mean
X 5% or 95% Level

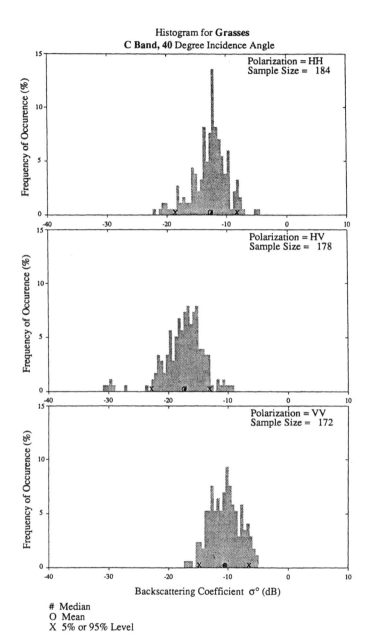

Histogram for **Grasses**
C Band, 40 Degree Incidence Angle

Median
O Mean
X 5% or 95% Level

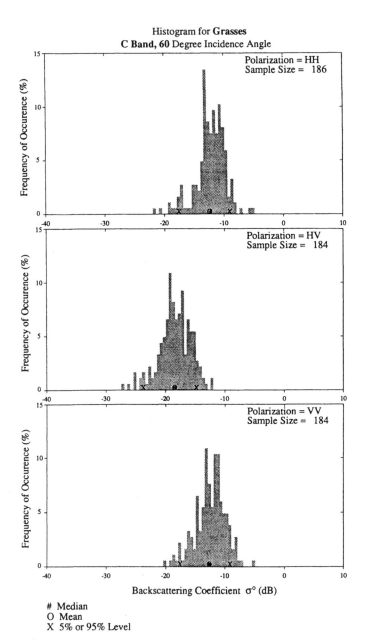

Histogram for **Grasses**
C Band, 60 Degree Incidence Angle

Median
O Mean
X 5% or 95% Level

C-5   X-Band Data:

Statistical Distribution Table for Grasses

X Band, HH Polarization

Angle	N	σ^o_{max}	σ^o_{5}	σ^o_{25}	Median	σ^o_{75}	σ^o_{95}	σ^o_{min}	Mean	Std. Dev.
0°	238	8.6	6.9	3.6	1.7	-1.4	-8.8	-15.0	0.8	4.4
10°	345	10.6	7.2	0.5	-2.3	-4.3	-10.1	-17.1	-1.8	4.8
20°	377	1.4	-0.4	-4.8	-6.2	-7.4	-12.1	-18.3	-6.3	3.1
30°	516	-1.2	-4.9	-7.2	-8.6	-10.2	-13.5	-18.6	-8.8	2.6
40°	931	-4.4	-7.5	-9.5	-11.2	-13.4	-17.4	-24.4	-11.7	3.1
45°	41	-5.3	-6.7	-8.7	-9.9	-12.2	-15.3	-15.7	-10.4	2.8
50°	890	-6.1	-8.6	-10.7	-12.4	-14.5	-18.4	-24.4	-12.8	3.1
60°	1033	-4.6	-9.5	-11.3	-12.8	-14.5	-18.5	-24.6	-13.2	2.8
70°	728	-3.4	-10.6	-12.7	-14.5	-16.5	-19.9	-28.5	-14.7	3.1
80°	242	-2.3	-12.9	-13.9	-14.6	-15.8	-24.3	-33.5	-15.5	3.5

X Band, HV Polarization

Angle	N	σ^o_{max}	σ^o_{5}	σ^o_{25}	Median	σ^o_{75}	σ^o_{95}	σ^o_{min}	Mean	Std. Dev.
0°	10	0.0						-21.5	-14.4	7.2
10°	10	-12.3						-21.6	-17.4	3.8
20°	153	-9.3	-11.8	-14.3	-16.0	-18.0	-22.8	-27.0	-16.5	3.2
30°	201	-11.2	-13.6	-18.3	-20.4	-22.8	-25.3	-29.5	-20.3	3.5
40°	607	-7.8	-12.2	-15.1	-17.3	-19.3	-22.6	-27.4	-17.3	3.3
45°	54	-10.9	-15.6	-17.6	-20.3	-22.6	-24.8	-28.1	-20.0	3.2
50°	1244	-7.9	-12.7	-16.6	-18.9	-21.1	-24.9	-32.4	-18.9	3.7
60°	671	-10.0	-12.9	-16.3	-19.1	-21.0	-23.8	-33.2	-18.7	3.3
70°	643	-6.5	-13.9	-16.7	-19.4	-21.4	-24.1	-36.7	-19.1	3.3

X Band, VV Polarization

Angle	N	σ^o_{max}	σ^o_{5}	σ^o_{25}	Median	σ^o_{75}	σ^o_{95}	σ^o_{min}	Mean	Std. Dev.
0°	237	10.1	7.0	4.0	2.2	-0.6	-8.8	-12.7	1.2	4.4
10°	343	10.8	7.6	-0.5	-3.2	-5.6	-9.6	-14.1	-2.5	5.0
20°	466	1.3	-1.6	-5.8	-8.0	-9.7	-12.6	-17.5	-7.7	3.1
30°	510	-3.2	-6.5	-9.4	-10.9	-13.0	-15.6	-18.8	-11.1	2.8
40°	928	-4.7	-8.1	-11.9	-14.2	-16.8	-20.2	-25.2	-14.2	3.6
45°	40	-5.2	-6.5	-8.7	-10.9	-14.0	-16.6	-17.1	-11.2	3.3
50°	1366	-2.7	-7.9	-10.9	-13.7	-16.6	-20.1	-23.0	-13.8	3.8
60°	979	-4.9	-8.7	-11.8	-14.6	-17.0	-20.2	-23.5	-14.5	3.5
70°	729	-3.4	-8.8	-11.9	-15.0	-17.4	-20.4	-26.5	-14.8	3.7
80°	236	-2.3	-11.0	-12.7	-13.9	-15.3	-22.3	-33.5	-14.6	3.8

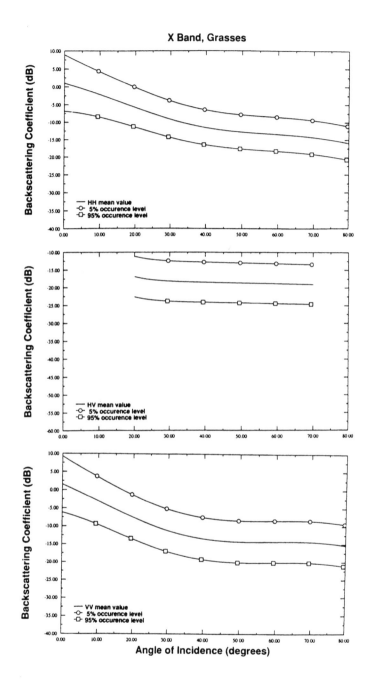

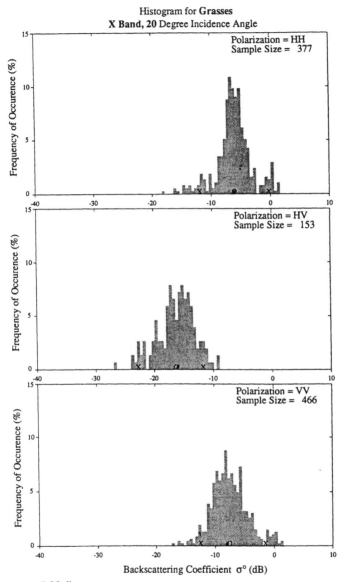

Histogram for **Grasses**
X Band, 20 Degree Incidence Angle

Median
O Mean
X 5% or 95% Level

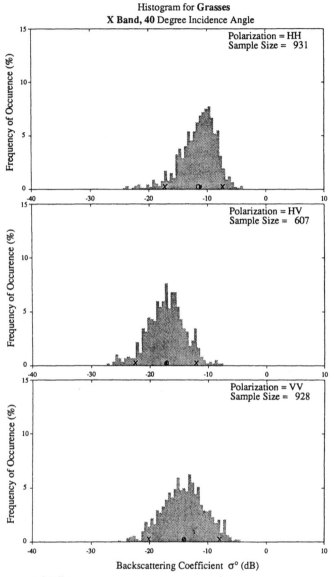

Histogram for **Grasses**
X Band, 40 Degree Incidence Angle

Median
O Mean
X 5% or 95% Level

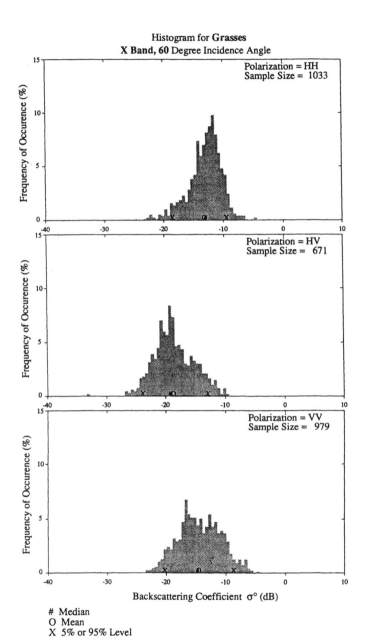

Histogram for **Grasses**
X Band, 60 Degree Incidence Angle

Median
O Mean
X 5% or 95% Level

C-6 Ku-Band Data:

Statistical Distribution Table for Grasses

Ku Band, HH Polarization

Angle	N	σ^o_{max}	σ^o_5	σ^o_{25}	Median	σ^o_{75}	σ^o_{95}	σ^o_{min}	Mean	Std. Dev.
0°	530	14.0	5.4	2.6	0.2	-2.7	-8.3	-9.8	-0.1	3.8
10°	655	11.2	3.6	-1.2	-3.9	-5.9	-9.5	-13.6	-3.4	4.2
20°	664	2.5	-2.8	-5.2	-6.6	-8.0	-10.4	-21.1	-6.6	2.7
30°	666	0.4	-5.0	-7.2	-8.5	-10.6	-13.1	-20.5	-8.8	2.6
40°	1334	-2.3	-6.9	-9.5	-11.3	-13.2	-16.9	-22.9	-11.5	3.1
50°	841	-5.0	-8.0	-10.9	-12.8	-14.9	-17.8	-23.9	-12.9	3.1
60°	1325	-5.6	-9.0	-11.3	-13.1	-14.7	-17.6	-21.5	-13.1	2.6
70°	764	-5.7	-10.2	-12.0	-13.8	-15.8	-18.7	-21.3	-14.0	2.6
80°	503	-7.4	-11.9	-14.1	-15.3	-16.4	-18.3	-24.3	-15.3	2.1

Ku Band, HV Polarization

Angle	N	σ^o_{max}	σ^o_5	σ^o_{25}	Median	σ^o_{75}	σ^o_{95}	σ^o_{min}	Mean	Std. Dev.
0°	1	-13.4						-13.4	-13.4	0.0
5°	1	-12.0						-12.0	-12.0	0.0
10°	2	-11.7						-13.3	-12.5	1.1
15°	2	-13.1						-14.9	-14.0	1.3
20°	2	-13.1						-17.0	-15.1	2.8
30°	2	-11.2						-18.7	-15.0	5.3
40°	662	-7.2	-11.0	-14.3	-15.9	-18.2	-21.3	-26.4	-16.2	3.2
50°	648	-8.7	-11.4	-14.8	-17.0	-18.8	-21.9	-33.0	-16.9	3.3
60°	601	-9.1	-11.4	-14.3	-17.5	-19.4	-22.7	-25.6	-17.0	3.4
70°	544	-10.2	-11.6	-14.4	-17.2	-19.5	-23.1	-25.3	-17.1	3.4

Ku Band, VV Polarization

Angle	N	σ^o_{max}	σ^o_5	σ^o_{25}	Median	σ^o_{75}	σ^o_{95}	σ^o_{min}	Mean	Std. Dev.
0°	531	13.5	7.1	3.5	1.0	-1.6	-7.2	-10.8	0.8	4.0
10°	657	9.3	3.6	-1.0	-3.7	-6.1	-9.1	-13.4	-3.3	4.0
20°	668	2.3	-3.1	-6.0	-7.9	-9.3	-11.7	-22.1	-7.6	2.9
30°	672	-2.3	-5.1	-8.1	-10.2	-12.4	-14.9	-21.3	-10.2	3.0
40°	1339	-4.3	-7.1	-10.8	-13.2	-15.6	-19.4	-24.0	-13.2	3.7
50°	849	-3.8	-8.2	-12.1	-15.1	-17.1	-20.2	-23.7	-14.6	3.7
60°	1311	-2.6	-8.1	-11.3	-14.1	-16.3	-19.6	-22.4	-13.9	3.5
70°	767	-1.6	-8.1	-11.5	-14.5	-16.6	-20.3	-22.2	-14.1	3.6
80°	512	-3.3	-10.0	-12.2	-13.7	-15.2	-17.4	-24.1	-13.7	2.4

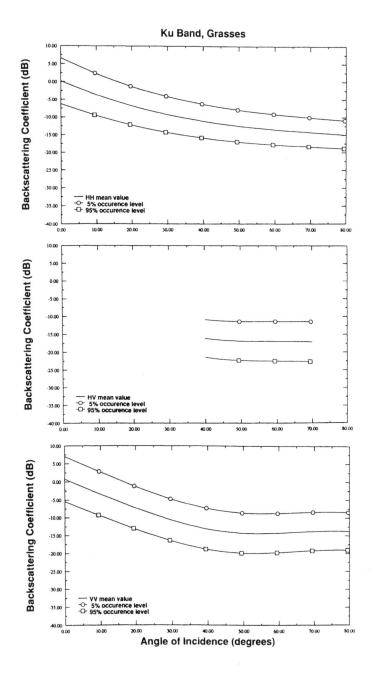

Ku Band, Grasses

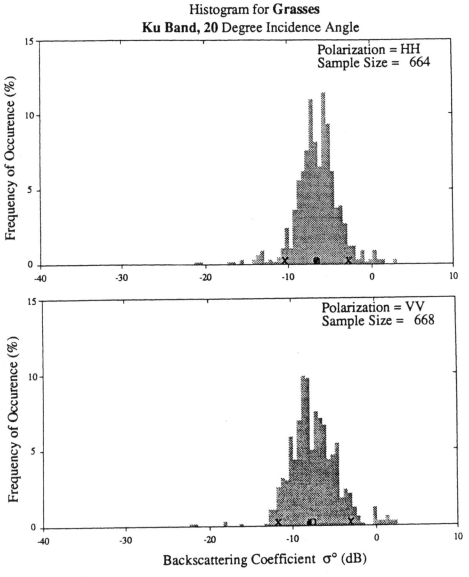

Histogram for **Grasses**
Ku Band, 20 Degree Incidence Angle

Polarization = HH
Sample Size = 664

Polarization = VV
Sample Size = 668

Backscattering Coefficient σ° (dB)

\# Median
O Mean
X 5% or 95% Level

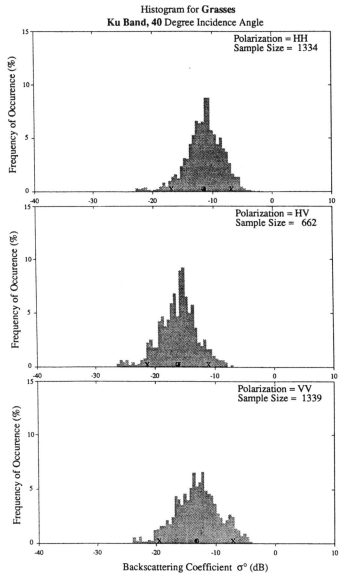

Histogram for **Grasses**
Ku Band, 40 Degree Incidence Angle

Median
O Mean
X 5% or 95% Level

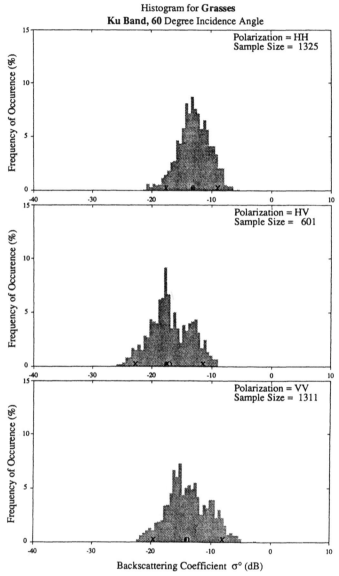

Histogram for **Grasses**
Ku Band, 60 Degree Incidence Angle

Polarization = HH
Sample Size = 1325

Polarization = HV
Sample Size = 601

Polarization = VV
Sample Size = 1311

Backscattering Coefficient σ° (dB)

\# Median
O Mean
X 5% or 95% Level

C-7 Ka-Band Data:

Statistical Distribution Table for **Grasses**

Ka Band, HH Polarization

Angle	N	σ^o_{max}	σ^o_5	σ^o_{25}	Median	σ^o_{75}	σ^o_{95}	σ^o_{min}	Mean	Std. Dev.
0°	1	-8.4						-8.4	-8.4	0.0
5°	1	-5.8						-5.8	-5.8	0.0
10°	19	-2.0	-3.6	-5.1	-7.3	-9.1	-10.3	-10.5	-7.0	2.5
15°	1	-6.7						-6.7	-6.7	0.0
20°	38	0.0	-1.2	-4.3	-6.5	-9.1	-11.3	-11.3	-6.4	3.0
30°	38	-3.0	-3.5	-5.6	-7.6	-10.1	-12.7	-13.1	-7.6	2.8
40°	32	-4.0	-4.9	-6.6	-8.4	-10.8	-14.6	-15.2	-8.8	2.9
45°	6	-3.4						-10.3	-6.6	3.0
50°	52	-5.4	-6.4	-8.5	-10.2	-13.8	-17.0	-20.2	-11.0	3.4
60°	38	-6.4	-6.8	-8.0	-10.0	-11.9	-18.9	-19.5	-10.5	3.1
70°	38	-5.0	-7.5	-10.1	-11.3	-13.3	-20.0	-21.7	-11.7	3.3
80°	15	-10.5	-11.1	-12.0	-14.8	-16.6	-21.4	-25.0	-14.8	3.8

Ka Band, HV Polarization

Angle	N	σ^o_{max}	σ^o_5	σ^o_{25}	Median	σ^o_{75}	σ^o_{95}	σ^o_{min}	Mean	Std. Dev.
0°	2	-15.3						-16.5	-15.9	0.8
5°	2	-13.8						-15.8	-14.8	1.4
10°	2	-13.6						-17.6	-15.6	2.8
15°	2	-15.7						-17.4	-16.5	1.2
20°	2	-13.8						-18.4	-16.1	3.3
30°	2	-13.8						-17.8	-15.8	2.8
40°	2	-13.0						-18.9	-15.9	4.2
50°	2	-13.4						-20.8	-17.1	5.2
60°	2	-17.3						-21.4	-19.3	2.9
70°	1	-23.8						-23.8	-23.8	0.0

Ka Band, VV Polarization

Angle	N	σ^o_{max}	σ^o_5	σ^o_{25}	Median	σ^o_{75}	σ^o_{95}	σ^o_{min}	Mean	Std. Dev.
0°	3	-0.8						-11.1	-6.7	5.3
5°	4	-4.2						-10.2	-7.3	2.7
10°	19	-3.1	-5.6	-6.4	-7.8	-9.3	-10.7	-11.1	-7.7	1.9
15°	4	-6.7						-9.6	-7.9	1.4
20°	38	-1.6	-2.4	-5.5	-7.3	-8.9	-11.3	-11.3	-7.1	2.6
30°	42	-2.3	-3.5	-6.5	-8.0	-10.5	-12.0	-12.1	-7.9	2.7
40°	32	-4.7	-5.7	-7.3	-9.2	-11.6	-14.5	-15.1	-9.3	2.6
45°	6	-4.5						-8.5	-6.3	1.8
50°	51	-5.4	-7.3	-9.0	-10.4	-13.2	-18.6	-20.4	-11.4	3.5
60°	38	-5.7	-6.9	-9.0	-10.6	-12.7	-17.2	-18.0	-10.8	2.8
70°	35	-8.4	-9.0	-10.6	-11.6	-15.0	-17.6	-18.0	-12.4	2.7
80°	12	-11.4						-20.1	-15.4	3.0

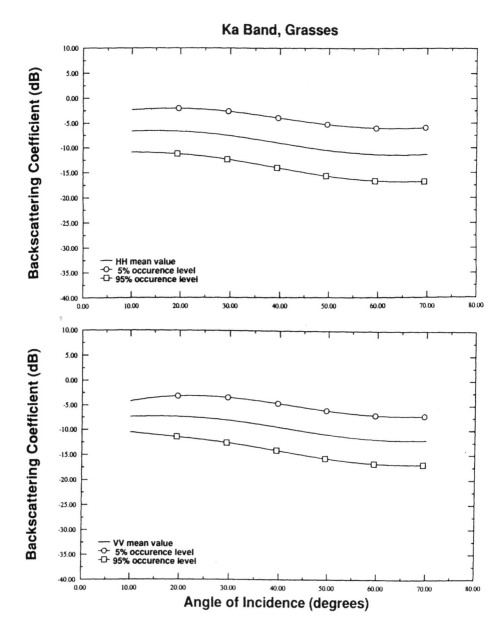

Ka Band, Grasses

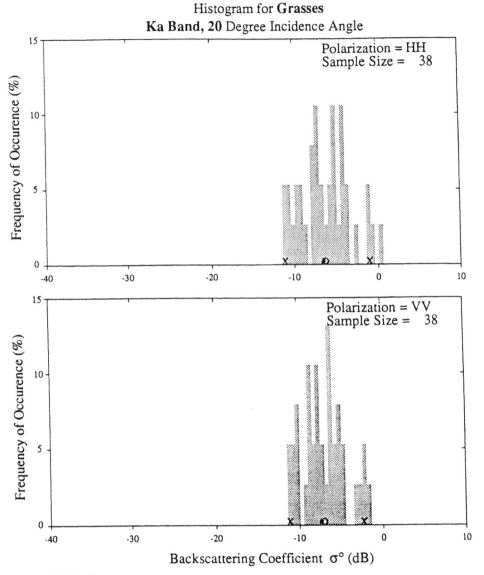

Histogram for **Grasses**
Ka Band, 20 Degree Incidence Angle

Polarization = HH
Sample Size =   38

Polarization = VV
Sample Size =   38

Backscattering Coefficient σ° (dB)

Median
O Mean
X 5% or 95% Level

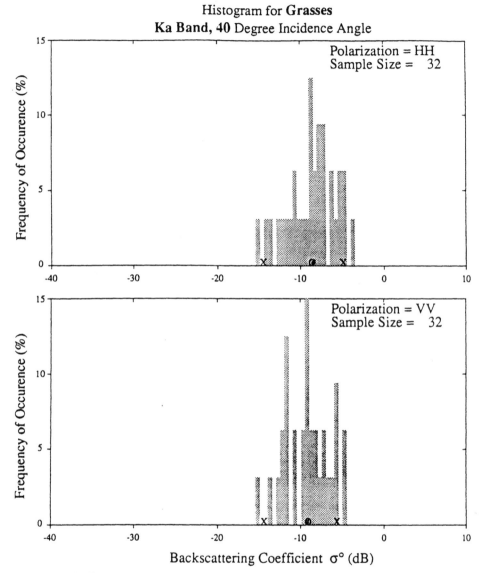

Histogram for **Grasses**
Ka Band, 40 Degree Incidence Angle

Median
O Mean
X 5% or 95% Level

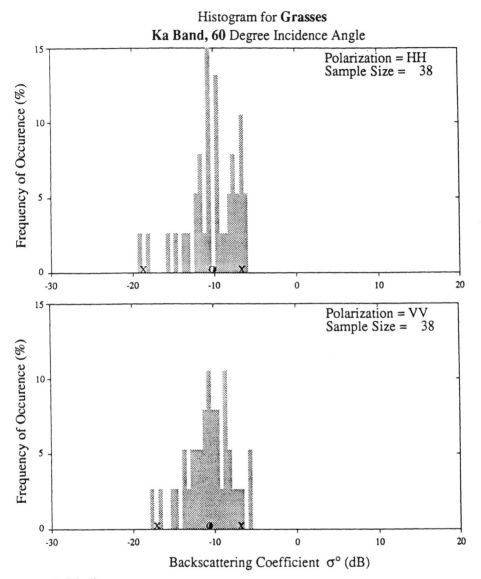

Histogram for **Grasses**
Ka Band, 60 Degree Incidence Angle

Median
O Mean
X 5% or 95% Level

C-8   W-Band Data:

Statistical Distribution Table for Grasses

W Band, VV Polarization

Angle	N	σ^o_{max}	σ^o_5	σ^o_{25}	Median	σ^o_{75}	σ^o_{95}	σ^o_{min}	Mean	Std. Dev.
0°	3	3.8						-2.1	0.7	3.0
5°	4	4.0						-2.7	-0.3	3.1
10°	4	0.7						-1.7	-0.9	1.1
15°	3	-1.5						-2.1	-1.8	0.3
20°	4	-0.3						-4.1	-1.7	1.7
30°	4	0.1						-4.8	-3.1	2.2
40°	4	-1.4						-4.3	-3.0	1.2
50°	4	-1.0						-4.7	-3.0	1.9
60°	3	-2.7						-7.5	-5.3	2.4
70°	2	-6.8						-8.3	-7.6	1.1

APPENDIX D
BACKSCATTERING DATA FOR SHRUBS

D-1 Data Sources and Parameter Loadings

Table D.1   Data sources for shrubs.

Band	References
L	25, 26, 28, 46, 96, 130
S	26, 46, 130
C	6, 19, 26, 28, 46, 89, 90, 107, 130
X	6, 34, 37, 45, 46, 48, 50, 56, 89, 90, 96
Ku	25, 26, 28, 34, 37, 46, 48, 50, 96
Ka	37, 45, 50, 96
W	

Table D.2   Mean and standard deviation parameter loadings for shrubs.

Band	Pol.	Angular Range θ_{min}	θ_{max}	P_1	P_2	P_3	P_4	P_5	P_6	M_1	M_2	M_3
L	HH	0	80	-26.688	29.454	1.814	0.873	4.135	-3.142	-9.0	14.931	0.092
	HV	0	80	-99.0	99.0	0.086	-21.298	0.0	0.0	4.747	-0.044	-2.826
	VV	0	80	-81.371	99.0	0.567	16.200	-1.948	3.142	-9.0	13.808	0.053
S	HH	0	80	-21.202	21.177	2.058	-0.132	-5.0	-3.142	1.713	3.205	1.729
	HV	0	80	-89.222	44.939	.253	30.0	-0.355	0.526	12.735	-9.0	-0.159
	VV	0	80	-20.566	20.079	1.776	-1.332	5.0	-1.983	2.475	2.308	3.858
C	HH	0	80	-91.950	99.0	0.270	6.980	1.922	-3.142	1.723	3.376	1.975
	HV	0	80	-99.0	91.003	0.156	3.948	2.239	-3.142	13.237	-9.0	-0.178
	VV	0	80	-91.133	99.0	0.294	8.107	2.112	-3.142	1.684	3.422	2.376
X	HH	0	80	-99.0	97.280	0.107	-0.538	5.0	-2.688	2.038	4.238	2.997
	HV	20	70	-28.057	0.0	0.0	13.575	1.0	-0.573	3.301	-0.001	-4.934
	VV	0	80	-99.0	97.682	0.113	-0.779	5.0	-2.076	2.081	4.025	2.997
Ku	HH	0	80	-99.0	98.254	0.098	-0.710	5.0	-2.225	1.941	4.096	2.930
	HV	40	70	-30.403	0.0	0.0	19.378	1.0	-0.590	-9.0	11.516	0.020
	VV	0	80	-99.0	98.741	0.103	-0.579	5.0	-2.210	2.192	3.646	3.320
Ka	HH	20	70	-41.170	27.831	0.076	-8.728	0.869	3.142	2.171	4.391	4.618
	HV											
	VV	20	70	-43.899	41.594	0.215	-0.794	5.0	-1.372	2.117	2.880	4.388
W	HH											
	HV		Insufficient Data									
	VV											

$$\partial^{\circ} = P_1 + P_2 \exp(-P_3\theta) + P_4 \cos(P_5\theta + P_6)$$
$$SD(\theta) = M_1 + M_2 \exp(-M_3\theta)$$

where θ is the angle of incidence in radians.

D-2 L-Band Data:

Statistical Distribution Table for **Shrubs**

L Band, HH Polarization

Angle	N	σ^o_{max}	σ^o_{5}	σ^o_{25}	Median	σ^o_{75}	σ^o_{95}	σ^o_{min}	Mean	Std. Dev.
0°	396	13.7	10.5	5.6	-0.4	-2.6	-5.2	-10.8	1.3	5.1
5°	204	12.4	8.0	3.9	0.9	-1.9	-6.7	-14.1	0.8	4.6
10°	427	6.7	1.4	-3.2	-7.9	-12.5	-15.7	-18.4	-7.5	5.5
15°	205	4.1	1.2	-2.1	-5.2	-8.7	-14.0	-18.5	-5.6	4.6
20°	441	5.5	-2.2	-6.6	-12.9	-18.8	-21.9	-24.8	-12.5	6.7
25°	43	2.3	-2.3	-4.2	-7.5	-11.0	-16.9	-22.0	-7.9	4.9
30°	440	0.7	-4.7	-9.7	-15.7	-20.7	-24.2	-26.8	-15.0	6.5
35°	43	-1.7	-3.4	-8.0	-10.1	-13.9	-19.0	-25.5	-10.5	4.9
40°	279	-2.2	-7.7	-17.1	-21.0	-23.1	-26.6	-29.1	-19.4	5.7
45°	48	-7.8	-9.3	-11.0	-14.5	-16.3	-22.1	-27.3	-14.3	4.0
50°	78	-5.4	-7.7	-11.0	-13.8	-18.0	-24.8	-26.2	-14.5	4.9
60°	242	-6.7	-15.4	-22.2	-24.5	-26.3	-29.6	-31.4	-23.8	4.5
70°	19	-8.6	-9.0	-9.7	-12.6	-20.7	-23.8	-23.9	-14.3	5.4
80°	223	-17.2	-21.4	-23.8	-25.3	-26.7	-29.0	-33.6	-25.2	2.4

L Band, HV Polarization

Angle	N	σ^o_{max}	σ^o_{5}	σ^o_{25}	Median	σ^o_{75}	σ^o_{95}	σ^o_{min}	Mean	Std. Dev.
0°	244	-9.4	-16.1	-20.9	-23.9	-25.3	-27.2	-31.2	-22.9	3.6
5°	74	-12.1	-14.0	-15.8	-17.0	-18.8	-24.0	-25.4	-17.5	2.8
10°	351	-11.1	-15.8	-20.5	-24.7	-28.1	-30.6	-33.1	-24.1	4.8
15°	147	-10.8	-13.3	-16.9	-19.4	-22.1	-26.1	-32.2	-19.6	3.9
20°	378	-11.1	-16.1	-20.0	-25.1	-29.5	-34.3	-36.0	-24.8	5.7
25°	46	-14.3	-17.0	-18.9	-21.1	-23.4	-29.1	-31.4	-21.6	3.7
30°	378	-14.6	-16.8	-20.5	-25.4	-29.6	-34.8	-36.5	-25.4	5.5
35°	46	-17.3	-18.6	-19.6	-22.0	-24.2	-29.7	-31.6	-22.3	3.3
40°	269	-18.4	-21.2	-25.1	-28.5	-31.1	-35.6	-38.5	-28.3	4.4
45°	46	-20.6	-21.2	-22.6	-24.7	-26.5	-31.2	-31.8	-24.9	2.9
50°	63	-19.3	-20.2	-22.0	-24.2	-26.6	-31.3	-32.1	-24.6	3.4
60°	223	-21.7	-26.2	-28.5	-30.8	-32.9	-37.0	-38.5	-30.9	3.5
80°	221	-23.1	-29.4	-30.9	-32.4	-34.7	-36.7	-42.7	-32.8	2.6

L Band, VV Polarization

Angle	N	σ^o_{max}	σ^o_{5}	σ^o_{25}	Median	σ^o_{75}	σ^o_{95}	σ^o_{min}	Mean	Std. Dev.
0°	396	13.4	8.8	4.6	-0.6	-2.4	-5.2	-13.4	0.8	4.6
5°	160	11.8	7.6	3.3	0.7	-2.4	-7.4	-15.3	0.4	4.6
10°	382	5.1	0.0	-4.4	-7.6	-12.1	-15.6	-19.2	-7.9	5.0
15°	161	1.5	-2.7	-5.9	-9.2	-11.5	-15.6	-19.7	-8.9	4.1
20°	398	-0.5	-5.4	-10.3	-14.1	-17.1	-20.7	-23.0	-13.5	4.6
30°	397	-4.1	-8.2	-12.5	-16.2	-19.1	-22.6	-27.4	-15.8	4.4
40°	236	-4.6	-10.6	-16.6	-18.9	-21.7	-25.2	-28.6	-18.7	4.5
45°	5	-16.5						-19.1	-17.8	1.0
50°	36	-5.4	-7.0	-9.5	-15.3	-18.4	-23.0	-23.1	-14.5	5.2
60°	241	-7.4	-12.5	-17.5	-20.1	-23.6	-26.8	-29.3	-20.1	4.4
70°	19	-7.0	-8.2	-9.7	-12.2	-20.5	-22.5	-22.6	-14.2	5.7
80°	223	-14.1	-16.6	-19.5	-22.0	-23.9	-27.3	-31.1	-21.9	3.2

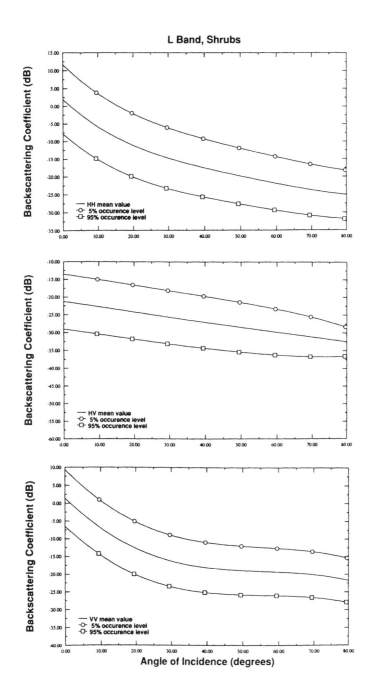

L Band, Shrubs

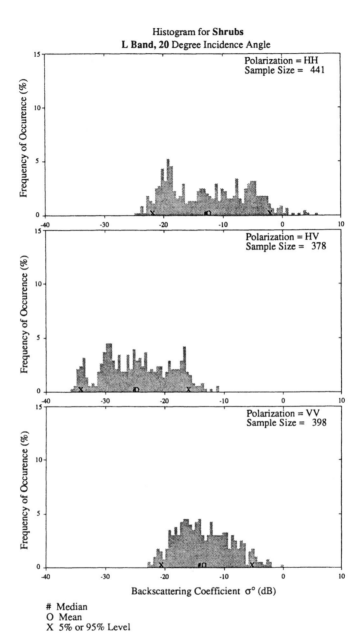

Histogram for **Shrubs**
L Band, 20 Degree Incidence Angle

Polarization = HH
Sample Size = 441

Polarization = HV
Sample Size = 378

Polarization = VV
Sample Size = 398

Backscattering Coefficient σ° (dB)

Frequency of Occurence (%)

Median
O Mean
X 5% or 95% Level

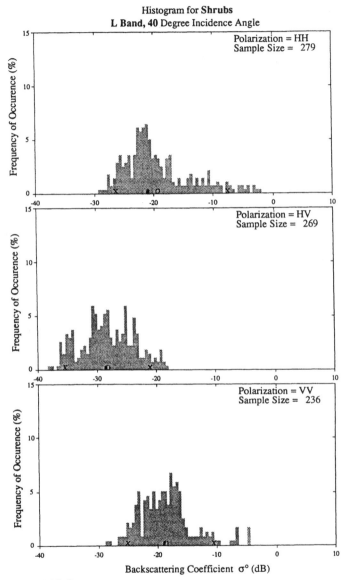

Histogram for **Shrubs**
L Band, 40 Degree Incidence Angle

Median
O Mean
X 5% or 95% Level

Histogram for **Shrubs**
L Band, 60 Degree Incidence Angle

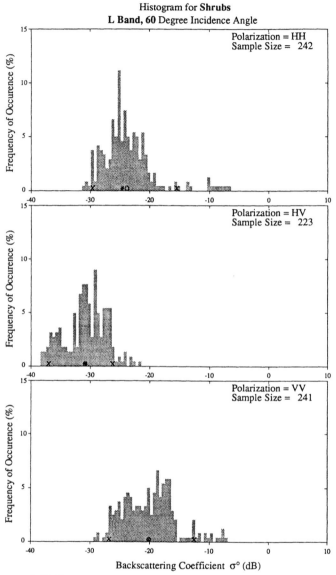

Median
O Mean
X 5% or 95% Level

D-3   S-Band Data:

Statistical Distribution Table for Shrubs

S Band, HH Polarization

Angle	N	σ^o_{max}	σ^o_5	σ^o_{25}	Median	σ^o_{75}	σ^o_{95}	σ^o_{min}	Mean	Std. Dev.
0°	397	13.4	9.3	3.6	-1.8	-3.8	-6.5	-12.9	-0.2	4.9
5°	146	9.2	5.4	2.3	-1.1	-3.7	-7.7	-14.4	-0.9	4.5
10°	397	4.3	0.0	-4.2	-7.3	-10.5	-13.2	-16.9	-7.2	4.2
15°	146	0.2	-2.2	-4.6	-6.9	-9.1	-12.0	-18.4	-7.0	3.2
20°	397	0.1	-5.9	-9.1	-11.8	-14.4	-16.8	-20.0	-11.7	3.5
30°	397	-5.5	-8.6	-11.0	-13.4	-16.1	-18.2	-26.9	-13.5	3.2
40°	251	-9.8	-12.8	-15.0	-16.9	-18.6	-20.3	-22.6	-16.7	2.4
60°	251	-11.9	-14.8	-16.9	-18.4	-20.2	-22.1	-24.9	-18.5	2.3
80°	251	-13.9	-16.8	-18.8	-20.1	-21.5	-23.3	-25.0	-20.0	2.0

S Band, HV Polarization

Angle	N	σ^o_{max}	σ^o_5	σ^o_{25}	Median	σ^o_{75}	σ^o_{95}	σ^o_{min}	Mean	Std. Dev.
0°	315	-3.6	-10.8	-16.5	-20.0	-21.7	-23.6	-26.7	-18.8	4.0
5°	107	-8.6	-12.1	-14.1	-16.0	-18.0	-21.9	-26.1	-16.2	3.1
10°	391	-10.3	-13.8	-17.3	-20.6	-23.1	-25.2	-28.7	-20.1	3.7
15°	140	-11.5	-13.7	-15.6	-17.3	-19.4	-21.8	-27.4	-17.6	2.7
20°	391	-12.8	-15.7	-18.2	-21.6	-23.8	-25.5	-29.6	-21.0	3.3
30°	386	-14.5	-16.7	-18.7	-21.7	-24.0	-25.9	-27.4	-21.4	3.0
40°	251	-17.3	-19.3	-22.0	-23.6	-25.0	-26.9	-28.5	-23.4	2.3
60°	251	-18.1	-21.6	-23.6	-24.7	-25.9	-27.6	-30.7	-24.7	1.9
80°	251	-23.0	-25.0	-26.6	-27.7	-29.0	-31.0	-32.9	-27.9	1.8

S Band, VV Polarization

Angle	N	σ^o_{max}	σ^o_5	σ^o_{25}	Median	σ^o_{75}	σ^o_{95}	σ^o_{min}	Mean	Std. Dev.
0°	395	13.2	8.3	3.4	-1.7	-3.6	-6.2	-12.0	-0.3	4.6
5°	146	9.5	5.2	2.1	-1.4	-4.5	-9.1	-14.0	-1.3	4.6
10°	397	6.1	0.0	-4.1	-7.3	-9.9	-12.7	-16.1	-6.8	4.0
15°	146	-0.5	-3.9	-6.5	-8.9	-10.2	-13.0	-19.0	-8.5	3.0
20°	397	-1.5	-6.3	-9.6	-11.9	-13.7	-15.6	-21.9	-11.6	2.9
30°	395	-4.8	-9.0	-11.6	-13.6	-15.1	-16.7	-21.2	-13.3	2.5
40°	251	-7.8	-9.9	-13.4	-15.0	-16.7	-19.1	-21.7	-15.0	2.7
60°	251	-8.5	-11.9	-14.4	-16.2	-18.1	-20.7	-23.2	-16.2	2.7
80°	251	-10.1	-14.7	-17.6	-19.4	-21.0	-23.2	-24.8	-19.2	2.6

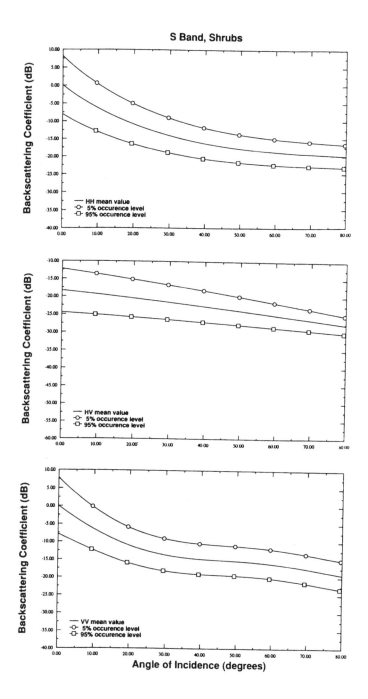

S Band, Shrubs

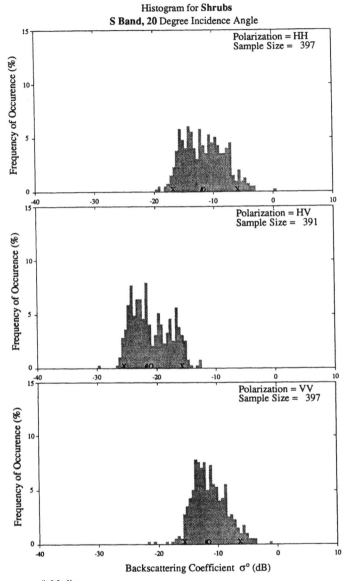

Histogram for **Shrubs**
S Band, 20 Degree Incidence Angle

Median
O Mean
X 5% or 95% Level

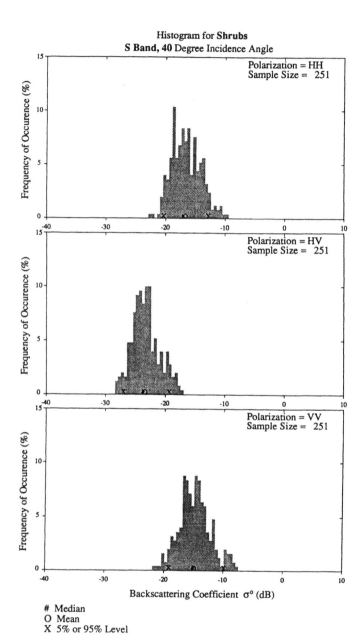

Histogram for **Shrubs**
S Band, 40 Degree Incidence Angle

Median
O Mean
X 5% or 95% Level

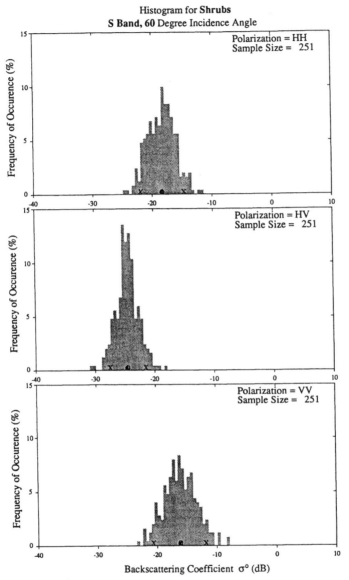

Histogram for **Shrubs**
S Band, 60 Degree Incidence Angle

Polarization = HH
Sample Size = 251

Polarization = HV
Sample Size = 251

Polarization = VV
Sample Size = 251

Backscattering Coefficient σ° (dB)

Median
O Mean
X 5% or 95% Level

D-4 C-Band Data:

Statistical Distribution Table for Shrubs

C Band, HH Polarization

Angle	N	σ°_{max}	σ°_{5}	σ°_{25}	Median	σ°_{75}	σ°_{95}	σ°_{min}	Mean	Std. Dev.
0°	805	15.4	8.6	2.9	-2.5	-4.8	-7.3	-9.6	-0.9	5.1
5°	352	15.1	8.1	3.0	-0.4	-3.0	-6.1	-12.6	0.1	4.5
10°	1176	11.4	2.1	0.0	-3.5	-7.0	-10.7	-16.2	-3.7	4.3
15°	352	5.7	-0.1	-2.7	-4.9	-6.9	-9.8	-15.8	-4.9	3.2
20°	1040	3.6	-2.4	-5.8	-8.1	-10.5	-13.3	-17.6	-8.1	3.3
25°	48	-5.8	-7.3	-8.5	-9.5	-10.5	-13.4	-17.6	-9.6	1.9
30°	939	-0.9	-4.3	-7.7	-9.7	-11.7	-14.7	-21.1	-9.7	3.1
35°	48	-7.6	-8.6	-9.6	-10.5	-11.2	-14.8	-18.8	-10.7	1.9
40°	553	-5.9	-8.1	-10.3	-12.0	-13.7	-15.9	-18.3	-12.0	2.4
45°	57	-9.3	-9.5	-10.2	-11.4	-12.4	-16.1	-18.8	-11.5	2.0
50°	643	0.7	-9.1	-11.0	-12.8	-14.6	-18.3	-22.9	-13.0	2.9
60°	524	-6.3	-10.6	-12.6	-13.7	-15.1	-17.0	-19.0	-13.8	1.9
80°	505	-3.7	-15.3	-16.9	-18.0	-19.1	-20.5	-28.5	-18.0	1.8

C Band, HV Polarization

Angle	N	σ°_{max}	σ°_{5}	σ°_{25}	Median	σ°_{75}	σ°_{95}	σ°_{min}	Mean	Std. Dev.
0°	733	3.4	-2.5	-9.8	-13.4	-16.0	-18.6	-20.6	-12.5	4.8
5°	298	3.4	-2.4	-8.5	-11.8	-13.5	-15.9	-21.4	-10.8	4.3
10°	824	0.2	-6.9	-12.8	-15.1	-17.3	-19.7	-22.8	-14.6	3.8
15°	321	0.0	-6.7	-11.7	-13.8	-15.9	-19.2	-22.9	-13.3	3.9
20°	993	0.0	-10.2	-14.2	-16.2	-18.3	-20.8	-27.8	-16.0	3.4
25°	46	-12.3	-14.9	-15.5	-16.5	-17.6	-20.7	-22.8	-16.7	2.0
30°	886	-0.4	-11.0	-14.7	-16.6	-18.8	-21.6	-26.6	-16.5	3.4
35°	46	-14.7	-16.2	-16.9	-17.6	-19.6	-21.9	-24.4	-18.2	2.0
40°	549	-11.3	-14.1	-16.2	-17.6	-19.3	-21.4	-23.8	-17.7	2.3
45°	57	-12.1	-15.1	-17.8	-18.9	-20.9	-26.3	-27.3	-19.2	3.1
50°	638	-7.5	-12.7	-15.9	-17.8	-20.0	-24.9	-30.5	-18.1	3.4
60°	520	-11.9	-15.3	-17.2	-18.4	-19.7	-21.7	-23.6	-18.5	1.9
80°	503	-9.6	-18.5	-20.6	-22.1	-23.5	-25.2	-26.9	-22.0	2.1

C Band, VV Polarization

Angle	N	σ°_{max}	σ°_{5}	σ°_{25}	Median	σ°_{75}	σ°_{95}	σ°_{min}	Mean	Std. Dev.
0°	798	17.0	8.5	2.9	-2.4	-4.8	-7.3	-10.7	-0.9	5.1
5°	302	15.9	8.8	3.4	0.0	-2.8	-6.4	-12.7	0.5	4.8
10°	802	12.4	1.6	-2.8	-5.4	-7.9	-10.5	-14.8	-5.1	3.8
15°	301	5.2	-0.1	-3.1	-5.2	-7.5	-10.8	-16.0	-5.3	3.4
20°	804	3.5	-2.8	-6.4	-8.7	-10.6	-12.8	-17.3	-8.4	3.1
30°	887	0.5	-4.8	-7.8	-9.8	-11.7	-14.1	-22.0	-9.7	2.9
40°	503	-5.6	-7.5	-10.1	-11.6	-13.1	-15.0	-18.3	-11.5	2.2
45°	11	-7.9						-14.1	-11.0	2.2
50°	111	-7.1	-9.0	-11.3	-13.0	-14.6	-16.5	-17.9	-12.9	2.3
60°	508	-7.3	-10.0	-11.9	-13.2	-14.5	-16.3	-19.2	-13.2	1.9
80°	503	-10.1	-14.6	-16.4	-17.7	-18.8	-20.3	-22.3	-17.6	1.8

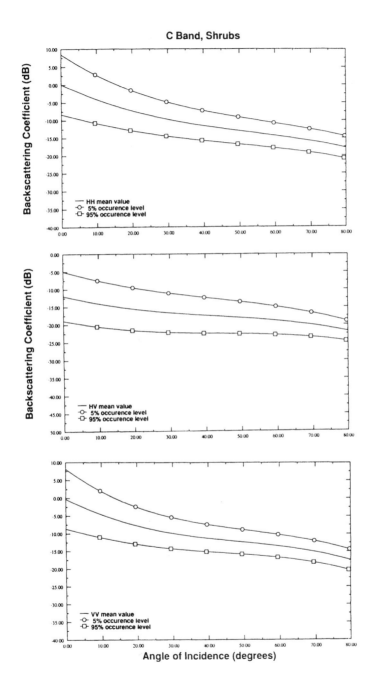

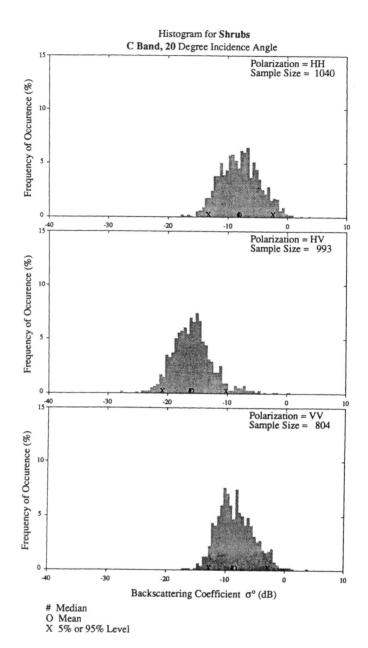

Histogram for **Shrubs**
C Band, 20 Degree Incidence Angle

Polarization = HH
Sample Size = 1040

Polarization = HV
Sample Size = 993

Polarization = VV
Sample Size = 804

Backscattering Coefficient σ° (dB)

\# Median
O Mean
X 5% or 95% Level

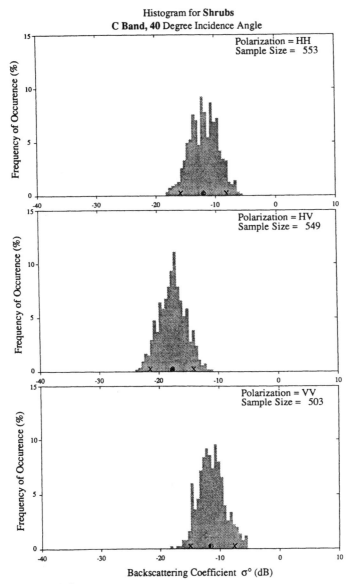

Histogram for **Shrubs**
C Band, 40 Degree Incidence Angle

Median
O Mean
X 5% or 95% Level

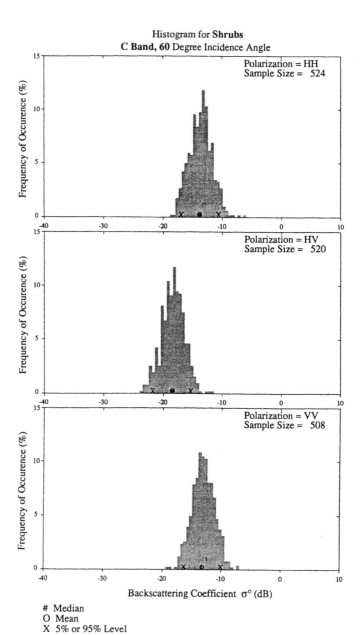

Histogram for **Shrubs**
C Band, 60 Degree Incidence Angle

Median
O Mean
X 5% or 95% Level

D-5 X-Band Data:

Statistical Distribution Table for Shrubs

X Band, HH Polarization

Angle	N	σ^o_{max}	σ^o_5	σ^o_{25}	Median	σ^o_{75}	σ^o_{95}	σ^o_{min}	Mean	Std. Dev.
0°	606	16.1	7.7	-1.6	-3.9	-6.5	-9.0	-11.4	-2.9	5.3
10°	1075	16.2	10.9	1.9	-4.1	-6.5	-8.8	-11.2	-2.1	6.1
20°	1111	9.1	-0.7	-4.4	-6.6	-8.1	-10.0	-12.1	-6.0	3.1
30°	1202	-0.5	-3.3	-5.9	-7.5	-9.0	-10.6	-14.6	-7.3	2.3
40°	2523	-1.7	-5.9	-7.7	-9.1	-10.5	-12.7	-16.8	-9.2	2.1
50°	2569	-2.6	-6.8	-8.7	-10.0	-12.1	-14.8	-22.8	-10.4	2.5
60°	2616	0.0	-7.6	-9.8	-11.1	-12.7	-15.4	-18.8	-11.2	2.5
70°	1930	-2.2	-9.8	-11.6	-12.9	-14.5	-17.7	-22.2	-13.2	2.4
80°	605	-10.5	-13.3	-14.6	-15.6	-16.8	-18.1	-20.8	-15.7	1.6

X Band, HV Polarization

Angle	N	σ^o_{max}	σ^o_5	σ^o_{25}	Median	σ^o_{75}	σ^o_{95}	σ^o_{min}	Mean	Std. Dev.
0°	4	-9.0						-14.0	-11.5	2.9
10°	6	-9.8						-15.4	-12.9	2.4
20°	171	-8.3	-9.9	-13.6	-15.5	-17.3	-20.4	-27.0	-15.5	3.1
30°	160	-9.1	-11.1	-14.3	-17.0	-19.6	-24.0	-29.1	-17.2	4.0
40°	1478	-7.5	-10.1	-11.7	-13.0	-14.7	-17.2	-21.7	-13.2	2.2
50°	2690	-6.6	-10.9	-12.9	-15.0	-18.2	-24.6	-35.3	-16.0	4.2
60°	1528	-6.5	-12.1	-13.6	-14.9	-16.7	-20.0	-25.2	-15.3	2.4
70°	1492	-4.8	-13.9	-15.7	-16.9	-18.6	-22.5	-25.9	-17.3	2.5

X Band, VV Polarization

Angle	N	σ^o_{max}	σ^o_5	σ^o_{25}	Median	σ^o_{75}	σ^o_{95}	σ^o_{min}	Mean	Std. Dev.
0°	610	15.6	7.3	-1.5	-3.9	-6.4	-8.6	-11.3	-2.7	5.2
10°	1074	15.1	10.3	1.5	-4.1	-6.7	-8.6	-11.6	-2.2	6.0
20°	1207	8.3	-0.6	-4.5	-6.6	-8.1	-9.7	-14.1	-6.0	3.0
30°	1198	0.5	-3.9	-5.9	-7.6	-9.1	-10.9	-17.2	-7.5	2.2
40°	2521	-3.1	-5.8	-7.8	-9.2	-10.6	-12.9	-18.4	-9.2	2.1
50°	3684	-1.1	-5.6	-7.7	-9.4	-11.4	-14.8	-20.8	-9.7	2.8
60°	2565	-2.5	-8.1	-10.0	-11.3	-12.9	-15.2	-18.5	-11.5	2.2
70°	1929	-2.2	-9.8	-11.7	-13.0	-14.6	-17.9	-22.3	-13.3	2.4
80°	601	-11.2	-13.9	-15.5	-16.7	-17.7	-19.5	-22.0	-16.6	1.7

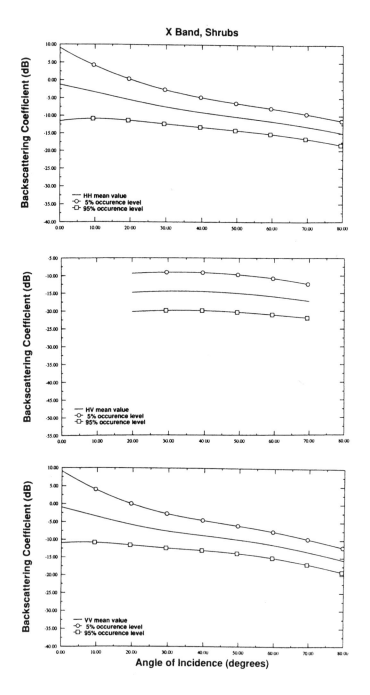

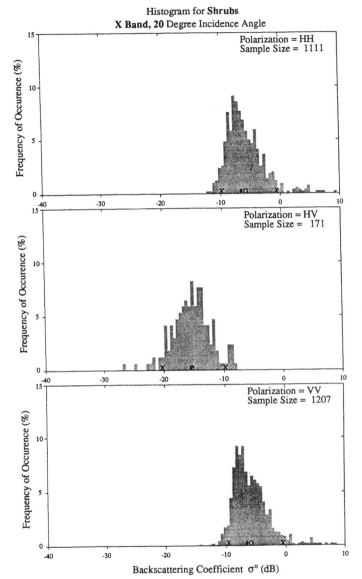

Histogram for **Shrubs**
X Band, 20 Degree Incidence Angle

Median
O Mean
X 5% or 95% Level

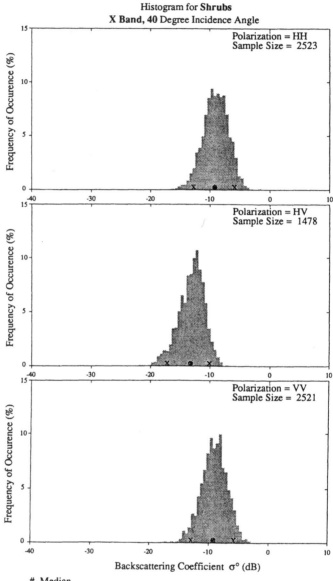

Histogram for **Shrubs**
X Band, 40 Degree Incidence Angle

Median
O Mean
X 5% or 95% Level

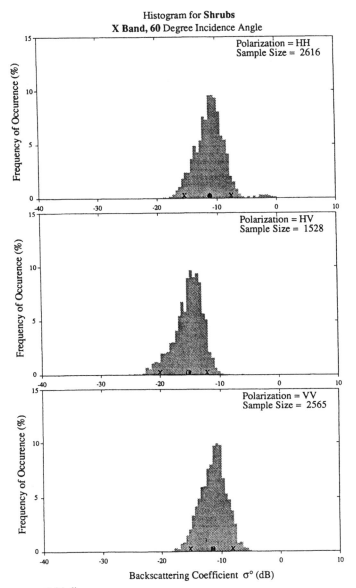

Histogram for **Shrubs**
X Band, 60 Degree Incidence Angle

Median
O Mean
X 5% or 95% Level

D-6 Ku-Band Data:

Statistical Distribution Table for Shrubs

Ku Band, HH Polarization

Angle	N	σ^o_{max}	σ^o_5	σ^o_{25}	Median	σ^o_{75}	σ^o_{95}	σ^o_{min}	Mean	Std. Dev.
0°	742	19.8	9.5	-0.2	-2.9	-5.3	-8.0	-10.7	-1.8	5.2
10°	1298	18.0	10.0	1.8	-3.1	-5.6	-8.1	-16.2	-1.5	5.7
20°	1303	11.2	0.2	-3.5	-5.5	-6.8	-8.8	-13.1	-4.9	3.1
30°	1304	0.8	-2.5	-4.7	-6.3	-7.7	-9.8	-15.2	-6.2	2.2
40°	2985	0.4	-4.0	-6.0	-7.5	-9.0	-11.5	-15.3	-7.6	2.2
50°	2715	-2.1	-5.0	-7.0	-8.3	-9.7	-12.6	-15.4	-8.4	2.2
60°	2991	-2.1	-5.9	-7.8	-9.4	-11.0	-13.8	-17.1	-9.5	2.4
70°	2231	-5.3	-7.8	-9.6	-11.0	-12.5	-15.4	-19.0	-11.2	2.2
80°	687	-9.1	-11.2	-13.1	-14.4	-15.8	-17.7	-20.0	-14.4	2.0

Ku Band, HV Polarization

Angle	N	σ^o_{max}	σ^o_5	σ^o_{25}	Median	σ^o_{75}	σ^o_{95}	σ^o_{min}	Mean	Std. Dev.
0°	15	0.0	-3.3	-9.9	-12.7	-15.7	-18.3	-19.1	-11.5	5.5
10°	16	-12.8	-12.9	-13.8	-15.3	-17.1	-20.7	-21.8	-15.9	2.7
20°	17	-15.0	-15.3	-16.1	-17.3	-19.0	-21.0	-22.1	-17.5	2.0
30°	16	-14.3	-17.0	-17.7	-18.3	-20.1	-22.0	-22.1	-18.7	2.0
40°	1702	-4.6	-7.6	-9.5	-10.9	-12.5	-15.5	-17.7	-11.1	2.3
50°	1715	-6.5	-8.4	-10.1	-11.7	-13.4	-16.1	-22.5	-11.9	2.4
60°	1677	-7.7	-9.6	-11.2	-12.7	-14.4	-17.2	-20.4	-12.9	2.3
70°	1549	-9.7	-11.7	-13.3	-14.6	-16.0	-19.5	-22.7	-14.8	2.2

Ku Band, VV Polarization

Angle	N	σ^o_{max}	σ^o_5	σ^o_{25}	Median	σ^o_{75}	σ^o_{95}	σ^o_{min}	Mean	Std. Dev.
0°	745	20.3	9.7	0.8	-2.3	-4.6	-7.2	-11.3	-1.0	5.3
5°	63	6.8	2.4	-0.1	-1.5	-2.8	-6.8	-10.7	-1.5	2.9
10°	1348	15.4	9.5	1.4	-2.6	-4.9	-7.4	-14.8	-1.3	5.3
15°	64	4.1	0.1	-1.5	-2.1	-7.6	-10.6	-15.2	-3.8	3.9
20°	1353	10.4	0.5	-3.1	-4.9	-6.3	-8.4	-14.8	-4.5	3.0
25°	46	-1.1	-2.9	-3.6	-4.2	-5.2	-7.7	-11.5	-4.6	1.7
30°	1354	1.3	-1.9	-4.3	-5.7	-7.2	-9.4	-17.5	-5.8	2.3
35°	46	-4.2	-4.9	-5.4	-6.1	-7.6	-9.6	-12.7	-6.7	1.7
40°	3034	0.1	-3.8	-5.7	-7.2	-8.9	-11.6	-15.8	-7.4	2.4
45°	50	-3.3	-4.6	-5.3	-5.7	-8.1	-10.7	-11.6	-6.7	2.1
50°	2763	-2.8	-4.8	-6.8	-8.2	-9.8	-12.9	-17.9	-8.4	2.4
60°	2995	-2.7	-5.7	-7.9	-9.4	-11.1	-13.8	-17.6	-9.6	2.5
70°	2240	-5.3	-7.8	-9.9	-11.2	-12.8	-16.1	-19.4	-11.5	2.4
80°	676	-8.3	-10.6	-12.6	-14.1	-15.6	-17.5	-26.4	-14.1	2.1

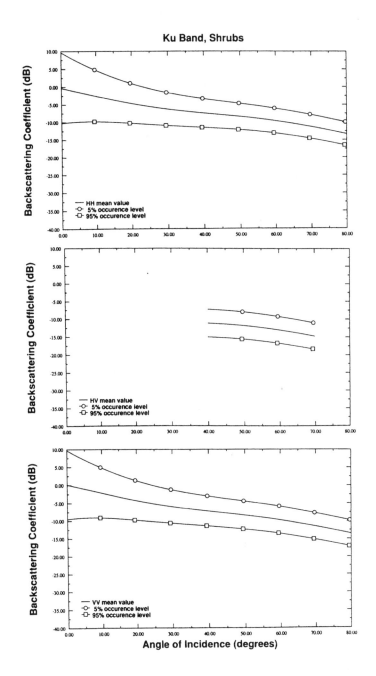

Ku Band, Shrubs

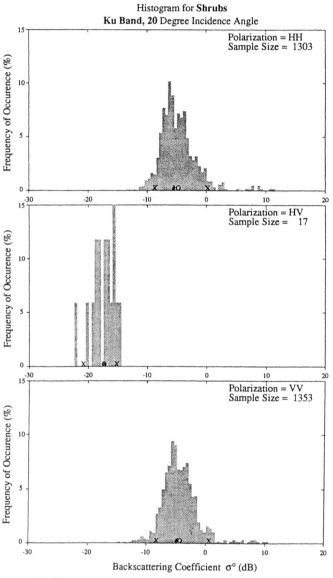

Histogram for **Shrubs**
Ku Band, 20 Degree Incidence Angle

Median
O Mean
X 5% or 95% Level

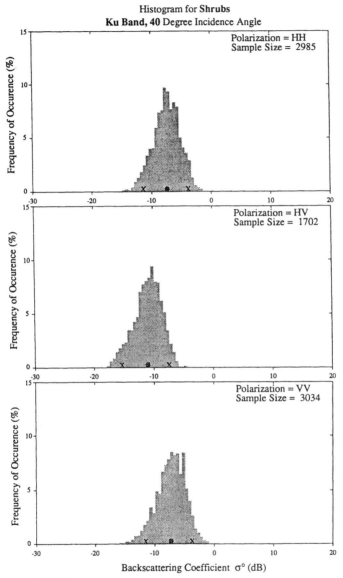

Histogram for **Shrubs**
Ku Band, 40 Degree Incidence Angle

Median
O Mean
X 5% or 95% Level

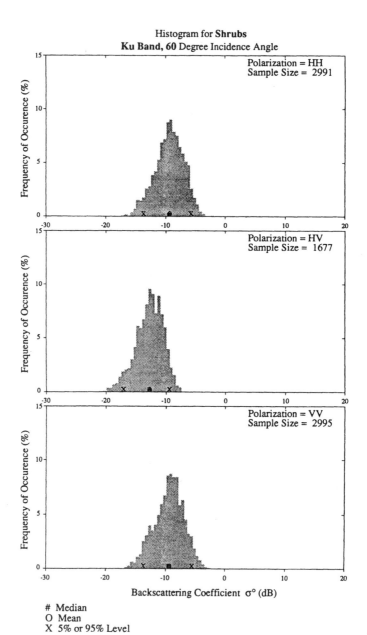

Histogram for **Shrubs**
Ku Band, 60 Degree Incidence Angle

Median
O Mean
X 5% or 95% Level

D-7 Ka-Band Data:

Statistical Distribution Table for Shrubs

Ka Band, HH Polarization

Angle	N	σ^o_{max}	σ^o_5	σ^o_{25}	Median	σ^o_{75}	σ^o_{95}	σ^o_{min}	Mean	Std. Dev.
10°	13	0.5						-11.1	-6.4	3.7
20°	29	0.0	-2.2	-3.5	-6.0	-7.5	-11.3	-11.3	-5.8	2.9
30°	29	-0.5	-2.9	-4.6	-6.6	-8.6	-11.6	-11.6	-6.4	3.0
40°	25	-1.0	-4.6	-6.1	-7.4	-9.8	-12.2	-12.2	-7.7	2.9
45°	4	-6.2						-7.8	-6.8	0.7
50°	230	-2.0	-6.4	-7.5	-8.4	-10.0	-12.5	-15.5	-8.8	2.0
60°	29	-3.0	-6.5	-8.1	-10.5	-11.8	-14.0	-14.1	-9.9	2.7
70°	29	-4.7	-7.9	-9.6	-12.0	-13.8	-15.7	-15.7	-11.6	2.9
80°	10	-6.6						-19.6	-15.7	4.1

Ka Band, VV Polarization

Angle	N	σ^o_{max}	σ^o_5	σ^o_{25}	Median	σ^o_{75}	σ^o_{95}	σ^o_{min}	Mean	Std. Dev.
10°	15	0.4	-2.3	-5.8	-8.1	-10.1	-10.8	-11.1	-6.9	3.4
20°	32	-0.8	-2.2	-4.1	-5.9	-8.5	-10.3	-10.3	-6.1	2.7
30°	31	-1.6	-3.9	-4.5	-6.9	-9.:	-10.6	-10.7	-6.8	2.5
40°	28	-2.6	-4.5	-6.2	-8.7	-10.2	-11.1	-11.6	-8.1	2.5
45°	5	-4.7						-7.5	-6.0	1.2
50°	233	-3.8	-6.1	-7.1	-8.1	-9.7	-12.8	-16.5	-8.6	2.1
60°	33	-5.0	-6.4	-9.6	-11.4	-12.0	-13.2	-13.9	-10.3	2.2
70°	30	-6.6	-8.5	-10.0	-12.7	-13.8	-15.2	-15.4	-11.9	2.4
80°	11	-13.1						-17.1	-15.5	1.1

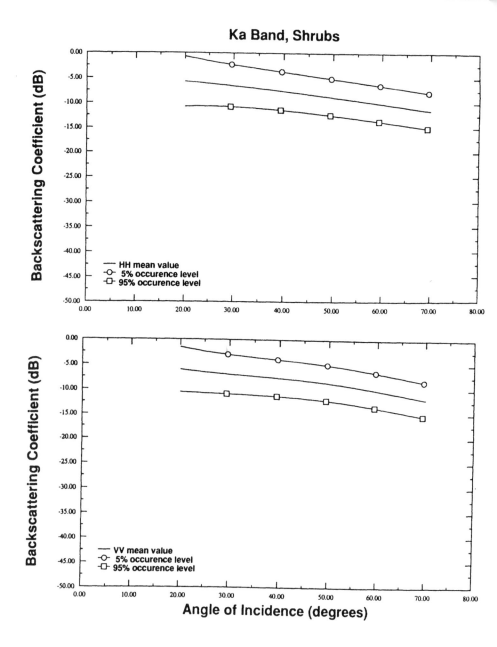

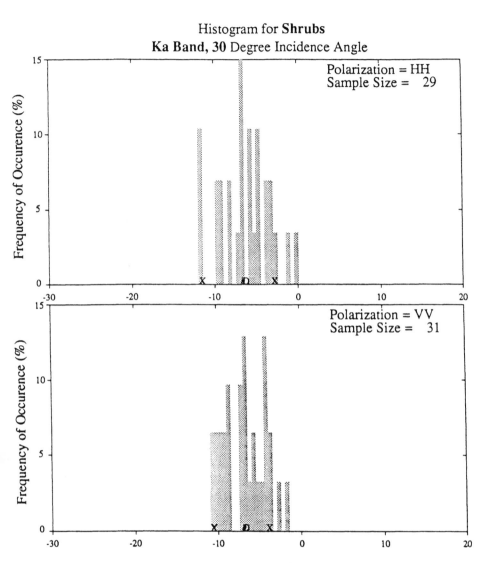

Histogram for **Shrubs**
Ka Band, 30 Degree Incidence Angle

Polarization = HH
Sample Size =   29

Polarization = VV
Sample Size =   31

Backscattering Coefficient σ° (dB)

\# Median
O Mean
X 5% or 95% Level

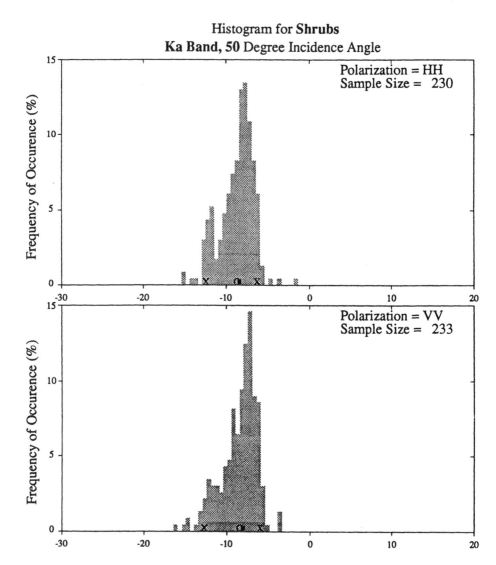

Histogram for **Shrubs**
Ka Band, 50 Degree Incidence Angle

Polarization = HH
Sample Size =   230

Polarization = VV
Sample Size =   233

Backscattering Coefficient σ° (dB)

\# Median
O Mean
X 5% or 95% Level

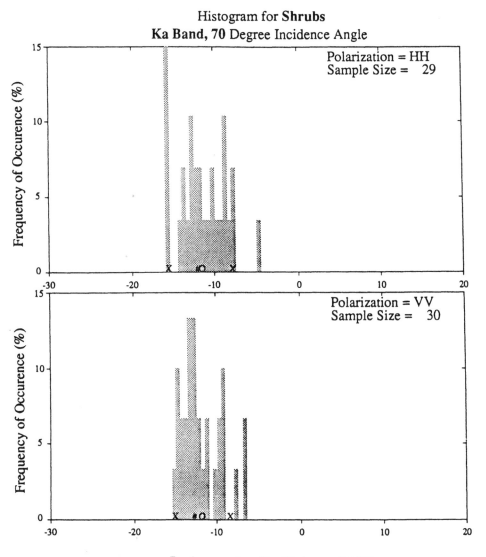

Histogram for **Shrubs**
Ka Band, 70 Degree Incidence Angle

Polarization = HH
Sample Size = 29

Polarization = VV
Sample Size = 30

Frequency of Occurence (%)

Backscattering Coefficient σ° (dB)

\# Median
O Mean
X 5% or 95% Level

APPENDIX E
BACKSCATTERING DATA FOR SHORT VEGETATION

E-1 Data Sources and Parameter Loadings

Table E.1 Data sources for short vegetation.

Band	References
L	25, 26, 28, 46, 71, 96, 99, 108, 130
S	26, 46, 68, 108, 130
C	6, 19, 26, 28, 46, 89, 90, 91, 107, 108, 130
X	6, 34, 37, 45, 46, 48, 50, 56, 89, 90, 91, 96, 99, 109
Ku	25, 26, 28, 34, 37, 45, 46, 48, 50, 96, 109
Ka	37, 45, 50, 96, 99, 109, 144
W	144

Table E.2 Mean and standard deviation parameter loadings for short vegetation.

Band	Pol.	Angular Range θ_{min}	Angular Range θ_{max}	$\bar{\sigma}^\circ$ Parameters P_1	P_2	P_3	P_4	P_5	P_6	SD(θ) Parameters M_1	M_2	M_3
L	HH	0	80	-27.265	32.390	2.133	1.438	-3.847	3.142	1.593	4.246	0.063
	HV	0	80	-41.60	22.872	0.689	-1.238	0.0	0.0	0.590	4.864	0.098
	VV	0	80	-24.614	27.398	2.265	-1.080	5.0	-1.999	4.918	0.819	15.0
S	HH	0	80	-20.779	21.867	2.434	0.347	-0.013	-0.393	2.527	3.273	3.001
	HV	0	80	-99.0	85.852	0.179	3.687	2.121	-3.142	13.195	-9.0	-0.148
	VV	0	80	-20.367	21.499	2.151	-1.069	5.0	-1.950	2.963	2.881	4.740
C	HH	0	80	-87.727	99.0	0.322	10.188	-1.747	3.142	2.586	2.946	2.740
	HV	0	80	-99.0	93.293	0.181	5.359	1.948	-3.142	13.717	-9.0	-0.169
	VV	0	80	-88.593	99.0	0.326	9.574	1.969	-3.142	2.287	3.330	2.674
X	HH	0	80	-99.0	97.417	0.114	-0.837	5.0	-2.984	2.490	3.514	3.217
	HV	10	70	-16.716	10.247	10.0	-1.045	5.0	-0.159	-9.0	13.278	0.066
	VV	0	80	-99.0	97.370	0.119	-1.171	5.0	-2.728	2.946	2.834	2.953
Ku	HH	0	80	-99.0	97.863	0.105	-0.893	5.0	-2.657	2.538	2.691	2.364
	HV	0	70	-14.234	3.468	10.0	-1.552	5.0	-0.562	-9.0	13.349	0.090
	VV	0	80	-99.0	97.788	0.105	-1.017	5.0	-3.142	1.628	3.117	0.566
Ka	HH	10	80	-99.0	79.050	0.263	-30.0	0.730	2.059	2.80	3.139	15.0
	HV											
	VV	10	80	-99.0	80.325	0.282	-30.0	0.833	1.970	2.686	-0.002	-2.853
W	HH											
	HV			Insufficient Data								
	VV											

$$\bar{\sigma}^\circ = P_1 + P_2 \exp(-P_3\theta) + P_4 \cos(P_5\theta + P_6)$$
$$SD(\theta) = M_1 + M_2 \exp(-M_3\theta)$$
where θ is the angle of incidence in radians.

E-2   L-Band Data:

Statistical Distribution Table for Short Vegetation

L Band, HH Polarization

Angle	N	σ^o_{max}	σ^o_5	σ^o_{25}	Median	σ^o_{75}	σ^o_{95}	σ^o_{min}	Mean	Std. Dev.
0°	537	17.8	12.4	8.2	2.4	-2.1	-5.0	-10.8	3.1	5.9
5°	263	12.8	9.1	4.3	0.9	-1.7	-6.5	-16.0	1.1	4.8
10°	605	6.7	1.2	-3.2	-6.8	-11.0	-15.7	-19.0	-7.1	5.2
15°	264	4.1	1.0	-3.1	-6.5	-10.6	-15.6	-21.0	-6.9	5.2
20°	611	5.5	-3.0	-7.9	-14.2	-18.9	-22.6	-28.7	-13.4	6.5
25°	53	2.3	-2.3	-5.0	-9.0	-15.6	-25.6	-27.5	-10.4	7.0
30°	606	0.7	-5.2	-10.9	-17.6	-21.6	-24.9	-30.0	-16.2	6.6
35°	53	-1.7	-3.4	-8.3	-12.1	-16.9	-28.8	-32.5	-13.1	7.3
40°	392	-2.2	-8.3	-17.3	-21.2	-24.0	-27.1	-35.0	-19.9	5.9
45°	58	-7.8	-9.3	-12.1	-15.3	-19.6	-32.3	-36.0	-16.7	6.7
50°	115	-5.4	-7.4	-11.0	-14.5	-21.1	-31.3	-35.0	-16.6	7.5
60°	347	-6.7	-13.5	-22.3	-24.9	-26.9	-29.9	-36.8	-23.9	4.9
80°	309	-17.2	-21.6	-24.4	-26.1	-27.9	-32.3	-38.1	-26.3	3.2

L Band, HV Polarization

Angle	N	σ^o_{max}	σ^o_5	σ^o_{25}	Median	σ^o_{75}	σ^o_{95}	σ^o_{min}	Mean	Std. Dev.
0°	348	1.7	-9.9	-16.8	-21.8	-24.8	-27.0	-31.2	-20.3	5.6
5°	92	-3.3	-8.7	-15.2	-17.0	-18.9	-23.0	-25.4	-16.6	4.1
10°	475	-9.7	-15.5	-20.5	-23.8	-27.4	-30.5	-35.3	-23.7	4.7
15°	178	-10.8	-13.4	-17.3	-19.9	-23.3	-29.1	-32.2	-20.4	4.4
20°	500	-11.1	-16.5	-21.3	-26.6	-30.6	-34.5	-41.0	-26.1	5.9
25°	52	-14.3	-17.0	-19.1	-22.1	-24.6	-34.0	-35.5	-23.0	5.2
30°	500	-14.6	-17.2	-22.1	-27.0	-31.1	-36.2	-42.3	-26.9	6.0
35°	52	-17.3	-18.6	-19.7	-22.3	-25.5	-35.8	-36.5	-23.8	5.2
40°	361	-18.4	-21.6	-26.0	-29.9	-33.1	-37.7	-40.8	-29.7	4.9
45°	52	-20.6	-21.2	-23.0	-25.1	-27.4	-36.6	-39.7	-26.2	4.6
50°	75	-19.3	-20.2	-22.4	-25.3	-29.4	-42.2	-44.2	-26.9	6.4
60°	309	-21.7	-26.7	-29.3	-31.8	-35.5	-39.5	-45.2	-32.3	4.2
80°	307	-22.7	-29.5	-31.4	-33.6	-36.2	-40.5	-48.6	-34.1	3.6

L Band, VV Polarization

Angle	N	σ^o_{max}	σ^o_5	σ^o_{25}	Median	σ^o_{75}	σ^o_{95}	σ^o_{min}	Mean	Std. Dev.
0°	537	18.1	13.2	7.1	1.8	-2.1	-5.0	-13.4	2.7	5.8
5°	208	12.5	8.7	3.7	0.8	-2.2	-6.4	-15.3	0.9	4.7
10°	523	5.1	0.7	-3.6	-6.8	-10.7	-15.2	-19.2	-7.2	4.8
15°	210	1.5	-2.7	-6.3	-9.6	-12.5	-16.8	-21.7	-9.5	4.4
20°	560	-0.5	-5.6	-10.4	-14.7	-17.6	-20.9	-26.6	-14.0	4.8
30°	559	-2.1	-8.6	-12.8	-17.0	-20.3	-24.3	-29.2	-16.6	5.0
40°	341	-3.9	-9.9	-13.5	-19.9	-23.6	-27.2	-36.5	-19.6	5.5
50°	65	-4.6	-7.0	-11.8	-15.0	-21.1	-32.4	-33.5	-16.4	7.5
60°	347	-6.2	-12.2	-17.8	-21.9	-25.8	-30.3	-37.3	-21.6	5.5
80°	309	-14.1	-17.2	-20.6	-23.3	-26.3	-29.5	-38.0	-23.5	4.0

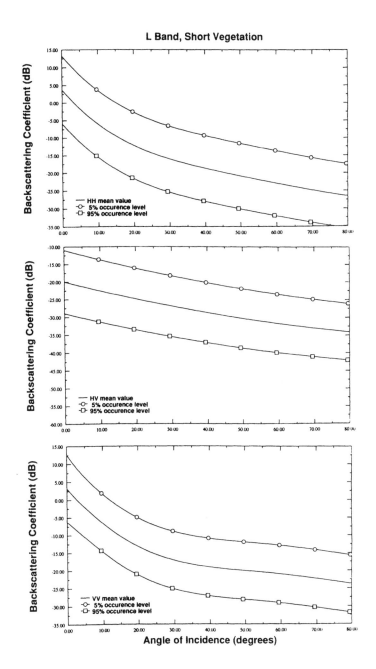

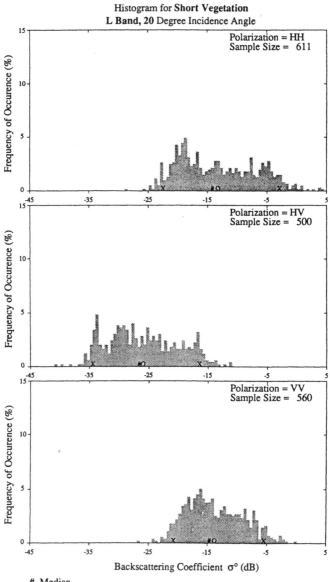

Histogram for **Short Vegetation**
L Band, 20 Degree Incidence Angle

Median
O Mean
X 5% or 95% Level

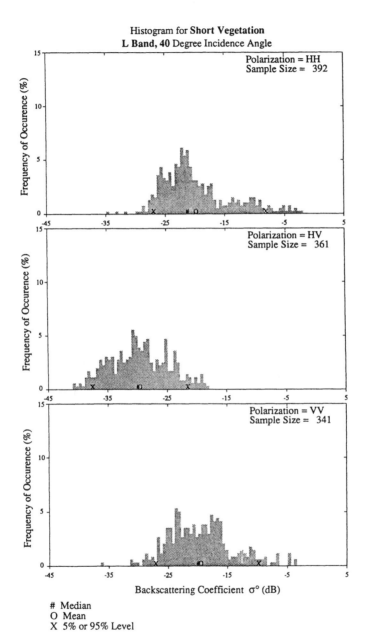

Histogram for **Short Vegetation**
L Band, 40 Degree Incidence Angle

Median
O Mean
X 5% or 95% Level

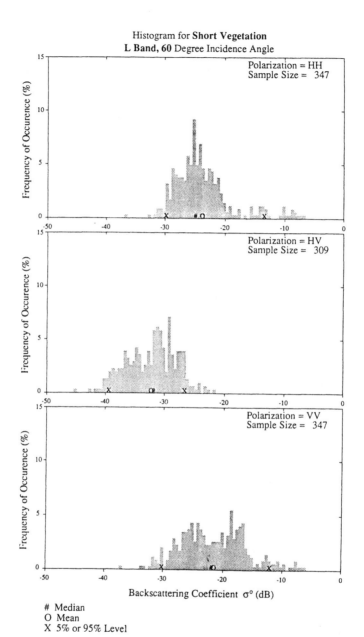

Histogram for **Short Vegetation**
L Band, 60 Degree Incidence Angle

Median
O Mean
X 5% or 95% Level

E-3 S-Band Data:

Statistical Distribution Table for Short Vegetation

S Band, HH Polarization

Angle	N	σ^o_{max}	σ^o_5	σ^o_{25}	Median	σ^o_{75}	σ^o_{95}	σ^o_{min}	Mean	Std. Dev.
0°	535	16.9	11.0	5.8	0.0	-3.3	-6.3	-25.0	1.3	5.9
5°	192	9.2	5.4	2.2	-0.7	-3.4	-7.7	-23.9	-0.9	4.7
10°	534	5.3	0.5	-3.7	-6.5	-9.9	-13.3	-23.5	-6.7	4.4
15°	192	0.2	-1.6	-4.6	-7.2	-9.8	-13.6	-24.4	-7.5	4.0
20°	535	0.1	-5.8	-9.3	-12.2	-14.7	-17.5	-24.1	-12.0	3.7
30°	535	-5.5	-8.8	-11.4	-14.2	-16.7	-19.5	-26.9	-14.1	3.4
40°	337	-8.4	-12.4	-14.9	-16.9	-18.7	-20.9	-23.5	-16.7	2.7
50°	24	-12.1	-13.2	-13.7	-15.9	-23.6	-27.7	-28.9	-18.3	5.3
60°	337	-10.3	-13.9	-16.8	-18.4	-20.3	-22.7	-26.1	-18.5	2.6
80°	337	-12.1	-15.6	-18.7	-20.2	-21.7	-24.1	-30.0	-20.1	2.5

S Band, HV Polarization

Angle	N	σ^o_{max}	σ^o_5	σ^o_{25}	Median	σ^o_{75}	σ^o_{95}	σ^o_{min}	Mean	Std. Dev.
0°	437	-2.1	-8.5	-13.3	-17.8	-21.2	-23.3	-26.7	-17.0	4.9
5°	134	-8.5	-11.4	-14.2	-15.9	-18.0	-22.3	-26.1	-16.2	3.2
10°	521	-9.1	-13.6	-17.2	-20.3	-22.9	-25.3	-32.9	-19.9	3.9
15°	179	-11.5	-13.8	-16.0	-17.9	-20.2	-24.6	-35.5	-18.5	3.6
20°	523	-10.5	-15.8	-18.7	-21.9	-24.1	-26.4	-38.4	-21.5	3.7
30°	518	-14.5	-16.8	-19.4	-22.2	-24.4	-27.4	-37.0	-22.1	3.4
40°	337	-17.3	-19.3	-22.3	-23.8	-25.5	-28.1	-33.2	-23.8	2.6
50°	22	-20.0	-20.7	-21.2	-23.6	-34.1	-36.7	-37.2	-26.7	6.4
60°	337	-18.1	-21.7	-24.0	-25.1	-26.9	-30.4	-35.4	-25.4	2.5
80°	337	-17.4	-25.0	-26.9	-28.3	-30.0	-32.4	-39.1	-28.5	2.5

S Band, VV Polarization

Angle	N	σ^o_{max}	σ^o_5	σ^o_{25}	Median	σ^o_{75}	σ^o_{95}	σ^o_{min}	Mean	Std. Dev.
0°	533	17.7	10.7	5.4	-0.5	-3.2	-6.0	-24.0	1.3	5.8
5°	192	10.4	5.0	2.1	-1.2	-4.1	-9.1	-24.1	-1.3	4.9
10°	535	7.3	1.5	-3.2	-6.4	-9.2	-12.8	-23.7	-6.2	4.4
15°	192	1.1	-2.9	-6.4	-9.0	-10.6	-14.5	-24.1	-8.9	3.7
20°	534	-1.1	-5.9	-9.3	-11.9	-13.9	-16.7	-25.0	-11.7	3.4
30°	533	-3.1	-9.0	-11.8	-13.9	-15.8	-19.2	-23.2	-13.9	3.1
40°	337	-7.8	-10.3	-13.6	-15.5	-17.4	-21.0	-23.4	-15.6	3.0
50°	24	-13.2	-15.8	-16.3	-17.5	-21.1	-28.4	-30.0	-19.2	4.3
60°	337	-8.5	-12.1	-14.9	-16.9	-19.0	-22.6	-27.3	-17.0	3.2
80°	337	-10.1	-14.8	-17.6	-19.5	-21.5	-24.1	-29.4	-19.6	2.9

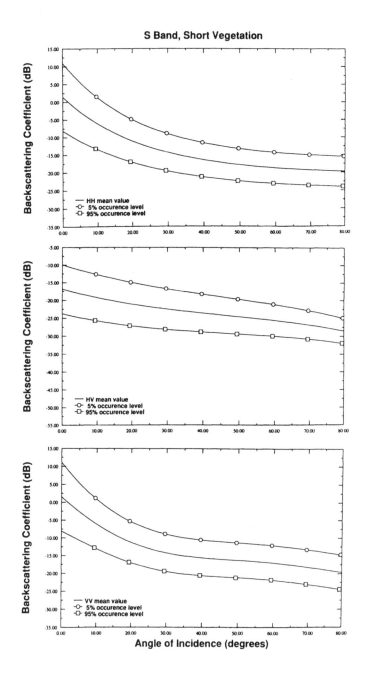

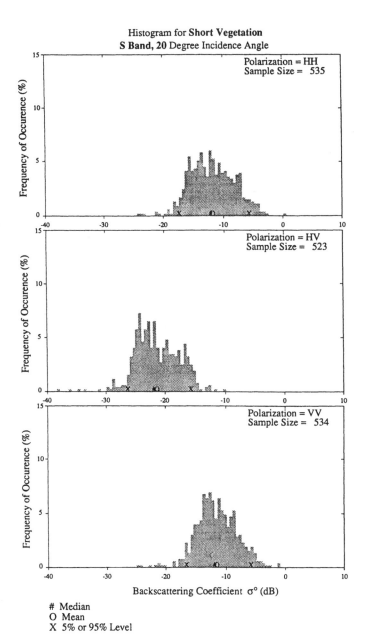

Histogram for **Short Vegetation**
S Band, 20 Degree Incidence Angle

Median
O Mean
X 5% or 95% Level

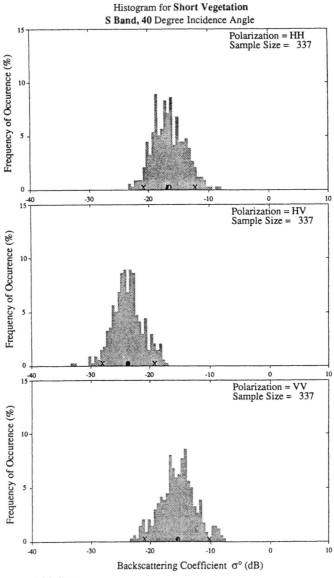

Histogram for **Short Vegetation**
S Band, 40 Degree Incidence Angle

Median
O Mean
X 5% or 95% Level

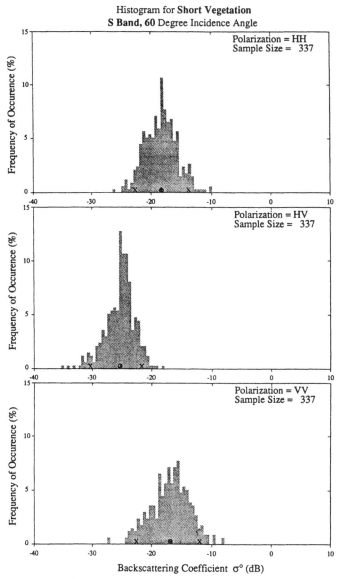

Histogram for **Short Vegetation**
S Band, 60 Degree Incidence Angle

Median
O Mean
X 5% or 95% Level

E-4 C-Band Data:

Statistical Distribution Table for **Short Vegetation**

C Band, **HH** Polarization

Angle	N	σ^o_{max}	σ^o_5	σ^o_{25}	Median	σ^o_{75}	σ^o_{95}	σ^o_{min}	Mean	Std. Dev.
0°	1086	15.7	10.3	4.9	-0.6	-4.4	-7.3	-12.0	0.3	5.6
5°	457	16.7	8.1	2.8	-0.3	-2.6	-5.9	-12.6	0.3	4.4
10°	1489	11.4	2.9	0.3	-3.3	-6.8	-10.9	-18.6	-3.5	4.6
15°	458	5.7	0.8	-2.5	-4.9	-7.0	-11.7	-18.5	-5.0	3.7
20°	1332	4.6	-2.1	-5.8	-8.3	-10.9	-13.9	-23.6	-8.3	3.7
25°	57	-5.8	-7.3	-8.9	-9.7	-11.1	-17.4	-20.9	-10.6	3.1
30°	1359	-0.3	-4.4	-7.9	-10.0	-12.2	-15.6	-24.4	-10.1	3.4
35°	59	-7.6	-8.6	-9.8	-11.0	-13.7	-20.9	-23.2	-12.3	3.9
40°	737	-4.7	-8.1	-10.4	-12.2	-13.8	-16.7	-22.5	-12.2	2.6
45°	89	-9.3	-9.6	-10.6	-12.2	-14.7	-21.4	-24.0	-13.1	3.5
50°	986	0.7	-9.6	-11.9	-13.7	-15.7	-19.7	-30.9	-14.0	3.3
55°	22	-9.3	-10.8	-12.2	-13.9	-16.0	-17.1	-17.3	-13.8	2.3
60°	710	-5.4	-10.0	-11.9	-13.4	-14.8	-17.2	-22.0	-13.4	2.2
80°	677	-3.7	-11.6	-15.8	-17.6	-18.8	-20.4	-28.5	-17.0	2.8

C Band, **HV** Polarization

Angle	N	σ^o_{max}	σ^o_5	σ^o_{25}	Median	σ^o_{75}	σ^o_{95}	σ^o_{min}	Mean	Std. Dev.
0°	992	3.4	-2.2	-8.0	-12.1	-15.3	-18.5	-23.4	-11.5	5.0
5°	381	3.4	-2.1	-8.0	-11.6	-13.4	-15.8	-21.4	-10.5	4.3
10°	1093	0.7	-6.1	-11.8	-14.5	-16.9	-20.0	-28.8	-14.1	4.3
15°	405	0.0	-5.2	-10.8	-13.6	-16.0	-20.8	-31.7	-13.3	4.8
20°	1260	0.0	-8.8	-13.8	-15.9	-18.2	-21.4	-32.9	-15.9	4.1
25°	52	-12.3	-14.9	-15.5	-16.6	-18.6	-29.7	-30.7	-18.1	4.3
30°	1281	0.4	-10.0	-14.7	-16.6	-18.9	-22.7	-33.7	-16.7	4.0
35°	52	-14.7	-16.2	-17.0	-17.9	-20.2	-30.6	-32.3	-19.6	4.3
40°	727	-9.5	-13.8	-16.0	-17.5	-19.3	-21.8	-30.6	-17.7	2.6
45°	85	-12.1	-16.2	-18.3	-19.8	-22.9	-30.3	-32.8	-20.8	4.2
50°	971	-7.5	-13.7	-16.5	-18.8	-21.1	-26.4	-35.7	-19.1	3.9
55°	22	-14.4	-15.3	-18.6	-21.1	-24.6	-30.0	-30.0	-21.3	4.8
60°	704	-11.9	-15.1	-17.1	-18.4	-19.8	-22.1	-27.4	-18.5	2.2
80°	675	-9.6	-17.2	-20.3	-21.9	-23.5	-25.7	-33.8	-21.8	2.7

C Band, **VV** Polarization

Angle	N	σ^o_{max}	σ^o_5	σ^o_{25}	Median	σ^o_{75}	σ^o_{95}	σ^o_{min}	Mean	Std. Dev.
0°	1078	17.0	9.8	4.9	-0.7	-4.4	-7.3	-12.5	0.3	5.6
5°	397	17.5	8.9	3.2	0.0	-2.8	-6.2	-12.7	0.4	4.8
10°	1082	12.4	4.4	-1.7	-4.6	-7.3	-10.6	-18.1	-4.2	4.5
15°	397	8.1	0.9	-2.7	-5.2	-7.7	-11.9	-19.8	-5.3	3.9
20°	1083	3.5	-2.4	-6.0	-8.4	-10.5	-13.2	-22.8	-8.2	3.5
30°	1283	1.7	-4.8	-7.9	-10.0	-12.1	-15.1	-23.2	-10.0	3.3
40°	675	-5.1	-7.1	-9.9	-11.4	-12.9	-15.0	-18.3	-11.3	2.3
45°	31	-7.9	-8.9	-11.0	-14.1	-15.6	-17.7	-18.3	-13.2	2.7
50°	251	-7.1	-9.9	-12.6	-14.5	-16.3	-22.4	-29.2	-14.7	3.6
55°	23	-7.1	-9.8	-11.8	-14.9	-15.7	-16.8	-17.0	-13.7	2.8
60°	692	-5.5	-9.7	-11.6	-13.0	-14.5	-16.5	-20.1	-13.1	2.1
80°	675	-7.9	-12.9	-15.5	-17.2	-18.7	-20.7	-27.0	-17.1	2.6

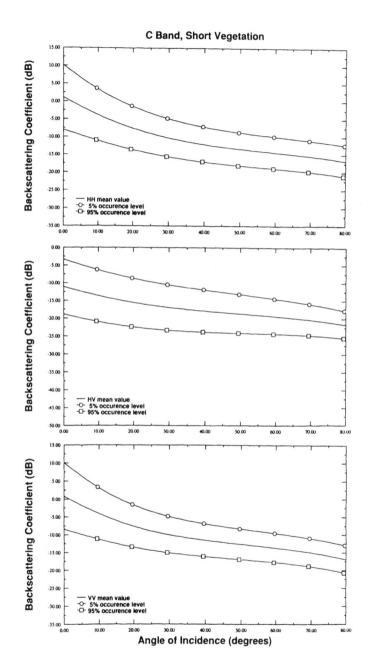

C Band, Short Vegetation

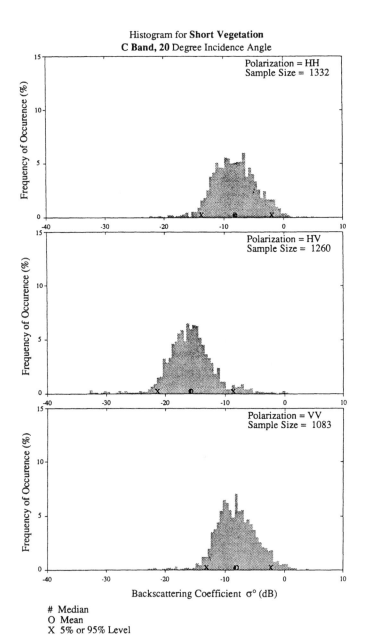

Histogram for **Short Vegetation**
C Band, 20 Degree Incidence Angle

Median
O Mean
X 5% or 95% Level

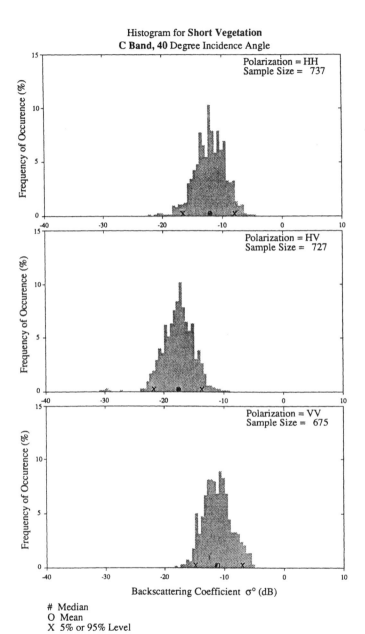

Histogram for **Short Vegetation**
C Band, 40 Degree Incidence Angle

Median
O Mean
X 5% or 95% Level

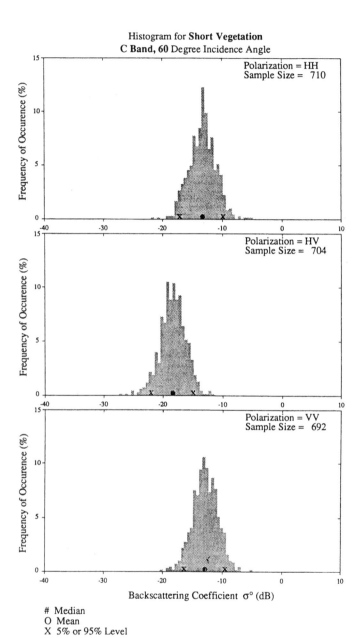

Histogram for **Short Vegetation**
C Band, 60 Degree Incidence Angle

Median
O Mean
X 5% or 95% Level

E-5 X-Band Data:

Statistical Distribution Table for **Short Vegetation**

X Band, HH Polarization

Angle	N	σ^o_{max}	σ^o_5	σ^o_{25}	Median	σ^o_{75}	σ^o_{95}	σ^o_{min}	Mean	Std. Dev.
0°	844	16.1	7.3	2.2	-2.9	-5.8	-9.0	-15.0	-1.8	5.3
5°	2	-5.0						-5.0	-5.0	0.0
10°	1420	16.2	8.9	1.5	-3.6	-6.2	-8.9	-17.1	-2.1	5.8
15°	3	-6.7						-8.6	-7.5	1.0
20°	1488	9.1	-0.6	-4.5	-6.4	-8.0	-10.3	-18.3	-6.1	3.1
30°	1718	-0.5	-3.7	-6.1	-7.8	-9.3	-11.8	-18.6	-7.8	2.5
40°	3454	-1.7	-6.1	-8.1	-9.6	-11.2	-14.6	-24.4	-9.9	2.7
45°	65	-3.2	-5.7	-7.6	-9.1	-11.0	-15.1	-15.7	-9.3	2.8
50°	3459	-2.6	-7.1	-9.0	-10.6	-12.8	-16.0	-24.4	-11.0	2.9
60°	3649	0.0	-8.0	-10.1	-11.5	-13.4	-16.3	-24.6	-11.8	2.7
70°	2658	-2.2	-9.9	-11.8	-13.3	-15.1	-18.7	-28.5	-13.6	2.7
80°	847	-2.3	-13.1	-14.3	-15.3	-16.7	-18.6	-33.5	-15.6	2.3

X Band, HV Polarization

Angle	N	σ^o_{max}	σ^o_5	σ^o_{25}	Median	σ^o_{75}	σ^o_{95}	σ^o_{min}	Mean	Std. Dev.
0°	14	0.0						-21.5	-13.6	6.3
5°	2	-14.9						-15.0	-14.9	0.1
10°	16	-9.8	-11.0	-12.9	-15.1	-19.2	-21.6	-21.6	-15.7	4.0
15°	3	-15.5						-16.2	-15.8	0.4
20°	324	-8.3	-11.4	-13.9	-15.7	-17.9	-21.9	-27.0	-16.0	3.2
30°	361	-9.1	-12.4	-16.2	-19.2	-21.9	-25.1	-29.5	-18.9	4.0
40°	2085	-7.5	-10.2	-12.2	-13.8	-16.2	-20.3	-27.4	-14.4	3.1
45°	81	-10.9	-14.6	-16.8	-18.6	-21.6	-24.0	-28.1	-19.0	3.3
50°	3967	-6.6	-11.2	-13.5	-16.5	-19.8	-24.7	-35.3	-16.9	4.2
60°	2199	-6.5	-12.2	-13.9	-15.7	-18.6	-22.1	-33.2	-16.3	3.2
70°	2135	-4.8	-13.9	-15.9	-17.4	-19.6	-23.1	-36.7	-17.9	2.9

X Band, VV Polarization

Angle	N	σ^o_{max}	σ^o_5	σ^o_{25}	Median	σ^o_{75}	σ^o_{95}	σ^o_{min}	Mean	Std. Dev.
0°	847	15.6	7.2	2.5	-2.7	-5.7	-8.7	-12.7	-1.6	5.3
5°	2	-5.0						-5.0	-5.0	0.0
10°	1417	15.1	9.4	1.0	-3.9	-6.5	-8.7	-14.1	-2.3	5.8
15°	3	-6.7						-9.1	-7.6	1.3
20°	1673	8.3	-1.0	-4.8	-7.0	-8.5	-10.7	-17.5	-6.5	3.1
30°	1708	0.5	-4.2	-6.5	-8.5	-10.1	-14.0	-18.8	-8.6	2.9
40°	3449	-3.1	-6.1	-8.2	-9.9	-12.2	-17.6	-25.2	-10.6	3.4
45°	63	-3.4	-5.4	-7.5	-9.0	-13.5	-16.0	-17.1	-9.9	3.5
50°	5083	-1.1	-5.9	-8.3	-10.2	-13.0	-17.7	-23.0	-10.8	3.6
60°	3544	-2.5	-8.1	-10.3	-11.8	-14.0	-17.8	-23.5	-12.3	2.9
70°	2658	-2.2	-9.5	-11.8	-13.4	-15.4	-19.0	-26.5	-13.7	2.9
80°	837	-2.3	-12.1	-14.6	-16.1	-17.5	-19.8	-33.5	-16.1	2.6

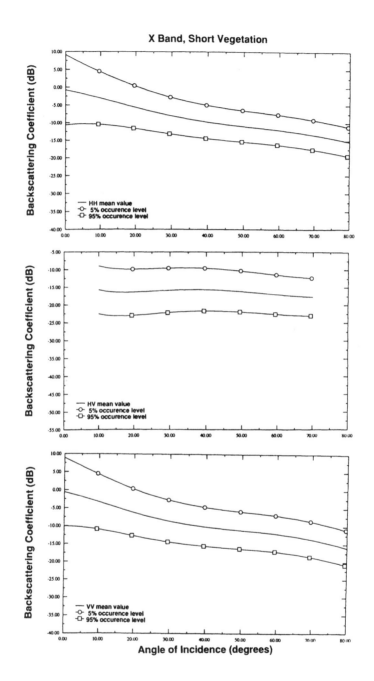

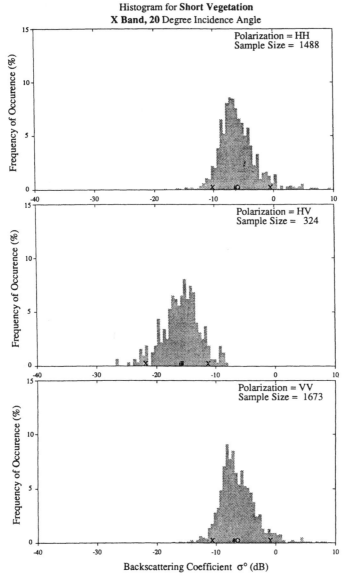

Histogram for **Short Vegetation**
X Band, 20 Degree Incidence Angle

Polarization = HH
Sample Size = 1488

Polarization = HV
Sample Size = 324

Polarization = VV
Sample Size = 1673

Backscattering Coefficient σ° (dB)

Frequency of Occurence (%)

Median
O Mean
X 5% or 95% Level

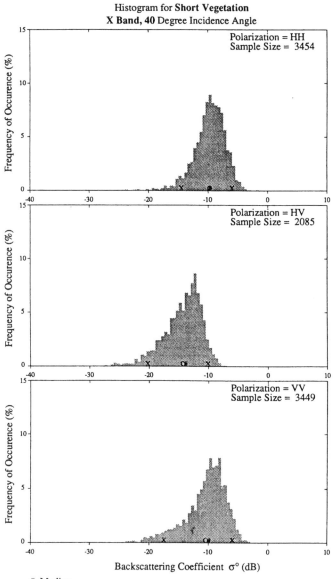

Histogram for **Short Vegetation**
X Band, 40 Degree Incidence Angle

Median
O Mean
X 5% or 95% Level

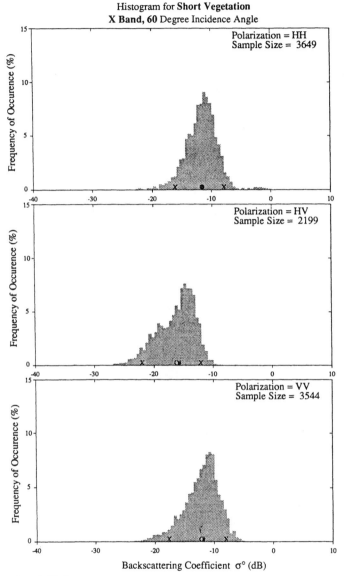

Histogram for **Short Vegetation**
X Band, 60 Degree Incidence Angle

Median
O Mean
X 5% or 95% Level

E-6   Ku-Band Data:

Statistical Distribution Table for **Short Vegetation**

Ku Band, HH Polarization

Angle	N	σ^o_{max}	σ^o_{5}	σ^o_{25}	Median	σ^o_{75}	σ^o_{95}	σ^o_{min}	Mean	Std. Dev.
0°	1272	19.8	7.6	1.9	-1.9	-4.5	-8.1	-10.7	-1.1	4.8
5°	19	3.3	1.8	0.4	-1.1	-3.3	-9.7	-10.1	-1.6	3.5
10°	1953	18.0	8.6	0.3	-3.4	-5.7	-8.6	-16.2	-2.1	5.3
20°	1967	11.2	-0.4	-4.0	-5.8	-7.3	-9.4	-21.1	-5.4	3.1
30°	1970	0.8	-2.8	-5.3	-7.0	-8.6	-11.6	-20.5	-7.1	2.7
40°	4319	0.4	-4.3	-6.6	-8.4	-10.7	-14.4	-22.9	-8.8	3.1
50°	3556	-2.1	-5.3	-7.3	-9.0	-11.3	-15.4	-23.9	-9.5	3.1
60°	4316	-2.1	-6.2	-8.5	-10.3	-12.7	-15.7	-21.5	-10.6	3.0
70°	2995	-5.3	-8.1	-10.0	-11.6	-13.4	-16.8	-21.3	-11.9	2.6
80°	1190	-7.4	-11.5	-13.5	-14.9	-16.0	-18.1	-24.3	-14.8	2.1

Ku Band, HV Polarization

Angle	N	σ^o_{max}	σ^o_{5}	σ^o_{25}	Median	σ^o_{75}	σ^o_{95}	σ^o_{min}	Mean	Std. Dev.
0°	16	0.0	-3.3	-9.9	-12.6	-14.7	-18.3	-19.1	-11.6	5.3
5°	17	-10.6	-12.0	-13.0	-14.6	-15.4	-20.3	-21.4	-14.6	2.7
10°	18	-11.7	-12.8	-13.5	-15.2	-17.1	-20.7	-21.8	-15.5	2.8
15°	19	-13.1	-14.6	-15.3	-16.7	-19.4	-21.3	-22.2	-17.2	2.4
20°	19	-13.1	-15.0	-15.7	-17.1	-19.0	-21.0	-22.1	-17.3	2.1
30°	18	-11.2	-15.2	-17.7	-18.3	-20.1	-22.0	-22.1	-18.3	2.6
40°	2364	-4.6	-7.9	-10.0	-11.9	-14.8	-18.9	-26.4	-12.5	3.5
50°	2363	-6.5	-8.7	-10.7	-12.6	-15.5	-19.7	-33.0	-13.3	3.5
60°	2278	-7.7	-9.8	-11.6	-13.3	-15.8	-20.1	-25.6	-14.0	3.2
70°	2093	-9.7	-11.7	-13.4	-14.9	-16.9	-20.8	-25.3	-15.4	2.8

Ku Band, VV Polarization

Angle	N	σ^o_{max}	σ^o_{5}	σ^o_{25}	Median	σ^o_{75}	σ^o_{95}	σ^o_{min}	Mean	Std. Dev.
0°	1276	20.3	8.3	2.6	-0.8	-3.7	-7.2	-11.3	-0.3	4.9
5°	75	6.8	2.9	0.0	-1.6	-3.0	-7.0	-10.7	-1.5	3.1
10°	2005	15.4	8.2	0.3	-3.0	-5.3	-8.1	-14.8	-1.9	5.0
15°	76	4.1	0.1	-1.6	-2.6	-8.0	-14.3	-19.8	-4.8	4.6
20°	2021	10.4	0.0	-3.6	-5.6	-7.5	-10.5	-22.1	-5.5	3.3
25°	52	-1.1	-2.9	-3.7	-4.5	-5.8	-11.7	-13.6	-5.4	2.8
30°	2026	1.3	-2.5	-4.9	-6.8	-9.3	-13.6	-21.3	-7.2	3.3
35°	52	-4.2	-4.9	-5.4	-6.4	-8.5	-14.3	-15.1	-7.5	2.9
40°	4373	0.1	-4.2	-6.3	-8.3	-11.4	-16.8	-24.0	-9.2	3.9
45°	60	-3.3	-4.6	-5.4	-6.1	-10.0	-14.3	-15.8	-7.8	3.3
50°	3612	-2.8	-5.1	-7.2	-8.9	-12.0	-17.6	-23.7	-9.9	3.8
60°	4306	-2.6	-6.0	-8.4	-10.3	-13.1	-17.2	-22.4	-10.9	3.5
70°	3007	-1.6	-7.8	-10.1	-11.7	-13.9	-17.5	-22.2	-12.1	3.0
80°	1188	-3.3	-10.4	-12.4	-14.0	-15.4	-17.4	-26.4	-13.9	2.3

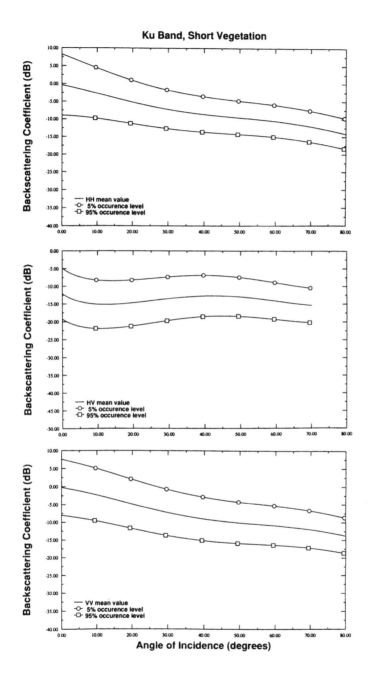

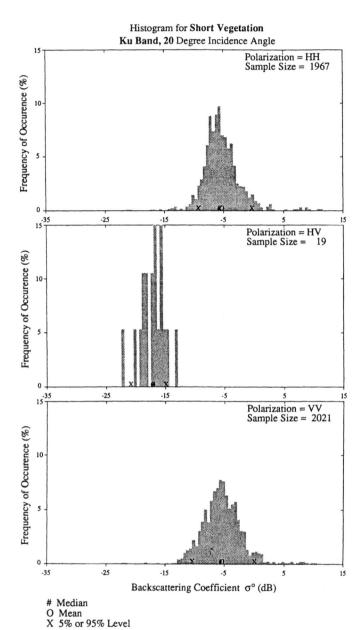

Histogram for **Short Vegetation**
Ku Band, 20 Degree Incidence Angle

Median
O Mean
X 5% or 95% Level

Histogram for **Short Vegetation**
Ku Band, 40 Degree Incidence Angle

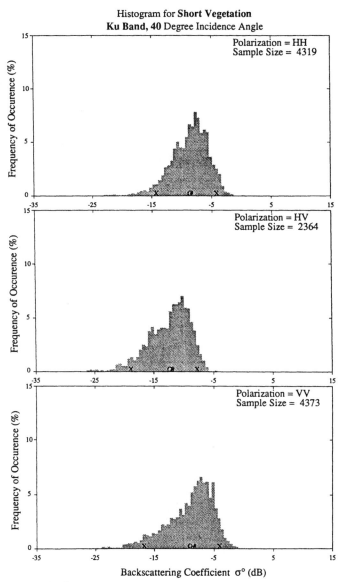

Median
O Mean
X 5% or 95% Level

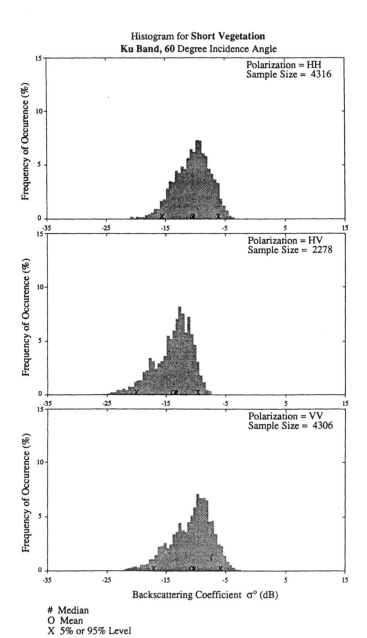

Histogram for **Short Vegetation**
Ku Band, 60 Degree Incidence Angle

\# Median
O Mean
X 5% or 95% Level

E-7 Ka-Band Data:

Statistical Distribution Table for Short Vegetation

Ka Band, HH Polarization

Angle	N	σ^o_{max}	σ^o_5	σ^o_{25}	Median	σ^o_{75}	σ^o_{95}	σ^o_{min}	Mean	Std. Dev.
0°	2	0.0						-8.4	-4.2	5.9
5°	1	-5.8						-5.8	-5.8	0.0
10°	32	0.5	-2.5	-4.8	-7.1	-8.8	-11.1	-11.1	-6.7	3.0
15°	1	-6.7						-6.7	-6.7	0.0
20°	67	0.0	-1.4	-4.1	-6.1	-7.9	-11.3	-11.3	-6.1	3.0
30°	67	-0.5	-3.2	-5.0	-6.6	-9.6	-11.9	-13.1	-7.0	2.9
40°	57	-1.0	-4.4	-6.1	-8.2	-10.2	-13.8	-15.2	-8.3	2.9
45°	10	-3.4						-10.3	-6.7	2.3
50°	282	-2.0	-6.4	-7.5	-8.6	-10.5	-14.0	-20.2	-9.2	2.5
60°	67	-3.0	-6.5	-8.0	-10.2	-11.9	-15.5	-19.5	-10.2	3.0
70°	67	-4.7	-7.7	-9.7	-11.6	-13.4	-16.4	-21.7	-11.6	3.1
80°	25	-6.6	-11.1	-12.5	-15.4	-17.8	-21.5	-25.0	-15.2	3.9

Ka Band, HV Polarization

Angle	N	σ^o_{max}	σ^o_5	σ^o_{25}	Median	σ^o_{75}	σ^o_{95}	σ^o_{min}	Mean	Std. Dev.
0°	3	0.0						-16.5	-10.6	9.2
5°	2	-13.8						-15.8	-14.8	1.4
10°	2	-13.6						-17.6	-15.6	2.8
15°	2	-15.7						-17.4	-16.5	1.2
20°	2	-13.8						-18.4	-16.1	3.3
30°	2	-13.8						-17.8	-15.8	2.8
40°	2	-13.0						-18.9	-15.9	4.2
50°	2	-13.4						-20.8	-17.1	5.2
60°	2	-17.3						-21.4	-19.3	2.9
70°	1	-23.8						-23.8	-23.8	0.0

Ka Band, VV Polarization

Angle	N	σ^o_{max}	σ^o_5	σ^o_{25}	Median	σ^o_{75}	σ^o_{95}	σ^o_{min}	Mean	Std. Dev.
0°	4	0.0						-11.1	-5.0	5.5
5°	4	-4.2						-10.2	-7.3	2.7
10°	34	0.4	-2.8	-6.1	-7.8	-9.3	-11.1	-11.1	-7.4	2.7
15°	4	-6.7						-9.6	-7.9	1.4
20°	70	-0.8	-2.2	-5.0	-6.4	-8.8	-10.7	-11.3	-6.7	2.7
30°	73	-1.6	-3.4	-5.9	-7.2	-9.3	-11.8	-12.1	-7.4	2.6
40°	60	-2.6	-4.8	-6.5	-9.2	-10.7	-12.4	-12.4	-8.7	2.6
45°	11	-4.5						-8.5	-6.2	1.5
50°	284	-3.8	-6.2	-7.2	-8.4	-10.4	-14.2	-20.4	-9.1	2.7
60°	71	-5.0	-6.3	-9.1	-10.8	-12.0	-15.1	-18.0	-10.6	2.5
70°	65	-6.6	-8.7	-10.5	-12.2	-13.8	-16.7	-18.0	-12.1	2.6
80°	23	-11.4	-12.8	-14.5	-15.5	-16.6	-19.8	-20.1	-15.4	2.3

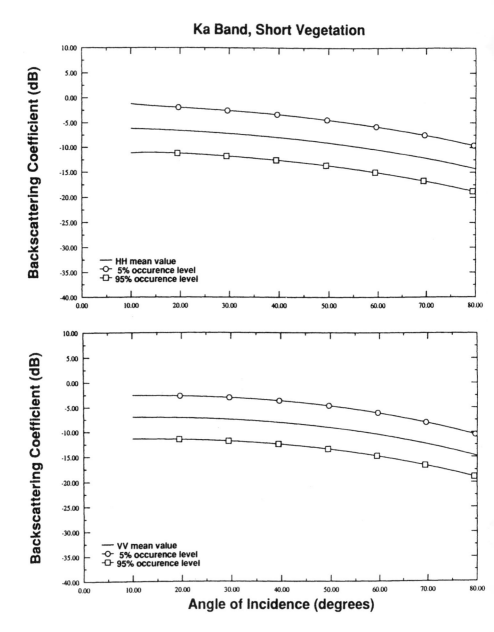

Ka Band, Short Vegetation

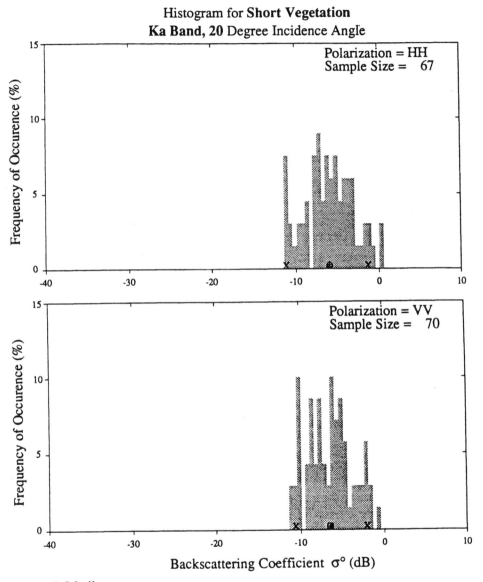

Histogram for **Short Vegetation**
Ka Band, 20 Degree Incidence Angle

Median
O Mean
X 5% or 95% Level

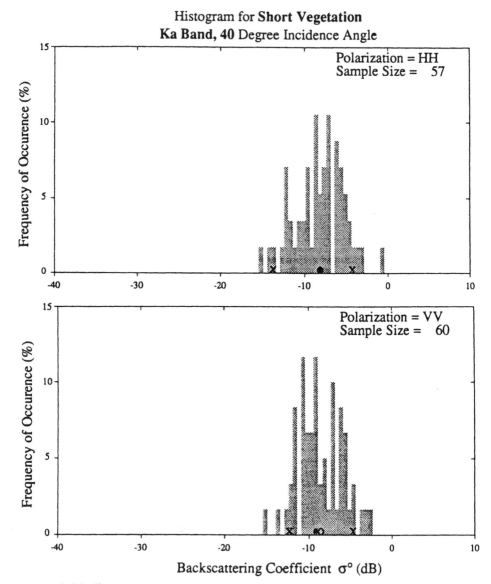

Histogram for **Short Vegetation**
Ka Band, 40 Degree Incidence Angle

Median
O Mean
X 5% or 95% Level

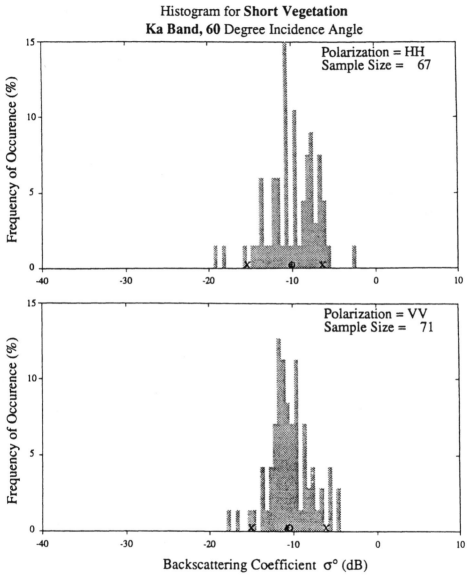

Histogram for **Short Vegetation**
Ka Band, 60 Degree Incidence Angle

Median
O Mean
X 5% or 95% Level

E-8 W-Band Data:

Statistical Distribution Table for **Short Vegetation**

W Band, VV Polarization

Angle	N	σ^{o}_{max}	σ^{o}_{5}	σ^{o}_{25}	Median	σ^{o}_{75}	σ^{o}_{95}	σ^{o}_{min}	Mean	Std. Dev.
0°	3	3.8						-2.1	0.7	3.0
5°	4	4.0						-2.7	-0.3	3.1
10°	4	0.7						-1.7	-0.9	1.1
15°	3	-1.5						-2.1	-1.8	0.3
20°	4	-0.3						-4.1	-1.7	1.7
30°	4	0.1						-4.8	-3.1	2.2
40°	4	-1.4						-4.3	-3.0	1.2
50°	4	-1.0						-4.7	-3.0	1.9
60°	3	-2.7						-7.5	-5.3	2.4
70°	2	-6.8						-8.3	-7.6	1.1

APPENDIX F
BACKSCATTERING DATA FOR ROAD SURFACES

F-1 Data Sources and Parameter Loadings

Table F.1   Data sources for road surfaces.

Band	References
L	99
S	
C	
X	37, 99, 109
Ku	109
Ka	37, 109, 144
W	144

Table F.2   Mean and standard deviation parameter loadings for road surfaces.

Band	Pol.	Angular Range θ_{min}	θ_{max}	$\sigma°$ Parameters P_1	P_2	P_3	P_4	P_5	P_6	SD(θ) Parameters M_1	M_2	M_3
L	HH											
	HV	Insufficient Data										
	VV											
S	HH											
	HV	Insufficient Data										
	VV											
C	HH											
	HV	Insufficient Data										
	VV											
X	HH	10	70	-94.472	99.0	0.892	30.0	1.562	-1.918	4.731	-0.007	-3.983
	HV											
	VV	10	70	-59.560	39.284	1.598	30.0	1.184	-1.178	4.260	-0.002	-4.807
Ku	HH	10	70	-90.341	82.900	0.030	1.651	5.0	0.038	5.490	0.001	-6.350
	HV											
	VV	10	70	-38.159	30.320	0.048	1.913	4.356	0.368	6.263	-0.840	0.064
Ka	HH	10	70	-94.900	99.0	0.694	30.0	1.342	-1.718	7.151	-5.201	0.778
	HV											
	VV	10	70	-84.761	99.0	0.797	-30.0	1.597	1.101	3.174	0.001	-0.095
W	HH											
	HV	Insufficient Data										
	VV											

$$\sigma° = P_1 + P_2 \exp(-P_3\theta) + P_4 \cos(P_5\theta + P_6)$$
$$SD(\theta) = M_1 + M_2 \exp(-M_3\theta)$$

where θ is the angle of incidence in radians.

F-2 L-Band Data:

Statistical Distribution Table for Road Surfaces

L Band, HH Polarization

Angle	N	σ^o_{max}	σ^o_5	σ^o_{25}	Median	σ^o_{75}	σ^o_{95}	σ^o_{min}	Mean	Std. Dev.
10°	1	-17.1						-17.1	-17.1	0.0
20°	1	-22.3						-22.3	-22.3	0.0
30°	1	-28.5						-28.5	-28.5	0.0
40°	1	-30.7						-30.7	-30.7	0.0
50°	1	-31.9						-31.9	-31.9	0.0
60°	1	-35.0						-35.0	-35.0	0.0
70°	1	-39.8						-39.8	-39.8	0.0

L Band, VV Polarization

Angle	N	σ^o_{max}	σ^o_5	σ^o_{25}	Median	σ^o_{75}	σ^o_{95}	σ^o_{min}	Mean	Std. Dev.
20°	1	-23.3						-23.3	-23.3	0.0
30°	1	-25.7						-25.7	-25.7	0.0
40°	1	-27.2						-27.2	-27.2	0.0
50°	1	-29.9						-29.9	-29.9	0.0
60°	1	-32.6						-32.6	-32.6	0.0
70°	1	-35.7						-35.7	-35.7	0.0

F-3 X-Band Data:

Statistical Distribution Table for Road Surfaces

X Band, HH Polarization

Angle	N	σ^o_{max}	σ^o_5	σ^o_{25}	Median	σ^o_{75}	σ^o_{95}	σ^o_{min}	Mean	Std. Dev.
0°	7	26.7						2.1	18.6	8.2
5°	3	9.8						0.2	6.6	5.5
10°	20	-4.0	-5.1	-7.6	-11.8	-16.3	-19.5	-21.6	-12.0	5.0
15°	8	-7.6						-18.8	-12.6	4.1
20°	20	-7.9	-11.3	-13.0	-15.7	-19.5	-22.8	-22.8	-16.1	4.2
30°	20	-10.5	-12.8	-14.3	-18.6	-23.2	-26.2	-26.6	-18.7	4.9
40°	20	-14.0	-15.6	-17.6	-21.2	-24.2	-28.3	-29.1	-21.1	4.5
50°	20	-16.9	-18.2	-19.9	-24.1	-25.7	-32.2	-32.7	-23.8	4.7
60°	18	-19.6	-22.2	-23.5	-25.8	-29.0	-34.2	-35.0	-26.4	4.1
70°	18	-24.4	-25.9	-29.0	-31.2	-33.8	-38.2	-38.2	-31.3	3.8
80°	14	-29.8						-47.7	-40.1	4.5

X Band, HV Polarization

Angle	N	σ^o_{max}	σ^o_5	σ^o_{25}	Median	σ^o_{75}	σ^o_{95}	σ^o_{min}	Mean	Std. Dev.
10°	6	-14.7						-24.6	-19.8	4.1
15°	8	-17.9						-27.9	-22.1	4.1
20°	8	-16.4						-27.5	-22.5	4.1
30°	8	-18.4						-32.2	-24.1	4.9
40°	8	-21.5						-33.3	-26.6	4.7
50°	8	-24.1						-38.5	-29.6	5.5
60°	7	-28.4						-40.6	-31.8	4.6
70°	6	-31.5						-38.5	-34.5	2.4
80°	3	-37.7						-45.8	-41.7	4.1

X Band, VV Polarization

Angle	N	σ^o_{max}	σ^o_5	σ^o_{25}	Median	σ^o_{75}	σ^o_{95}	σ^o_{min}	Mean	Std. Dev.
0°	7	26.7						3.2	19.1	7.9
5°	3	4.4						-3.3	0.6	3.9
10°	20	-2.8	-6.7	-11.1	-13.5	-15.9	-18.8	-20.6	-12.9	4.2
15°	8	-6.9						-19.3	-12.3	4.5
20°	20	-6.9	-11.1	-12.9	-15.2	-17.5	-22.8	-23.4	-15.5	4.0
30°	20	-10.1	-11.0	-13.4	-16.9	-20.2	-24.8	-25.1	-17.0	4.5
40°	20	-11.8	-13.5	-15.7	-18.2	-20.9	-26.0	-26.1	-18.5	4.2
50°	19	-14.9	-15.5	-17.4	-19.9	-24.2	-28.5	-29.8	-20.3	4.3
60°	19	-17.5	-18.3	-19.4	-21.8	-23.7	-29.9	-31.0	-22.1	3.7
70°	18	-21.7	-21.8	-22.2	-24.4	-26.2	-32.2	-34.7	-25.1	3.5
80°	14	-26.4						-43.1	-31.7	4.8

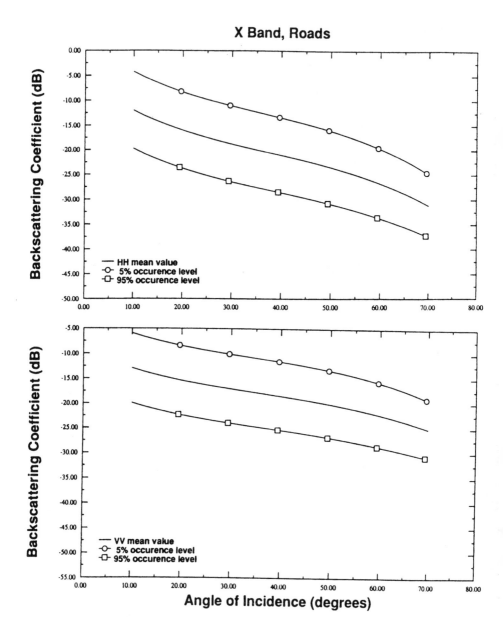

X Band, Roads

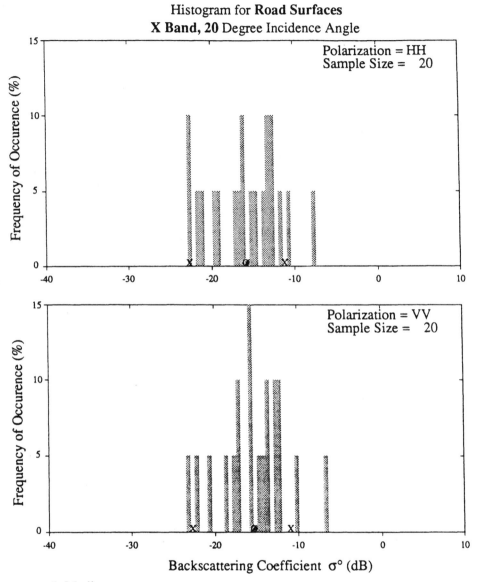

Histogram for **Road Surfaces**
X Band, 20 Degree Incidence Angle

Median
O Mean
X 5% or 95% Level

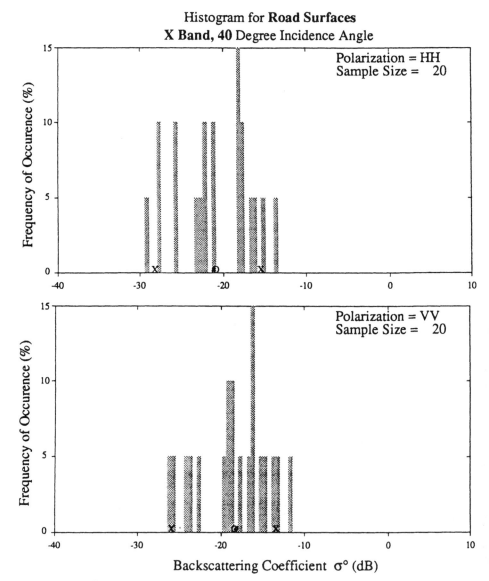

Histogram for **Road Surfaces**
X Band, 40 Degree Incidence Angle

Median
O Mean
X 5% or 95% Level

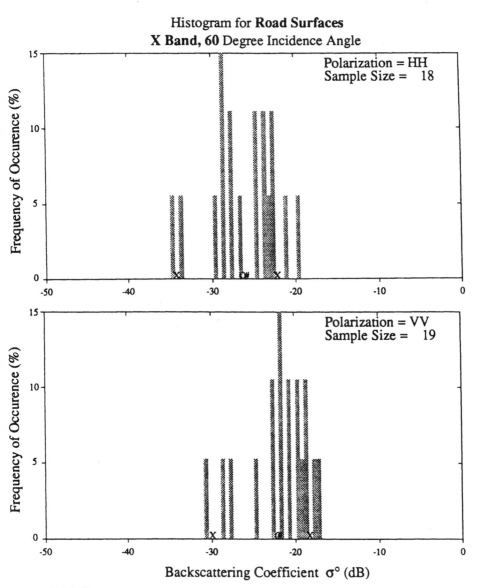

Histogram for **Road Surfaces**
X Band, 60 Degree Incidence Angle

Median
O Mean
X 5% or 95% Level

F-4 Ku-Band Data:

Statistical Distribution Table for Road Surfaces

Ku Band, HH Polarization

Angle	N	σ^{o}_{max}	σ^{o}_{5}	σ^{o}_{25}	Median	σ^{o}_{75}	σ^{o}_{95}	σ^{o}_{min}	Mean	Std. Dev.
0°	8	34.0						3.2	25.0	9.8
5°	8	9.4						-8.2	0.4	6.6
10°	18	3.9	0.8	-4.2	-7.5	-10.9	-13.0	-13.2	-6.8	5.1
15°	8	1.2						-14.7	-5.7	6.0
20°	18	1.6	-2.4	-5.2	-9.0	-14.3	-15.9	-16.1	-8.9	5.2
30°	18	-1.0	-3.8	-6.9	-9.1	-13.9	-20.7	-22.3	-10.0	5.7
40°	18	-3.2	-4.8	-7.2	-9.8	-11.9	-22.9	-24.6	-10.5	5.5
50°	18	-2.0	-3.8	-6.4	-10.0	-13.8	-25.5	-27.1	-10.7	6.6
60°	16	-0.8	-2.2	-3.7	-7.5	-13.4	-17.8	-18.1	-8.6	5.6
70°	16	0.4	-0.3	-1.0	-4.0	-17.6	-23.6	-24.2	-9.0	9.4
80°	12	2.6						-32.5	-6.0	13.9

Ku Band, HV Polarization

Angle	N	σ^{o}_{max}	σ^{o}_{5}	σ^{o}_{25}	Median	σ^{o}_{75}	σ^{o}_{95}	σ^{o}_{min}	Mean	Std. Dev.
0°	1	-8.4						-8.4	-8.4	0.0
5°	4	-6.9						-13.3	-9.5	2.7
10°	8	-7.5						-23.4	-13.5	5.6
15°	8	-10.7						-25.1	-15.3	5.9
20°	7	-10.0						-25.9	-16.5	6.1
30°	8	-11.5						-31.1	-18.0	7.6
40°	8	-13.8						-31.7	-19.8	7.1
50°	8	-13.2						-33.4	-22.0	7.5
60°	6	-18.0						-24.5	-21.1	2.7
70°	6	-21.9						-29.4	-25.7	3.2
80°	3	-27.6						-35.6	-32.0	4.1

Ku Band, VV Polarization

Angle	N	σ^{o}_{max}	σ^{o}_{5}	σ^{o}_{25}	Median	σ^{o}_{75}	σ^{o}_{95}	σ^{o}_{min}	Mean	Std. Dev.
0°	8	34.7						4.6	25.7	9.6
5°	8	8.4						-5.9	0.3	5.5
10°	18	3.9	1.0	-4.4	-8.9	-11.2	-14.5	-15.5	-7.2	5.5
15°	8	2.2						-16.6	-6.2	6.8
20°	18	0.8	-2.4	-5.3	-9.6	-13.9	-16.8	-17.3	-9.2	5.3
30°	18	-0.6	-3.0	-6.0	-9.3	-14.4	-20.3	-20.6	-10.0	5.7
40°	18	-3.3	-4.6	-7.3	-9.5	-13.2	-22.3	-22.5	-10.6	5.5
50°	18	-2.4	-5.3	-7.8	-8.8	-13.3	-24.2	-24.5	-10.5	5.8
60°	16	-1.5	-3.8	-6.5	-8.2	-10.8	-14.2	-16.8	-8.5	3.9
70°	16	-0.2	-1.7	-3.3	-5.8	-14.3	-17.6	-21.6	-8.1	6.5
80°	12	1.7						-25.1	-5.3	10.3

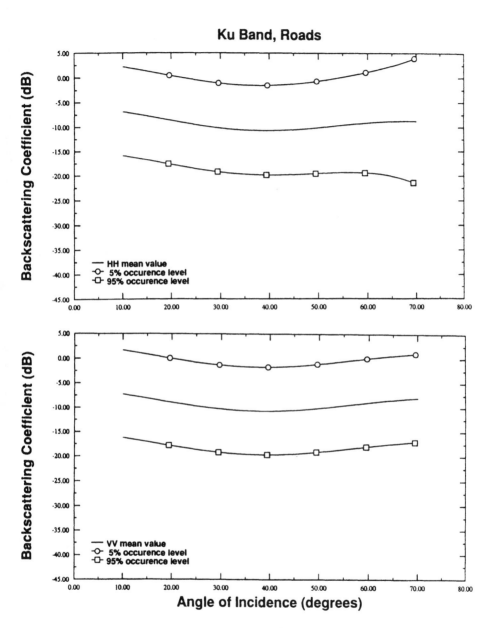

Ku Band, Roads

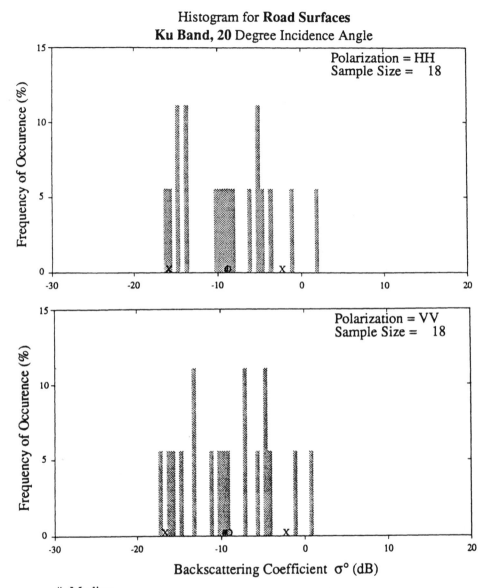

Histogram for **Road Surfaces**
Ku Band, 20 Degree Incidence Angle

Median
O Mean
X 5% or 95% Level

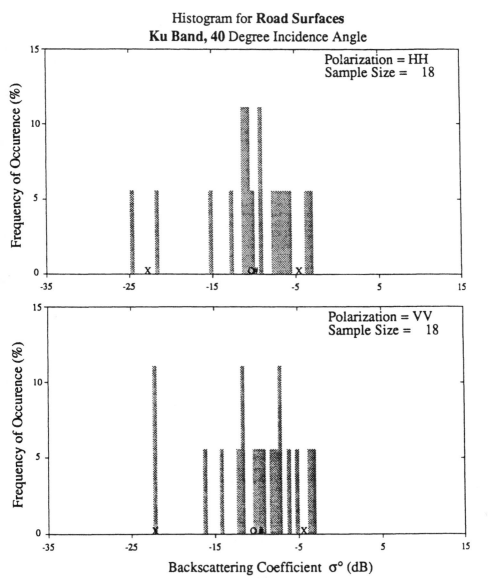

Histogram for **Road Surfaces**
Ku Band, 40 Degree Incidence Angle

Polarization = HH
Sample Size = 18

Polarization = VV
Sample Size = 18

Frequency of Occurence (%)

Backscattering Coefficient σ° (dB)

\# Median
O Mean
X 5% or 95% Level

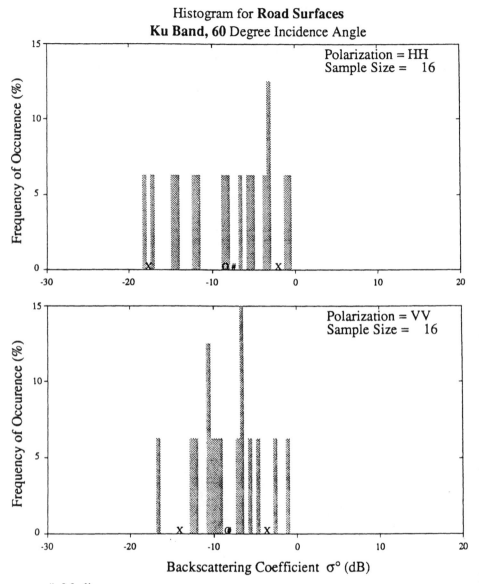

Histogram for **Road Surfaces**
Ku Band, 60 Degree Incidence Angle

Median
O Mean
X 5% or 95% Level

F-5   Ka-Band Data:

Statistical Distribution Table for Road Surfaces

Ka Band, HH Polarization

Angle	N	σ^o_{max}	σ^o_5	σ^o_{25}	Median	σ^o_{75}	σ^o_{95}	σ^o_{min}	Mean	Std. Dev.
0°	8	24.2						-4.8	17.3	10.0
5°	8	6.4						-4.6	-0.3	3.8
10°	26	1.7	-1.4	-2.8	-4.4	-6.6	-9.6	-10.9	-4.6	2.7
15°	8	-2.5						-11.6	-6.2	3.5
20°	26	-3.3	-4.5	-5.6	-6.8	-10.0	-13.8	-14.4	-7.7	3.1
30°	26	-5.1	-6.6	-7.9	-8.8	-12.9	-17.0	-17.1	-10.2	3.5
40°	26	-6.2	-8.0	-9.9	-11.3	-15.6	-20.8	-21.2	-12.6	4.2
50°	26	-7.9	-10.8	-13.1	-13.4	-19.4	-25.3	-25.9	-15.5	4.8
60°	25	-10.0	-13.0	-16.0	-17.3	-21.9	-26.9	-29.0	-18.2	4.7
70°	24	-15.1	-16.3	-19.8	-22.4	-25.4	-32.0	-33.8	-22.7	5.1
80°	13	-20.9						-39.6	-28.5	5.6

Ka Band, HV Polarization

Angle	N	σ^o_{max}	σ^o_5	σ^o_{25}	Median	σ^o_{75}	σ^o_{95}	σ^o_{min}	Mean	Std. Dev.
0°	3	2.5						-18.3	-8.8	10.5
5°	9	-4.2						-16.9	-11.5	3.5
10°	10	-6.5						-23.6	-15.0	4.9
15°	10	-8.5						-24.2	-16.1	4.9
20°	10	-9.6						-25.9	-16.2	4.8
30°	10	-12.8						-27.6	-17.3	4.8
40°	10	-13.2						-28.9	-18.6	5.4
50°	10	-13.9						-30.0	-20.1	5.4
60°	8	-17.3						-25.5	-20.2	2.5
70°	8	-22.0						-30.2	-24.2	2.8
80°	3	-26.9						-32.2	-29.9	2.7

Ka Band, VV Polarization

Angle	N	σ^o_{max}	σ^o_5	σ^o_{25}	Median	σ^o_{75}	σ^o_{95}	σ^o_{min}	Mean	Std. Dev.
0°	10	24.9						-4.1	16.5	10.0
5°	10	6.3						-5.1	0.2	3.5
10°	26	1.8	-0.7	-2.8	-4.4	-6.2	-8.9	-9.9	-4.3	2.8
15°	10	-1.2						-11.3	-5.8	3.2
20°	26	-1.7	-3.0	-5.0	-6.6	-10.3	-14.5	-16.7	-7.2	3.6
30°	26	-2.6	-5.9	-6.6	-7.3	-11.3	-14.6	-15.2	-8.6	3.1
40°	26	-4.7	-7.2	-7.7	-8.4	-12.7	-17.4	-18.3	-10.0	3.5
50°	26	-6.9	-8.6	-9.0	-10.7	-14.4	-19.6	-19.8	-11.7	3.6
60°	24	-9.0	-11.0	-11.0	-12.8	-15.7	-19.1	-19.6	-13.4	3.0
70°	24	-12.1	-14.2	-14.8	-16.2	-19.7	-24.8	-34.1	-17.6	4.5
80°	13	-17.7						-26.9	-22.0	3.2

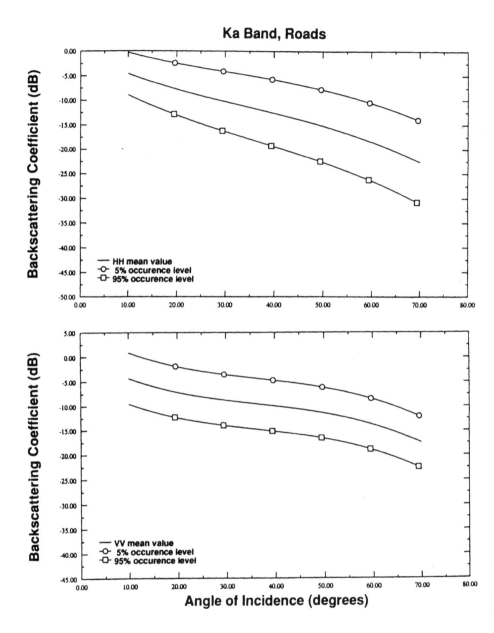

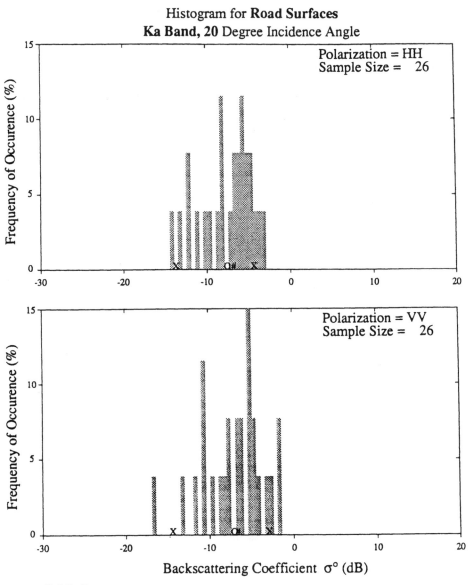

Histogram for **Road Surfaces**
Ka Band, 20 Degree Incidence Angle

Median
O Mean
X 5% or 95% Level

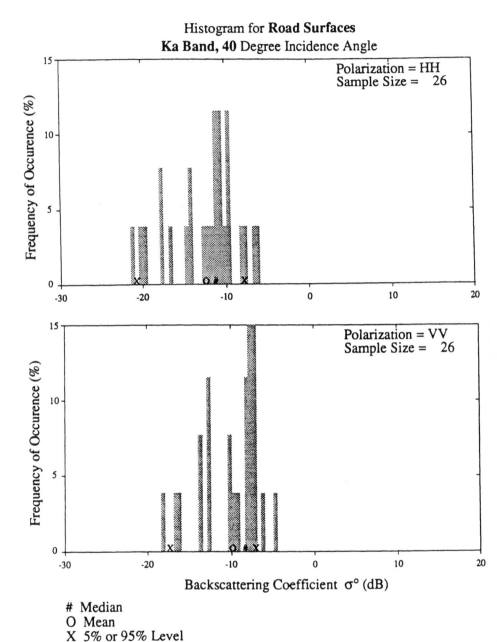

Histogram for **Road Surfaces**
Ka Band, 40 Degree Incidence Angle

Median
O Mean
X 5% or 95% Level

Histogram for **Road Surfaces**
Ka Band, 60 Degree Incidence Angle

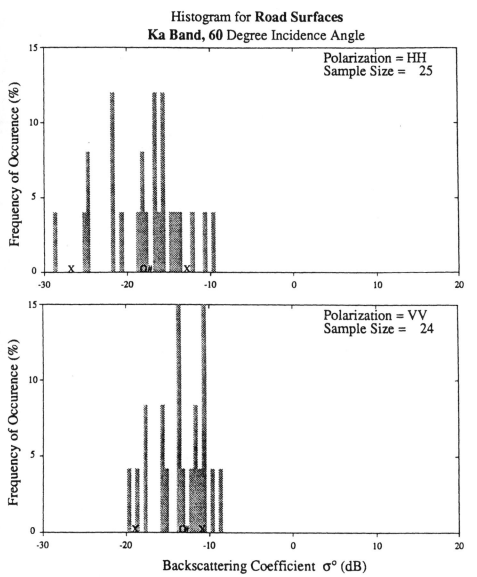

Median
O Mean
X 5% or 95% Level

F-6 W-Band Data:

Statistical Distribution Table for **Road Surfaces**

W Band, **VV** Polarization

Angle	N	σ^o_{max}	σ^o_5	σ^o_{25}	Median	σ^o_{75}	σ^o_{95}	σ^o_{min}	Mean	Std. Dev.
0°	3	17.2						4.6	9.5	6.8
5°	3	6.3						1.7	3.8	2.3
10°	3	3.6						0.3	1.4	1.9
15°	3	3.6						-0.4	1.0	2.3
20°	3	1.7						-1.5	0.3	1.6
30°	3	0.3						-3.0	-1.0	1.8
40°	3	-1.4						-4.7	-2.7	1.8
50°	2	-1.8						-6.2	-4.0	3.1
60°	2	-3.5						-6.0	-4.8	1.8
70°	2	-5.4						-7.0	-6.2	1.1

APPENDIX G
BACKSCATTERING DATA FOR URBAN AREAS

G-1 Data Sources and Parameter Loadings

Table G.1   Data sources for urban areas.

Band	References
L	4
S	4, 95
C	43
X	4, 51, 145, 151
Ku	
Ka	
W	

Table G.2   Mean and standard deviation parameter loadings for urban areas.

Band	Pol.	Angular Range θ_{min}	θ_{max}	P_1	P_2	P_3	P_4	P_5	P_6	M_1	M_2	M_3
L	HH											
	HV											
	VV											
S	HH											
	HV											
	VV											
C	HH											
	HV											
	VV											
X	HH											
	HV											
	VV											
Ku	HH											
	HV											
	VV											
Ka	HH											
	HV											
	VV											
W	HH											
	HV											
	VV											

(The θ° Parameters span P_1–P_6; the SD(θ) Parameters span M_1–M_3. The table is marked "Insufficient Data".)

$$\theta^\circ = P_1 + P_2\exp(-P_3\theta) + P_4\cos(P_5\theta + P_6)$$
$$SD(\theta) = M_1 + M_2\exp(-M_3\theta)$$
where θ is the angle of incidence in radians.

G-2   L-Band Data:

Statistical Distribution Table for **Urban Areas**

L Band, **HH** Polarization

Angle	N	σ^o_{max}	σ^o_5	σ^o_{25}	Median	σ^o_{75}	σ^o_{95}	σ^o_{min}	Mean	Std. Dev.
0°	2	7.0						1.5	4.3	3.9
10°	1	-1.4						-1.4	-1.4	0.0
20°	1	-5.8						-5.8	-5.8	0.0
30°	2	-8.9						-9.0	-8.9	0.1
40°	2	-9.0						-10.1	-9.6	0.8
50°	2	-11.0						-11.5	-11.3	0.4
60°	4	-4.8						-13.0	-9.5	3.4
70°	2	-12.1						-13.7	-12.9	1.1
75°	2	-6.0						-14.0	-10.0	5.7
80°	5	-8.0						-14.6	-12.4	2.8

L Band, **HV** Polarization

Angle	N	σ^o_{max}	σ^o_5	σ^o_{25}	Median	σ^o_{75}	σ^o_{95}	σ^o_{min}	Mean	Std. Dev.
0°	2	-10.4						-10.4	-10.4	0.0
10°	1	-15.9						-15.9	-15.9	0.0
20°	2	-17.0						-22.2	-19.6	3.7
30°	2	-17.6						-23.8	-20.7	4.4
40°	2	-18.4						-20.5	-19.5	1.5
50°	4	-15.2						-22.3	-18.8	3.5
60°	4	-16.7						-20.7	-18.2	1.7
70°	4	-15.0						-21.5	-19.2	3.0
80°	5	-15.9						-24.8	-19.6	3.7

L Band, **VV** Polarization

Angle	N	σ^o_{max}	σ^o_5	σ^o_{25}	Median	σ^o_{75}	σ^o_{95}	σ^o_{min}	Mean	Std. Dev.
0°	2	8.0						-0.1	4.0	5.7
10°	1	-3.5						-3.5	-3.5	0.0
20°	1	-10.4						-10.4	-10.4	0.0
30°	1	-11.5						-11.5	-11.5	0.0
40°	1	-14.5						-14.5	-14.5	0.0
50°	1	-13.8						-13.8	-13.8	0.0
60°	2	-14.2						-15.0	-14.6	0.6
70°	2	-13.9						-19.7	-16.8	4.1
80°	2	-18.5						-20.0	-19.3	1.1

G-3   S-Band Data:

Statistical Distribution Table for Urban Areas

S Band, HH Polarization

Angle	N	σ^o_{max}	σ^o_5	σ^o_{25}	Median	σ^o_{75}	σ^o_{95}	σ^o_{min}	Mean	Std. Dev.
0°	4	12.0						8.0	9.6	1.7
5°	3	11.0						7.5	9.3	1.8
10°	4	10.0						-2.8	5.4	5.7
15°	3	7.5						4.0	6.3	2.0
20°	4	5.0						-11.1	0.5	7.8
25°	3	2.0						-4.0	-1.7	3.2
30°	4	-5.0						-15.0	-10.6	4.4
40°	1	-10.0						-10.0	-10.0	0.0
50°	2	-13.5						-16.3	-14.9	2.0
60°	2	-11.9						-17.7	-14.8	4.1
70°	2	-17.1						-22.7	-19.9	4.0
80°	3	-20.1						-24.8	-23.0	2.6

S Band, HV Polarization

Angle	N	σ^o_{max}	σ^o_5	σ^o_{25}	Median	σ^o_{75}	σ^o_{95}	σ^o_{min}	Mean	Std. Dev.
0°	2	-1.7						-3.0	-2.3	0.9
10°	2	-13.0						-16.7	-14.9	2.6
20°	2	-13.3						-15.5	-14.4	1.6
30°	2	-12.9						-15.9	-14.4	2.1
40°	2	-12.5						-17.1	-14.8	3.3
50°	2	-17.7						-18.4	-18.0	0.5
60°	4	-18.6						-24.1	-21.0	2.7
70°	4	-23.1						-27.0	-24.6	1.8
80°	6	-23.4						-28.5	-24.9	2.0

S Band, VV Polarization

Angle	N	σ^o_{max}	σ^o_5	σ^o_{25}	Median	σ^o_{75}	σ^o_{95}	σ^o_{min}	Mean	Std. Dev.
0°	1	14.7						14.7	14.7	0.0
10°	1	-8.5						-8.5	-8.5	0.0
20°	1	-14.4						-14.4	-14.4	0.0
30°	1	-11.7						-11.7	-11.7	0.0
40°	1	-11.9						-11.9	-11.9	0.0
50°	1	-11.7						-11.7	-11.7	0.0
60°	2	-14.4						-16.6	-15.5	1.6
70°	2	-15.4						-19.0	-17.2	2.5
80°	3	-18.1						-19.7	-18.7	0.9

G-4 C-Band Data:

Statistical Distribution Table for **Urban Areas**

C Band, **HH** Polarization

Angle	N	σ^{o}_{max}	σ^{o}_{5}	σ^{o}_{25}	Median	σ^{o}_{75}	σ^{o}_{95}	σ^{o}_{min}	Mean	Std. Dev.
15°	1	-8.0						-8.0	-8.0	0.0
30°	2	-7.0						-9.0	-8.0	1.4
60°	2	-7.5						-10.0	-8.8	1.8
75°	1	-7.5						-7.5	-7.5	0.0
80°	2	-10.0						-11.0	-10.5	0.7

G-5 X-Band Data:

Statistical Distribution Table for **Urban Areas**

X Band, **HH** Polarization

Angle	N	σ^o_{max}	σ^o_5	σ^o_{25}	Median	σ^o_{75}	σ^o_{95}	σ^o_{min}	Mean	Std. Dev.
0°	4	20.0						-13.0	9.9	15.5
10°	4	3.1						-0.3	1.4	1.9
15°	2	-9.0						-10.2	-9.6	0.8
20°	4	-0.3						-10.3	-3.6	4.6
30°	9	13.0						-10.6	-4.1	7.7
35°	6	11.0						-7.0	2.9	7.9
40°	11	11.0						-12.2	-1.9	7.9
45°	6	13.5						-8.5	3.0	7.9
50°	11	15.5						-13.9	-2.1	8.9
55°	6	17.0						-9.0	3.8	9.1
60°	39	18.0	2.8	-9.5	-23.5	-27.0	-29.0	-29.0	-17.1	12.3
65°	6	18.5						-7.5	4.7	9.6
70°	11	18.0						-15.8	-2.9	11.3
75°	7	16.5						-12.5	1.8	10.8
80°	16	15.5	5.7	-2.2	-11.5	-15.7	-20.4	-21.0	-8.6	10.3

X Band, **HV** Polarization

Angle	N	σ^o_{max}	σ^o_5	σ^o_{25}	Median	σ^o_{75}	σ^o_{95}	σ^o_{min}	Mean	Std. Dev.
0°	3	-0.5						-5.0	-3.3	2.5
10°	4	-13.3						-18.4	-15.9	2.2
20°	2	-14.5						-15.0	-14.8	0.4
30°	2	-13.0						-17.0	-15.0	2.8
40°	2	-14.9						-17.2	-16.0	1.6
50°	2	-15.5						-16.6	-16.0	0.8
60°	4	-17.0						-21.4	-19.9	2.0
70°	2	-22.0						-23.3	-22.6	0.9
80°	5	-20.8						-26.9	-24.4	2.3

X Band, **VV** Polarization

Angle	N	σ^o_{max}	σ^o_5	σ^o_{25}	Median	σ^o_{75}	σ^o_{95}	σ^o_{min}	Mean	Std. Dev.
0°	2	19.5						10.5	15.0	6.4
10°	2	0.4						-4.0	-1.8	3.1
20°	1	-4.0						-4.0	-4.0	0.0
30°	4	-1.6						-12.6	-7.2	4.6
40°	4	-5.2						-13.7	-9.2	4.0
50°	4	-7.3						-15.9	-11.2	3.6
60°	6	-6.8						-23.0	-14.4	5.7
70°	5	-12.6						-17.6	-15.3	2.0
80°	5	-7.6						-19.0	-14.5	4.5

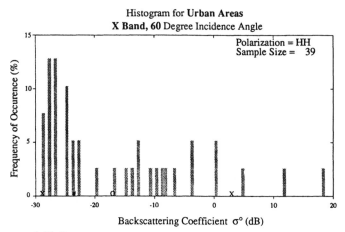

Histogram for **Urban Areas**
X Band, 60 Degree Incidence Angle

Median
O Mean
X 5% or 95% Level

APPENDIX H
BACKSCATTERING DATA FOR DRY SNOW

H-1   Data Sources and Parameter Loading

Table H.1   Data sources for dry snow.

Band	References
L	108, 111, 113, 136
S	68, 108, 111, 113, 136
C	68, 108, 111, 113, 136
X	37, 68, 93, 111, 113, 131, 136
Ku	37, 111, 113, 131, 136
Ka	37, 60, 111, 131, 136, 144
W	41, 52, 55, 57, 60, 68, 116, 144, 153, 154, 155

Table H.2   Mean and standard deviation parameter loadings for dry snow.

Band	Pol.	Angular Range θ_{min}	θ_{max}	σ° Parameters P_1	P_2	P_3	P_4	P_5	P_6	SD(θ) Parameters M_1	M_2	M_3
L	HH	0	70	-74.019	99.0	1.592	-30.0	1.928	0.905	-9.0	13.672	0.064
	HV	0	70	-91.341	99.0	1.202	30.0	1.790	-2.304	5.377	-0.571	3.695
	VV	0	70	-77.032	99.0	1.415	-30.0	1.720	0.997	4.487	-0.001	-5.725
S	HH	0	70	-47.055	30.164	5.788	30.0	1.188	-0.629	3.572	-2.0×10^{-5}	-9.0
	HV	0	70	-54.390	13.292	10.0	-30.0	-0.715	3.142	13.194	-9.0	-0.110
	VV	0	70	-40.652	18.826	9.211	30.0	0.690	0.214	-9.0	12.516	0.075
C	HH	0	70	-42.864	20.762	10.0	30.0	0.763	-0.147	4.398	0.0	0.0
	HV	0	70	-25.543	16.640	10.0	-2.959	3.116	2.085	13.903	-9.0	-0.085
	VV	0	70	-19.765	19.830	10.0	7.089	1.540	-0.012	13.370	-9.0	-0.016
X	HH	0	75	-13.298	20.048	10.0	4.529	2.927	-1.173	2.653	0.010	-2.457
	HV	20	75	-18.315	99.0	10.0	4.463	3.956	-2.128	2.460	1.0×10^{-5}	-8.314
	VV	0	70	-11.460	17.514	10.0	4.891	3.135	-0.888	12.339	-9.0	-0.072
Ku	HH	0	75	-36.188	15.340	10.0	30.0	0.716	-0.186	3.027	-0.033	0.055
	HV	0	75	-16.794	20.584	3.263	-2.243	5.0	0.096	12.434	-9.0	0.077
	VV	0	70	-10.038	13.975	10.0	-6.197	1.513	3.142	12.541	-9.0	-0.032
Ka	HH	0	75	-84.161	99.0	0.298	8.931	2.702	-3.142	-9.0	13.475	0.058
	HV											
	VV	0	70	-87.531	99.0	0.222	7.389	2.787	-3.142	-9.0	13.748	0.076
W	HH											
	HV											
	VV	0	75	-6.296	5.737	10.0	5.738	-2.356	1.065	3.364	0.0	0.0

$$\sigma^\circ = P_1 + P_2 \exp(-P_3\theta) + P_4 \cos(P_5\theta + P_6)$$
$$SD(\theta) = M_1 + M_2 \exp(-M_3\theta)$$
where θ is the angle of incidence in radians.

H-2 L-Band Data:

Statistical Distribution Table for Dry Snow

L Band, HH Polarization

Angle	N	σ°_{max}	σ°_{5}	σ°_{25}	Median	σ°_{75}	σ°_{95}	σ°_{min}	Mean	Std. Dev.
0°	126	12.5	11.1	9.2	7.4	4.2	-0.9	-9.3	6.4	4.1
10°	60	-0.1	-3.0	-5.6	-8.3	-11.2	-19.3	-21.0	-8.7	4.6
20°	88	-4.0	-7.6	-13.8	-16.3	-18.2	-27.7	-30.0	-16.5	5.5
30°	60	-15.7	-17.5	-18.9	-21.3	-23.7	-30.3	-32.1	-22.0	4.0
50°	128	-16.5	-17.6	-20.1	-24.1	-26.3	-31.4	-35.3	-23.6	4.1
55°	1	-22.5						-22.5	-22.5	0.0
60°	2	-25.5						-25.5	-25.5	0.0
65°	1	-27.7						-27.7	-27.7	0.0
70°	75	-23.8	-26.2	-28.5	-29.8	-31.1	-38.5	-38.7	-30.2	3.1
75°	1	-31.5						-31.5	-31.5	0.0
80°	2	-33.1						-33.1	-33.1	0.0

L Band, HV Polarization

Angle	N	σ°_{max}	σ°_{5}	σ°_{25}	Median	σ°_{75}	σ°_{95}	σ°_{min}	Mean	Std. Dev.
0°	47	-6.8	-7.7	-8.6	-11.8	-15.4	-24.2	-28.2	-12.6	4.9
10°	56	-16.0	-16.8	-20.3	-22.5	-25.0	-34.0	-37.1	-23.3	4.7
20°	77	-17.0	-20.2	-24.7	-28.8	-30.9	-41.8	-42.6	-28.7	5.8
30°	49	-27.8	-29.7	-31.6	-32.8	-35.9	-43.4	-44.4	-34.2	4.2
50°	51	-23.6	-25.7	-29.1	-31.9	-37.0	-48.0	-48.9	-33.8	6.6
70°	39	-31.4	-34.1	-36.6	-37.9	-40.7	-49.1	-49.4	-39.0	4.4
75°	1	-42.3						-42.3	-42.3	0.0
80°	2	-45.6						-45.6	-45.6	0.0

L Band, VV Polarization

Angle	N	σ°_{max}	σ°_{5}	σ°_{25}	Median	σ°_{75}	σ°_{95}	σ°_{min}	Mean	Std. Dev.
0°	106	12.9	11.2	8.3	6.2	3.8	-1.9	-7.2	5.7	4.0
10°	60	-0.7	-1.5	-5.3	-7.7	-10.4	-20.2	-22.0	-8.4	4.9
20°	102	-4.4	-8.6	-10.8	-14.4	-17.8	-25.3	-28.1	-14.7	5.2
30°	59	-15.8	-18.2	-19.8	-21.4	-22.9	-30.9	-31.2	-22.1	3.6
50°	98	-17.2	-19.0	-20.4	-23.4	-25.7	-36.2	-37.6	-23.8	4.4
55°	1	-22.8						-22.8	-22.8	0.0
60°	2	-24.0						-24.0	-24.0	0.0
65°	1	-26.5						-26.5	-26.5	0.0
70°	92	-23.0	-26.3	-27.7	-29.0	-30.2	-39.3	-41.3	-29.6	3.5
75°	1	-31.3						-31.3	-31.3	0.0
80°	2	-34.8						-34.8	-34.8	0.0

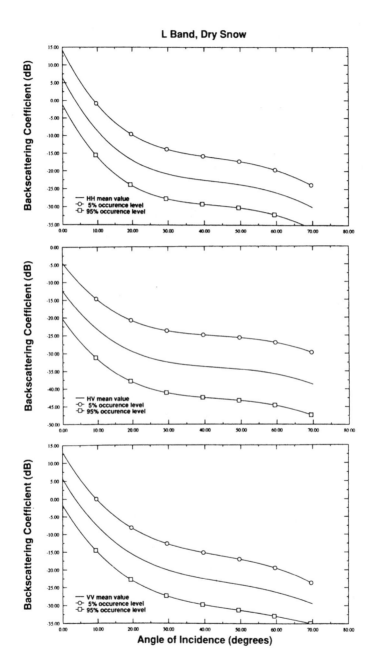

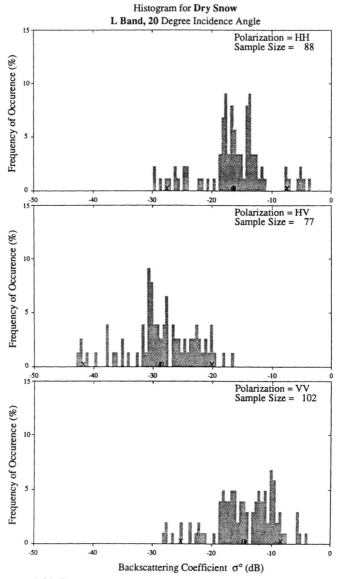

Histogram for **Dry Snow**
L Band, 20 Degree Incidence Angle

Median
O Mean
X 5% or 95% Level

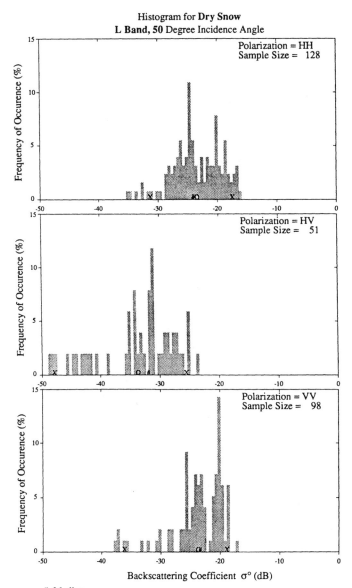

Histogram for **Dry Snow**
L Band, 50 Degree Incidence Angle

Median
O Mean
X 5% or 95% Level

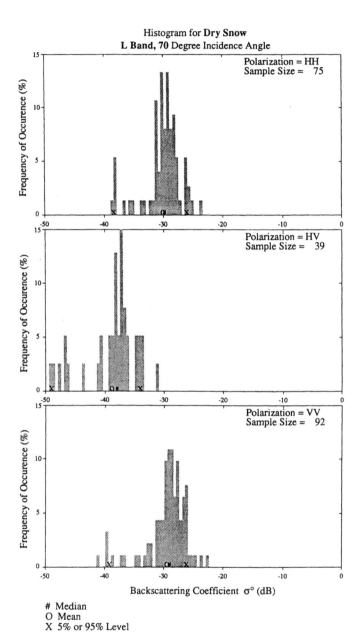

Histogram for **Dry Snow**
L Band, 70 Degree Incidence Angle

Median
O Mean
X 5% or 95% Level

H-3   S-Band Data:

Statistical Distribution Table for Dry Snow

S Band, HH Polarization

Angle	N	σ^o_{max}	σ^o_5	σ^o_{25}	Median	σ^o_{75}	σ^o_{95}	σ^o_{min}	Mean	Std. Dev.
0°	125	16.4	11.8	9.8	7.3	5.4	1.6	-0.8	7.4	3.2
10°	60	-2.6	-5.8	-7.0	-8.2	-11.4	-16.4	-19.3	-9.3	3.4
20°	127	-5.7	-8.0	-10.3	-13.3	-14.7	-19.4	-23.5	-13.0	3.4
30°	60	-10.0	-13.1	-14.9	-16.2	-18.7	-25.3	-25.8	-17.1	3.6
50°	167	-9.6	-11.7	-17.5	-19.4	-21.1	-25.7	-30.7	-19.1	4.0
55°	1	-21.7						-21.7	-21.7	0.0
60°	2	-22.5						-22.5	-22.5	0.0
65°	1	-24.1						-24.1	-24.1	0.0
70°	108	-21.7	-23.1	-25.2	-26.7	-28.4	-30.6	-32.4	-26.7	2.3
75°	1	-30.5						-30.5	-30.5	0.0
80°	2	-33.4						-33.4	-33.4	0.0

S Band, HV Polarization

Angle	N	σ^o_{max}	σ^o_5	σ^o_{25}	Median	σ^o_{75}	σ^o_{95}	σ^o_{min}	Mean	Std. Dev.
0°	71	-2.4	-5.7	-7.5	-9.9	-15.1	-19.1	-23.6	-11.1	4.5
10°	55	-16.7	-18.6	-19.7	-20.8	-23.8	-32.7	-34.1	-22.4	4.2
20°	106	-15.2	-19.5	-22.1	-24.1	-25.8	-34.0	-35.9	-24.5	3.7
30°	59	-22.7	-23.9	-25.4	-27.1	-28.6	-35.2	-35.5	-27.4	3.0
50°	116	-23.3	-24.0	-27.4	-29.8	-32.1	-37.5	-39.0	-29.8	3.4
55°	1	-32.4						-32.4	-32.4	0.0
60°	2	-35.0						-35.0	-35.0	0.0
65°	1	-33.0						-33.0	-33.0	0.0
70°	86	-29.3	-30.8	-32.8	-34.8	-38.0	-40.0	-41.4	-35.2	3.0
75°	1	-39.5						-39.5	-39.5	0.0
80°	2	-43.2						-43.2	-43.2	0.0

S Band, VV Polarization

Angle	N	σ^o_{max}	σ^o_5	σ^o_{25}	Median	σ^o_{75}	σ^o_{95}	σ^o_{min}	Mean	Std. Dev.
0°	105	16.7	11.8	10.0	7.3	5.3	1.7	-0.5	7.5	3.4
10°	60	-2.2	-5.0	-6.7	-7.8	-10.8	-15.0	-17.1	-8.8	3.3
20°	107	-5.2	-8.0	-10.2	-12.0	-14.3	-19.6	-20.6	-12.4	3.2
30°	60	-9.1	-13.1	-14.3	-15.6	-18.7	-22.3	-22.9	-16.5	3.1
50°	121	-9.6	-17.0	-18.5	-19.4	-20.8	-26.6	-31.0	-19.8	3.1
55°	1	-21.9						-21.9	-21.9	0.0
60°	2	-22.2						-22.2	-22.2	0.0
65°	1	-24.7						-24.7	-24.7	0.0
70°	108	-21.9	-23.3	-24.9	-25.6	-26.9	-31.3	-32.6	-26.0	2.1
75°	1	-28.7						-28.7	-28.7	0.0
80°	2	-33.4						-33.4	-33.4	0.0

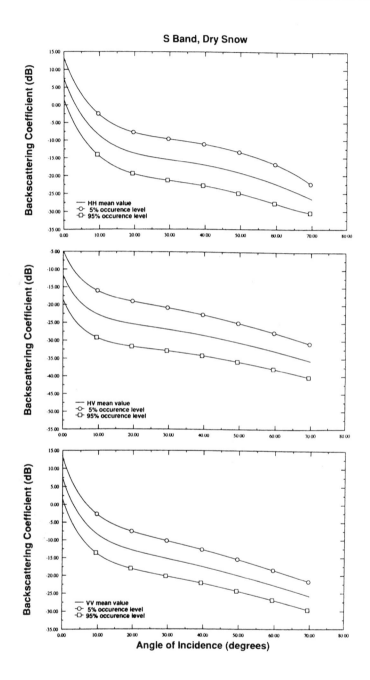

Histogram for **Dry Snow**
S Band, 20 Degree Incidence Angle

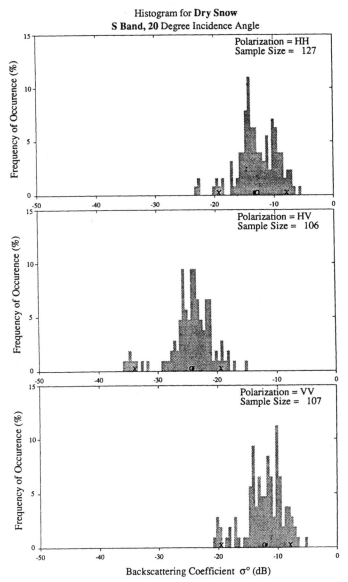

Backscattering Coefficient σ° (dB)

\# Median
O Mean
X 5% or 95% Level

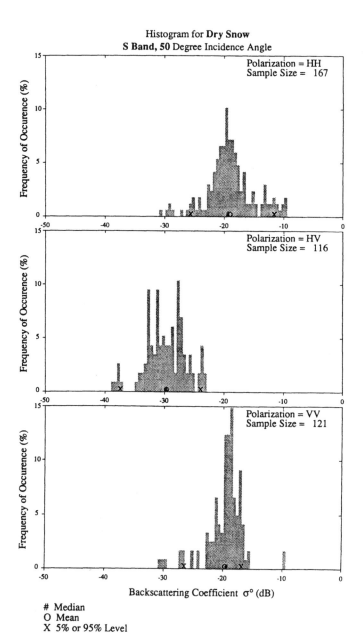

Histogram for **Dry Snow**
S Band, 50 Degree Incidence Angle

Median
O Mean
X 5% or 95% Level

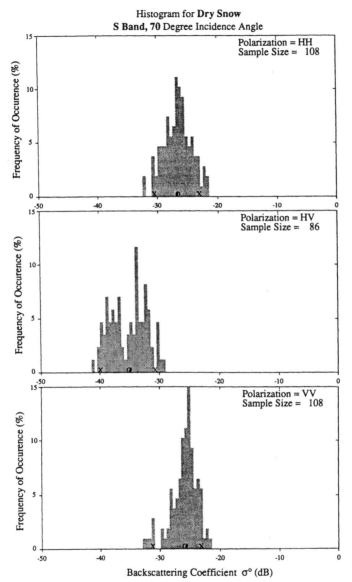

Histogram for **Dry Snow**
S Band, 70 Degree Incidence Angle

Median
O Mean
X 5% or 95% Level

H-4   C-Band Data:

Statistical Distribution Table for **Dry Snow**

C Band, **HH** Polarization

Angle	N	σ^o_{max}	σ^o_5	σ^o_{25}	Median	σ^o_{75}	σ^o_{95}	σ^o_{min}	Mean	Std. Dev.
0°	195	18.1	14.5	10.0	7.9	5.7	1.2	-16.1	7.6	4.4
10°	117	2.4	-2.5	-7.1	-9.3	-12.5	-17.7	-22.4	-9.7	4.4
20°	204	2.2	-6.2	-9.0	-11.2	-13.3	-19.7	-28.0	-11.6	4.2
30°	117	-5.7	-10.0	-12.7	-14.7	-17.9	-22.8	-24.9	-15.3	4.0
50°	268	-6.8	-9.6	-13.9	-15.8	-18.3	-27.6	-32.5	-16.5	4.9
55°	1	-18.6						-18.6	-18.6	0.0
60°	1	-17.2						-17.2	-17.2	0.0
65°	1	-19.3						-19.3	-19.3	0.0
70°	162	-10.6	-15.1	-19.6	-22.3	-24.3	-28.2	-32.7	-21.8	4.1
75°	1	-24.1						-24.1	-24.1	0.0
80°	1	-26.5						-26.5	-26.5	0.0

C Band, **HV** Polarization

Angle	N	σ^o_{max}	σ^o_5	σ^o_{25}	Median	σ^o_{75}	σ^o_{95}	σ^o_{min}	Mean	Std. Dev.
0°	100	4.4	2.4	-2.2	-7.1	-11.5	-21.7	-24.6	-7.4	6.7
10°	115	-10.6	-14.4	-17.8	-20.4	-22.4	-29.0	-32.2	-20.4	4.1
20°	142	-13.8	-15.5	-18.6	-21.3	-23.8	-28.9	-31.9	-21.5	4.1
30°	116	-15.6	-18.3	-21.4	-23.2	-25.9	-28.7	-33.0	-23.5	3.4
50°	150	-15.0	-19.8	-22.2	-25.2	-27.8	-34.7	-37.5	-25.7	4.6
55°	1	-27.9						-27.9	-27.9	0.0
65°	1	-25.9						-25.9	-25.9	0.0
70°	143	-16.8	-21.7	-25.1	-29.1	-31.5	-34.5	-36.2	-28.3	4.2
75°	1	-32.9						-32.9	-32.9	0.0

C Band, **VV** Polarization

Angle	N	σ^o_{max}	σ^o_5	σ^o_{25}	Median	σ^o_{75}	σ^o_{95}	σ^o_{min}	Mean	Std. Dev.
0°	159	17.8	15.4	9.9	6.9	5.0	-0.1	-14.1	7.2	4.7
10°	116	0.6	-3.9	-8.2	-9.6	-12.0	-17.9	-21.0	-10.0	3.9
20°	159	-2.4	-5.8	-8.8	-12.2	-14.6	-20.3	-25.0	-12.1	4.5
30°	117	-6.5	-10.8	-13.4	-15.2	-17.5	-22.7	-26.0	-15.7	3.6
50°	180	-9.6	-12.3	-15.2	-16.4	-18.9	-28.2	-31.3	-17.9	4.6
55°	1	-17.5						-17.5	-17.5	0.0
60°	1	-18.8						-18.8	-18.8	0.0
65°	1	-19.9						-19.9	-19.9	0.0
70°	120	-13.3	-15.5	-18.8	-22.2	-24.5	-28.9	-31.4	-21.9	4.1
80°	1	-28.0						-28.0	-28.0	0.0

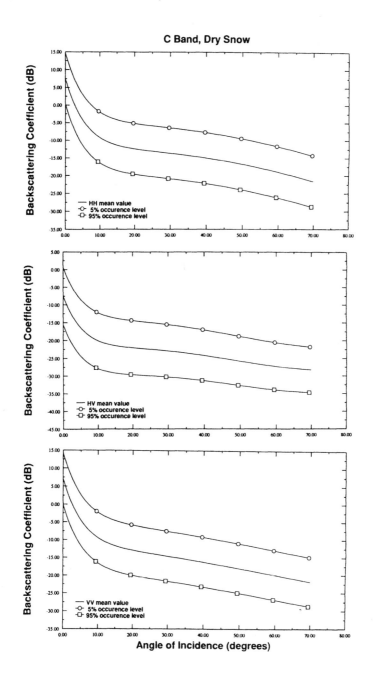

C Band, Dry Snow

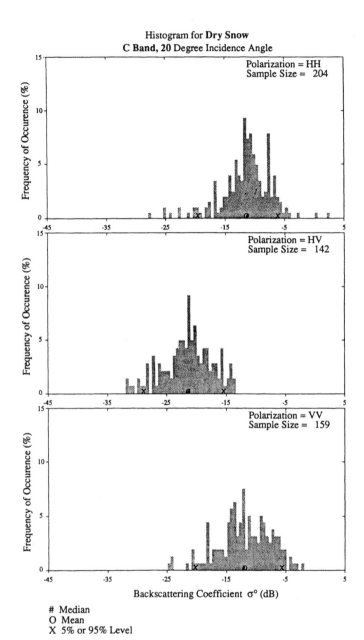

Histogram for **Dry Snow**
C Band, 20 Degree Incidence Angle

\# Median
O Mean
X 5% or 95% Level

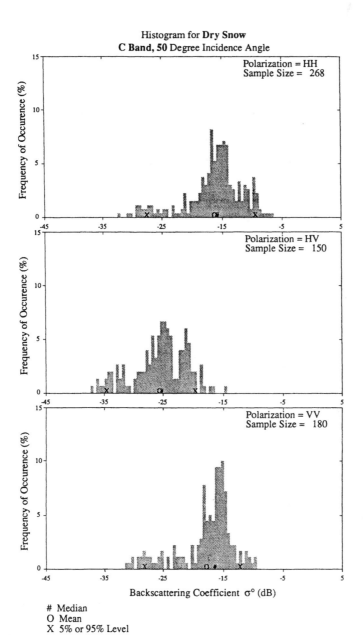

Histogram for **Dry Snow**
C Band, 50 Degree Incidence Angle

Median
O Mean
X 5% or 95% Level

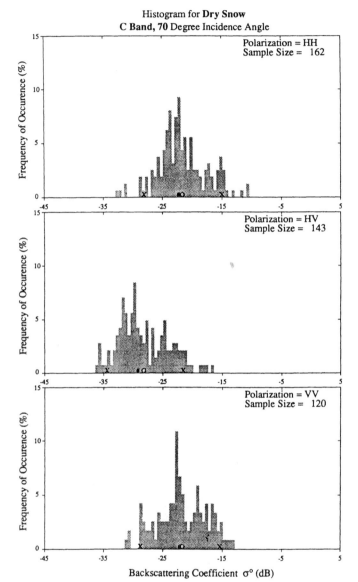

Histogram for **Dry Snow**
C Band, 70 Degree Incidence Angle

Median
O Mean
X 5% or 95% Level

H-5 X-Band Data:

Statistical Distribution Table for Dry Snow

X Band, HH Polarization

Angle	N	σ^o_{max}	σ^o_5	σ^o_{25}	Median	σ^o_{75}	σ^o_{95}	σ^o_{min}	Mean	Std. Dev.
0°	117	14.5	12.7	11.4	9.5	5.7	3.0	1.8	8.6	3.3
5°	52	5.1	2.3	-2.4	-3.4	-4.5	-8.0	-9.0	-3.2	2.9
10°	36	3.2	-2.5	-3.6	-4.8	-5.7	-9.8	-10.1	-4.5	2.2
20°	114	-2.5	-4.3	-6.3	-7.8	-8.7	-10.8	-15.2	-7.6	2.0
25°	48	-5.4	-8.2	-9.7	-10.5	-11.6	-16.2	-17.1	-10.9	2.3
30°	30	-6.9	-7.3	-8.4	-9.6	-11.8	-16.3	-16.3	-10.2	2.4
40°	55	-6.4	-6.5	-6.9	-9.1	-10.7	-15.3	-16.1	-9.4	2.7
50°	316	-4.4	-7.6	-10.2	-11.9	-14.0	-17.3	-20.5	-12.1	2.9
55°	53	-10.6	-12.5	-13.7	-14.9	-16.6	-19.3	-21.0	-15.2	2.1
60°	17	-5.5	-9.9	-14.6	-16.8	-18.7	-21.1	-21.1	-15.9	4.2
70°	82	-4.1	-14.3	-15.3	-16.8	-18.4	-21.6	-24.7	-16.9	3.0
75°	52	-13.6	-14.4	-14.7	-15.3	-19.2	-20.8	-25.7	-16.8	2.6
80°	11	-16.1						-30.6	-23.7	3.9

X Band, HV Polarization

Angle	N	σ^o_{max}	σ^o_5	σ^o_{25}	Median	σ^o_{75}	σ^o_{95}	σ^o_{min}	Mean	Std. Dev.
0°	9	3.6						-5.5	-0.8	3.1
20°	69	-6.6	-9.5	-10.3	-11.6	-12.8	-16.4	-21.5	-11.9	2.3
30°	20	-9.3	-10.4	-12.1	-14.0	-16.7	-21.2	-21.2	-14.7	3.4
40°	45	-11.4	-11.8	-12.2	-13.2	-14.1	-22.0	-22.0	-13.8	2.7
50°	120	-12.2	-14.5	-15.6	-17.3	-18.5	-21.9	-24.4	-17.4	2.3
55°	12	-15.1						-24.3	-19.5	2.8
60°	7	-19.5						-27.0	-23.8	3.3
70°	71	-14.6	-18.1	-21.1	-22.0	-23.5	-28.4	-30.1	-22.2	2.8
75°	49	-15.9	-19.9	-21.3	-21.7	-25.7	-27.3	-33.1	-22.9	3.0
80°	4	-25.1						-32.4	-30.0	3.4

X Band, VV Polarization

Angle	N	σ^o_{max}	σ^o_5	σ^o_{25}	Median	σ^o_{75}	σ^o_{95}	σ^o_{min}	Mean	Std. Dev.
0°	73	15.9	14.6	12.4	9.3	6.2	3.4	2.0	9.3	3.7
5°	17	6.4	6.3	0.8	-2.8	-6.7	-9.4	-10.2	-2.0	5.4
10°	29	3.4	-0.8	-2.8	-4.1	-5.7	-8.0	-10.1	-3.8	2.6
20°	74	-1.0	-1.4	-3.4	-4.8	-7.3	-10.7	-12.7	-5.3	2.6
30°	28	-4.8	-6.1	-8.3	-9.6	-11.7	-14.0	-14.0	-9.8	2.6
35°	11	-7.9						-19.3	-12.5	4.1
40°	13	-6.9						-16.7	-11.8	2.9
50°	92	-6.4	-8.9	-11.1	-12.4	-13.6	-17.5	-18.5	-12.6	2.4
55°	12	-9.0						-21.4	-15.1	3.7
60°	15	-5.5	-9.8	-13.0	-14.9	-19.9	-20.5	-20.5	-15.0	4.4
70°	78	-4.1	-12.9	-15.1	-15.9	-17.4	-21.5	-24.7	-16.3	3.0
75°	12	-15.0						-26.9	-17.8	3.2
80°	9	-18.6						-30.6	-23.6	4.2

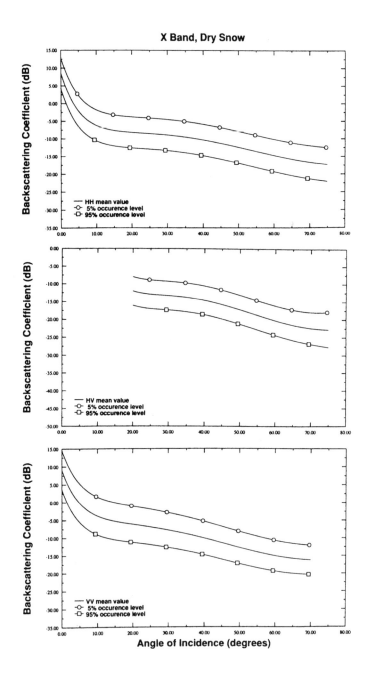

X Band, Dry Snow

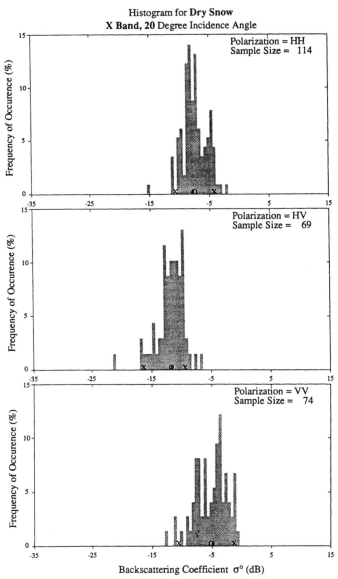

Histogram for **Dry Snow**
X Band, 20 Degree Incidence Angle

Median
O Mean
X 5% or 95% Level

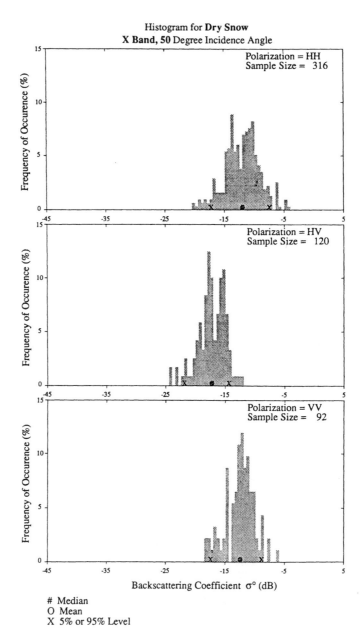

Histogram for **Dry Snow**
X Band, 50 Degree Incidence Angle

Median
O Mean
X 5% or 95% Level

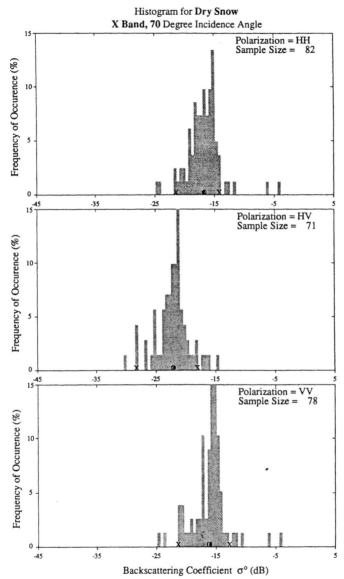

Histogram for **Dry Snow**
X Band, 70 Degree Incidence Angle

Median
O Mean
X 5% or 95% Level

H-6 Ku-Band Data:

Statistical Distribution Table for **Dry Snow**

Ku Band, **HH** Polarization

Angle	N	σ^o_{max}	σ^o_5	σ^o_{25}	Median	σ^o_{75}	σ^o_{95}	σ^o_{min}	Mean	Std. Dev.
0°	225	16.4	13.7	11.5	9.1	6.0	3.1	1.0	8.9	3.4
5°	60	2.1	-0.1	-1.8	-2.9	-4.1	-7.7	-9.8	-3.0	2.2
10°	86	3.4	1.2	-1.0	-3.0	-4.6	-6.6	-10.1	-2.8	2.6
20°	220	1.0	-1.5	-3.3	-5.1	-6.7	-10.8	-17.3	-5.2	2.8
25°	55	-2.9	-3.8	-4.9	-6.4	-7.5	-14.2	-14.5	-6.9	2.8
30°	80	0.6	-1.4	-3.2	-5.8	-7.9	-12.4	-15.5	-6.0	3.6
40°	90	-4.3	-5.2	-6.8	-8.6	-10.0	-14.1	-16.2	-8.7	2.6
50°	523	-0.9	-4.2	-7.1	-8.8	-11.1	-14.0	-19.2	-9.1	3.0
55°	57	-5.9	-7.1	-8.1	-9.2	-10.2	-16.5	-17.7	-9.7	2.7
60°	14	-8.1						-22.5	-14.9	4.7
70°	184	-6.2	-8.2	-9.8	-12.8	-14.5	-18.4	-25.7	-12.6	3.4
75°	94	-10.6	-10.9	-12.1	-15.0	-16.1	-19.8	-21.3	-14.7	2.7
80°	7	-13.5						-30.0	-22.0	6.9

Ku Band, **HV** Polarization

Angle	N	σ^o_{max}	σ^o_5	σ^o_{25}	Median	σ^o_{75}	σ^o_{95}	σ^o_{min}	Mean	Std. Dev.
0°	171	13.0	9.1	4.8	1.4	-1.8	-5.2	-8.4	1.5	4.4
5°	12	-2.6						-11.9	-6.8	3.8
10°	72	-0.1	-1.3	-3.3	-5.7	-7.5	-10.3	-11.4	-5.6	2.7
20°	173	-0.3	-4.1	-8.0	-10.4	-12.8	-16.9	-23.3	-10.4	3.8
25°	13	-5.4						-16.9	-11.5	4.6
30°	75	-3.2	-4.4	-6.0	-8.7	-10.7	-14.7	-18.2	-8.9	3.4
40°	85	-10.2	-10.6	-12.0	-13.9	-14.9	-17.9	-19.0	-13.7	2.0
50°	303	-5.8	-7.9	-13.2	-15.0	-17.3	-20.0	-27.2	-15.0	3.6
55°	15	-10.6	-11.1	-11.8	-15.7	-17.6	-18.4	-18.9	-14.8	3.0
60°	9	-15.2						-22.2	-18.2	2.8
70°	163	-9.6	-11.8	-15.9	-18.8	-21.8	-24.4	-30.2	-18.5	4.2
75°	91	34.0	-15.2	-16.3	-19.7	-21.6	-25.2	-26.8	-18.8	6.5
80°	2	-22.1						-23.8	-23.0	1.2

Ku Band, **VV** Polarization

Angle	N	σ^o_{max}	σ^o_5	σ^o_{25}	Median	σ^o_{75}	σ^o_{95}	σ^o_{min}	Mean	Std. Dev.
0°	172	18.7	16.3	12.8	10.7	8.0	3.4	1.1	10.3	3.8
5°	17	2.5	1.9	1.4	-0.5	-5.4	-7.1	-7.6	-1.7	3.6
10°	78	3.5	'2.8	0.2	-1.9	-4.3	-7.3	-9.6	-2.0	3.2
20°	172	2.1	0.1	-1.6	-3.2	-5.7	-9.8	-14.1	-3.8	3.0
25°	13	-1.2						-14.4	-7.3	5.2
30°	78	2.5	-0.4	-2.5	-5.4	-7.2	-13.3	-13.4	-5.6	4.0
40°	8	-9.7						-13.6	-12.3	1.4
50°	221	1.1	-3.3	-6.1	-9.1	-11.0	-13.8	-16.7	-8.6	3.3
55°	15	-4.5	-5.2	-6.1	-9.9	-14.6	-15.4	-15.7	-9.9	4.2
60°	12	-6.1						-19.7	-13.1	4.6
70°	177	-4.3	-7.1	-9.5	-11.7	-13.5	-17.0	-22.6	-11.7	3.2
75°	14	-10.8						-14.4	-12.7	1.2
80°	5	-14.5						-24.1	-20.0	4.0

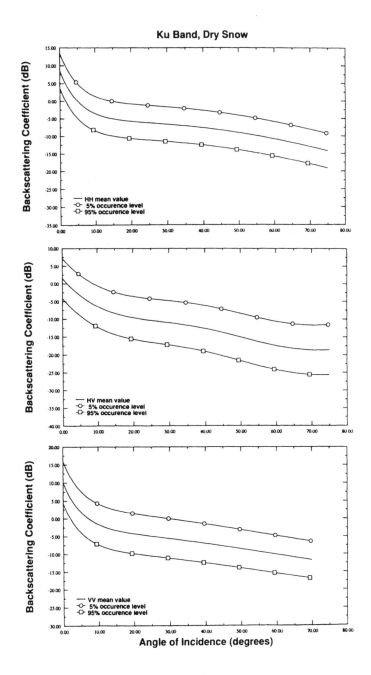

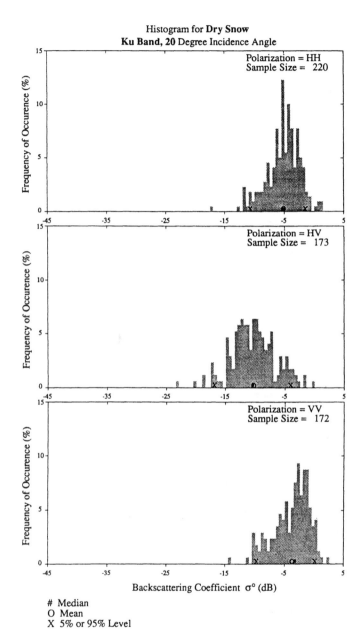

Histogram for **Dry Snow**
Ku Band, 20 Degree Incidence Angle

Polarization = HH
Sample Size = 220

Polarization = HV
Sample Size = 173

Polarization = VV
Sample Size = 172

Backscattering Coefficient σ° (dB)

\# Median
O Mean
X 5% or 95% Level

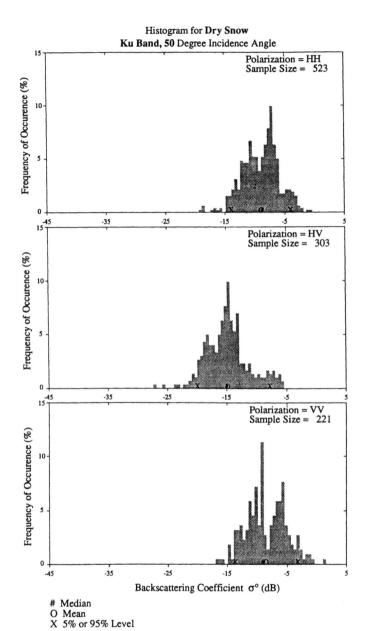

Histogram for **Dry Snow**
Ku Band, 50 Degree Incidence Angle

Median
O Mean
X 5% or 95% Level

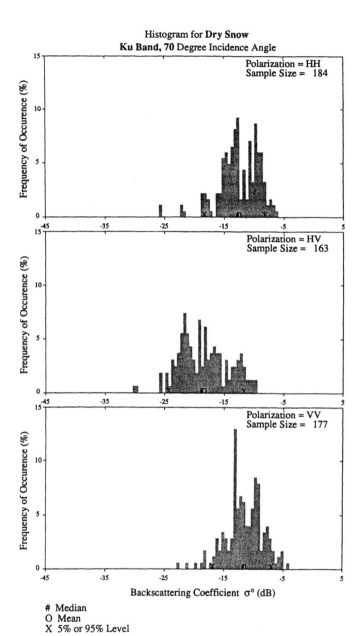

Histogram for **Dry Snow**
Ku Band, 70 Degree Incidence Angle

Median
O Mean
X 5% or 95% Level

H-7 Ka-Band Data:

Statistical Distribution Table for Dry Snow

Ka Band, HH Polarization

Angle	N	σ°_{max}	σ°_5	σ°_{25}	Median	σ°_{75}	σ°_{95}	σ°_{min}	Mean	Std. Dev.
0°	69	20.5	14.8	8.0	3.3	2.3	1.3	0.6	5.3	4.5
5°	10	6.5						-1.3	4.0	2.3
10°	21	8.0	6.8	5.6	4.5	1.2	-6.1	-6.1	3.0	4.5
20°	68	7.1	6.6	3.9	0.8	-0.3	-7.0	-8.0	1.1	3.5
25°	9	5.6						-5.6	2.5	3.2
30°	20	6.9	5.9	5.6	2.9	-1.2	-9.5	-9.6	1.1	5.4
40°	58	1.2	0.6	-4.1	-5.6	-7.0	-10.3	-16.6	-5.2	3.3
50°	157	7.1	4.9	2.0	-0.2	-5.6	-8.2	-13.3	-1.1	4.5
55°	16	4.2	3.7	3.5	3.2	2.8	-2.6	-6.6	2.4	2.6
60°	10	-0.2						-15.6	-6.3	5.9
70°	65	0.8	-0.5	-2.5	-4.2	-5.6	-15.5	-18.3	-4.8	4.0
75°	46	-3.8	-9.9	-10.6	-11.4	-12.9	-14.1	-14.5	-11.5	2.1
80°	2	-5.1						-25.1	-15.1	14.1

Ka Band, HV Polarization

Angle	N	σ°_{max}	σ°_5	σ°_{25}	Median	σ°_{75}	σ°_{95}	σ°_{min}	Mean	Std. Dev.
0°	10	2.8						-7.9	-3.3	3.4
5°	4	-9.3						-12.1	-10.4	1.2
10°	7	2.6						-14.1	-6.3	6.3
20°	7	0.6						-13.1	-6.1	5.6
25°	1	-9.5						-9.5	-9.5	0.0
30°	7	2.1						-13.6	-5.5	5.8
35°	1	-11.4						-11.4	-11.4	0.0
40°	74	-1.7	-3.0	-4.4	-8.7	-10.0	-14.1	-19.9	-7.9	3.5
45°	6	-7.1						-12.2	-9.5	1.9
50°	47	-0.7	-6.1	-9.0	-10.1	-11.4	-13.7	-14.3	-9.8	2.7
55°	1	-12.1						-12.1	-12.1	0.0
60°	6	-5.2						-15.8	-10.7	5.3
65°	1	-15.3						-15.3	-15.3	0.0
70°	6	-7.3						-21.2	-14.8	7.0
75°	46	-13.1	-13.8	-14.5	-15.6	-17.1	-19.2	-24.9	-16.0	2.1
80°	2	-26.8						-30.0	-28.4	2.3

Ka Band, VV Polarization

Angle	N	σ°_{max}	σ°_5	σ°_{25}	Median	σ°_{75}	σ°_{95}	σ°_{min}	Mean	Std. Dev.
0°	76	15.9	13.4	5.9	2.6	1.6	-2.2	-4.0	4.0	4.4
5°	15	3.9	3.8	3.3	1.2	-4.7	-10.4	-10.4	-0.6	5.1
10°	24	7.0	6.6	6.0	3.5	-3.1	-13.1	-13.1	1.1	6.1
20°	70	7.1	5.8	2.6	0.2	-1.0	-7.6	-13.6	0.3	4.1
25°	10	3.8						-7.9	0.2	4.3
30°	23	6.4	5.5	5.0	1.2	-5.8	-10.3	-10.3	-0.2	5.7
40°	38	3.0	1.9	1.4	0.1	-4.2	-10.7	-10.7	-1.7	4.2
45°	6	1.0						-8.3	-2.8	3.5
50°	120	5.7	4.3	2.7	0.2	-1.1	-8.1	-11.5	0.0	3.6
55°	17	1.2	1.0	0.6	0.2	-0.1	-8.1	-8.6	-0.6	2.8
60°	13	0.1						-14.3	-7.0	5.8
70°	68	-0.4	-1.9	-2.8	-4.0	-5.1	-14.7	-16.7	-5.1	3.8
75°	8	-4.6						-20.7	-10.9	4.7
80°	4	-6.1						-24.5	-18.3	8.3

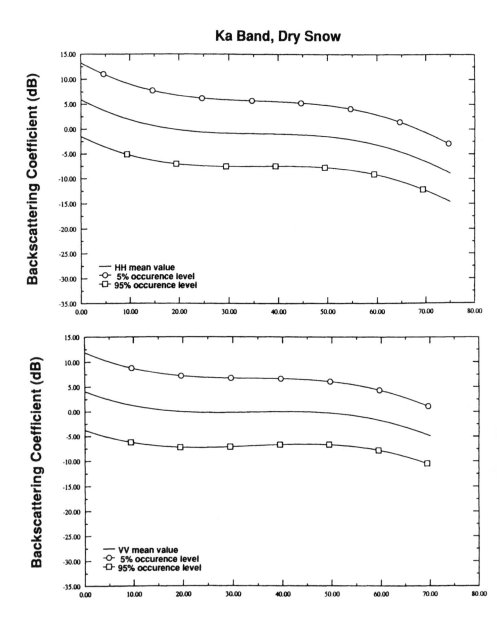

Ka Band, Dry Snow

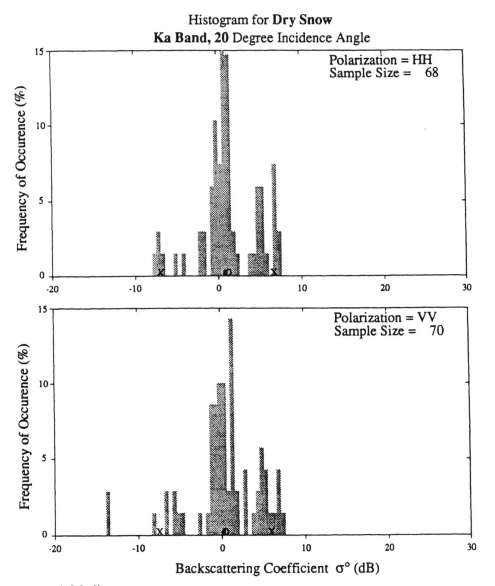

Histogram for **Dry Snow**
Ka Band, 20 Degree Incidence Angle

Median
O Mean
X 5% or 95% Level

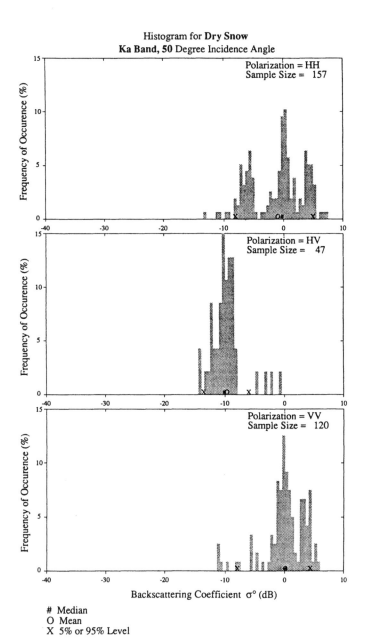

Histogram for **Dry Snow**
Ka Band, 50 Degree Incidence Angle

Polarization = HH
Sample Size = 157

Polarization = HV
Sample Size = 47

Polarization = VV
Sample Size = 120

Backscattering Coefficient σ° (dB)

\# Median
O Mean
X 5% or 95% Level

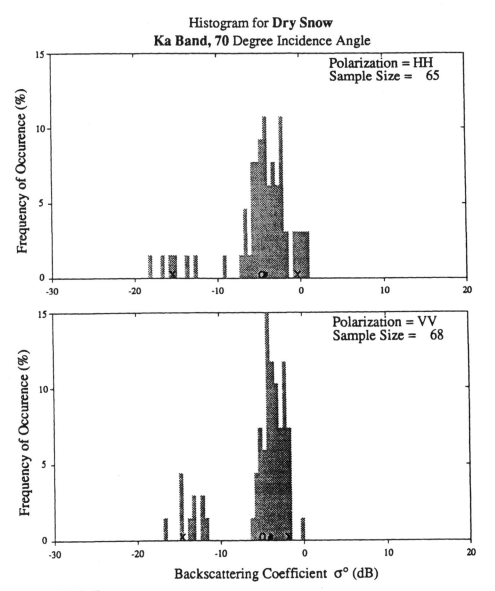

Histogram for **Dry Snow**
Ka Band, 70 Degree Incidence Angle

Median
O Mean
X 5% or 95% Level

H-8 W-Band Data:

Statistical Distribution Table for **Dry Snow**

W Band, **HH** Polarization

Angle	N	σ^o_{max}	σ^o_5	σ^o_{25}	Median	σ^o_{75}	σ^o_{95}	σ^o_{min}	Mean	Std. Dev.
0°	5	9.0						-0.6	4.1	4.3
10°	3	6.8						4.5	5.4	1.2
20°	3	7.9						4.9	6.0	1.7
30°	3	4.7						3.6	4.3	0.6
40°	36	4.9	4.8	4.0	3.2	2.5	-3.0	-8.5	2.8	2.4
45°	3	7.0						-11.0	-1.0	9.2
50°	11	4.7						-14.1	-7.5	7.7
55°	2	-1.0						-7.5	-4.3	4.6
60°	5	2.6						-7.5	-0.8	4.2
65°	2	-2.5						-7.5	-5.0	3.5
70°	5	0.1						-9.0	-3.3	3.8
75°	6	-9.0						-18.0	-12.6	4.2

W Band, **HV** Polarization

Angle	N	σ^o_{max}	σ^o_5	σ^o_{25}	Median	σ^o_{75}	σ^o_{95}	σ^o_{min}	Mean	Std. Dev.
0°	5	6.0						-3.9	-0.9	4.2
5°	5	-1.5						-3.0	-2.3	0.6
10°	7	2.3						-6.5	-1.6	3.4
15°	1	-4.0						-4.0	-4.0	0.0
20°	6	2.0						-5.0	-1.3	2.8
25°	4	-3.0						-5.5	-4.1	1.1
30°	4	1.9						-5.0	-0.5	3.1
35°	7	-3.5						-7.5	-5.4	1.3
40°	35	1.1	0.9	0.6	0.0	-2.2	-6.5	-7.0	-0.9	2.2
45°	12	-1.0						-9.0	-6.1	2.6
50°	3	-0.4						-1.2	-0.7	0.4
55°	13	0.0						-13.0	-8.2	4.0
60°	10	-1.0						-11.0	-5.7	4.1
65°	12	-2.5						-18.0	-11.8	4.7
70°	9	-3.0						-12.0	-7.6	3.8
75°	16	-9.0	-9.9	-12.5	-17.2	-18.5	-20.2	-21.0	-15.8	3.8

W Band, **VV** Polarization

Angle	N	σ^o_{max}	σ^o_5	σ^o_{25}	Median	σ^o_{75}	σ^o_{95}	σ^o_{min}	Mean	Std. Dev.
0°	18	9.0	6.3	3.1	1.5	0.3	-1.0	-1.3	2.2	2.7
5°	23	4.0	3.5	2.8	1.0	-3.9	-7.2	-7.5	-0.2	3.8
10°	22	7.4	3.7	2.0	-0.7	-3.2	-8.7	-9.5	-0.5	3.9
15°	9	3.0						-8.5	-1.6	3.2
20°	27	5.4	3.9	2.0	-1.5	-2.5	-6.9	-8.0	-0.5	3.2
25°	16	2.5	0.8	0.3	-1.5	-2.0	-2.2	-2.5	-0.9	1.4
30°	19	4.9	3.8	0.8	-2.0	-4.2	-9.2	-9.5	-1.7	3.9
35°	18	2.0	1.3	-1.2	-2.2	-3.5	-4.7	-5.0	-2.1	2.1
40°	54	4.8	4.1	3.2	1.4	-2.2	-8.0	-9.5	0.4	3.6
45°	23	4.0	-0.2	-1.7	-3.5	-5.2	-8.6	-10.0	-3.1	3.2
50°	24	3.9	1.7	-2.5	-4.7	-11.8	-13.5	-14.1	-6.1	5.5
55°	24	2.0	-0.7	-1.2	-3.2	-5.0	-6.5	-7.0	-3.3	2.2
60°	33	2.2	0.9	-3.7	-6.0	-7.0	-12.2	-12.5	-5.2	3.4
65°	25	0.0	-3.2	-5.0	-6.7	-8.0	-12.5	-13.0	-6.6	3.0
70°	18	-1.1	-1.5	-4.5	-8.0	-9.5	-10.0	-10.0	-6.5	3.3
75°	21	2.0	-6.2	-8.0	-10.0	-12.0	-15.5	-16.0	-9.4	4.1
80°	4	-13.0						-13.5	-13.3	0.3

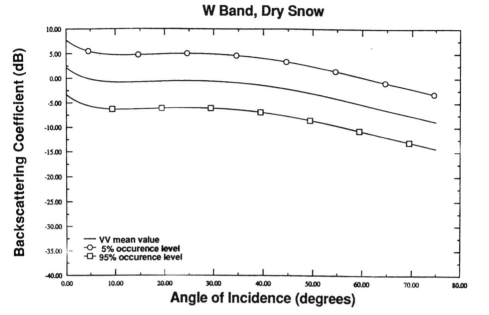

Histogram for **Dry Snow**
W Band, 20 Degree Incidence Angle

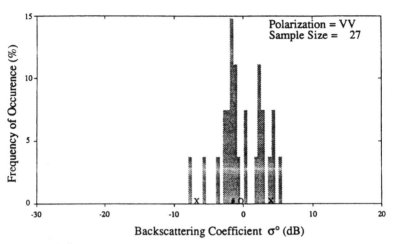

Median
O Mean
X 5% or 95% Level

Histogram for **Dry Snow**
W Band, 60 Degree Incidence Angle

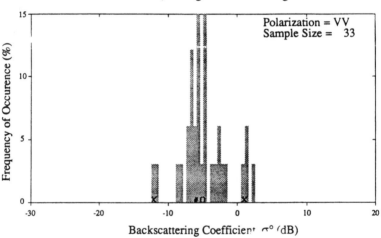

Median
O Mean
X 5% or 95% Level

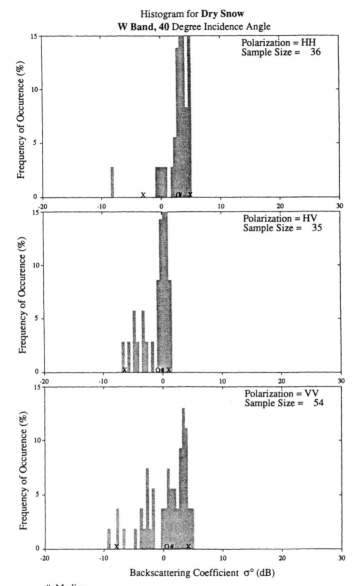

Histogram for **Dry Snow**
W Band, 40 Degree Incidence Angle

Polarization = HH
Sample Size = 36

Polarization = HV
Sample Size = 35

Polarization = VV
Sample Size = 54

Frequency of Occurence (%)

Backscattering Coefficient σ° (dB)

\# Median
O Mean
X 5% or 95% Level

APPENDIX I
BACKSCATTERING DATA FOR WET SNOW

I-1 Data Sources and Parameter Loadings

Table I.1   Data sources for wet snow.

Band	References
L	108, 111, 113, 136
S	68, 108, 111, 113, 136
C	68, 108, 111, 113, 136
X	37, 68, 93, 111, 113, 131, 136
Ku	111, 113, 131, 136
Ka	37, 60, 111, 131, 136, 144
W	52, 60, 144, 153, 154, 156

Table I.2   Mean and standard deviation parameter loadings for wet snow.

Band	Pol.	Angular Range θ_{min}	θ_{max}	$\sigma°$ Parameters P_1	P_2	P_3	P_4	P_5	P_6	SD(θ) Parameters M_1	M_2	M_3
L	HH	0	70	-73.069	95.221	1.548	30.0	1.795	-2.126	-9.0	14.416	0.109
	HV	0	70	-90.980	99.0	1.129	30.0	1.827	-2.308	4.879	0.349	15.0
	VV	0	70	-75.156	99.0	1.446	30.0	1.793	-2.179	5.230	-0.283	-1.557
S	HH	0	70	-45.772	25.160	5.942	30.0	0.929	-0.284	12.944	-9.0	-0.079
	HV	0	70	-42.940	9.935	15.0	30.0	0.438	0.712	3.276	1.027	8.958
	VV	0	70	-39.328	18.594	8.046	30.0	0.666	0.269	1.157	2.904	0.605
C	HH	0	70	-31.910	17.749	11.854	30.0	0.421	0.740	-9.0	13.0	-0.031
	HV	0	70	-24.622	15.102	15.0	-3.401	2.431	3.142	13.553	-9.0	-0.036
	VV	0	70	4.288	15.642	15.0	30.0	0.535	1.994	4.206	0.015	-2.804
X	HH	0	70	10.020	7.909	15.0	30.0	0.828	2.073	3.506	0.470	15.0
	HV	0	75	4.495	10.451	15.0	-30.0	-0.746	1.083	11.605	-9.0	0.104
	VV	0	70	10.952	6.473	15.0	30.0	0.777	2.081	4.159	0.150	1.291
Ku	HH	0	75	9.715	11.701	15.0	30.0	0.526	2.038	-9.0	13.066	-0.042
	HV	0	75	-79.693	99.0	0.981	30.0	-1.458	2.173	5.631	1.844	1.844
	VV	0	70	-9.080	13.312	15.0	-4.206	2.403	3.142	-9.0	14.014	0.043
Ka	HH	0	70	43.630	-13.027	-0.860	29.130	1.094	2.802	-8.198	15.0	-0.082
	HV											
	VV	0	70	-33.899	7.851	15.0	30.0	0.780	-0.374	5.488	1.413	0.552
W	HH											
	HV											
	VV	40	75	-22.126	99.0	2.466	0.0	0.0	0.0	4.134	15.0	3.991

$$\sigma° = P_1 + P_2 \exp(-P_3\theta) + P_4 \cos(P_5\theta + P_6)$$
$$SD(\theta) = M_1 + M_2 \exp(-M_3\theta)$$
where θ is the angle of incidence in radians.

I-2 L-Band Data:

Statistical Distribution Table for Wet Snow

L Band, HH Polarization

Angle	N	σ°_{max}	σ°_{5}	σ°_{25}	Median	σ°_{75}	σ°_{95}	σ°_{min}	Mean	Std. Dev.
0°	105	15.5	12.2	9.7	7.4	3.9	-5.2	-7.3	6.4	5.1
10°	55	3.3	-0.1	-5.1	-8.3	-11.4	-16.9	-17.7	-8.1	5.0
20°	79	-2.9	-6.6	-11.6	-15.6	-17.5	-26.1	-30.2	-14.8	5.4
30°	58	-7.2	-12.2	-17.9	-19.7	-22.7	-29.8	-31.6	-20.1	4.7
50°	120	-14.1	-16.4	-19.6	-22.5	-26.1	-29.6	-36.5	-22.9	4.4
60°	2	-25.2						-25.4	-25.3	0.1
70°	76	-20.2	-24.4	-27.4	-28.9	-30.0	-35.3	-39.1	-28.8	3.1
80°	2	-26.4						-31.7	-29.0	3.7

L Band, HV Polarization

Angle	N	σ°_{max}	σ°_{5}	σ°_{25}	Median	σ°_{75}	σ°_{95}	σ°_{min}	Mean	Std. Dev.
0°	41	0.0	-5.5	-9.0	-11.9	-14.0	-22.2	-24.9	-12.0	5.2
10°	50	-10.3	-15.0	-19.9	-23.0	-25.0	-32.5	-33.6	-22.8	5.1
20°	73	-8.3	-17.9	-20.8	-26.6	-30.3	-35.5	-41.9	-25.7	6.1
30°	47	-23.1	-26.1	-29.5	-31.6	-34.0	-37.4	-43.3	-31.6	3.8
50°	72	-20.7	-25.3	-28.2	-29.7	-33.7	-42.1	-47.4	-30.9	4.9
60°	1	-35.8						-35.8	-35.8	0.0
70°	41	-28.6	-32.5	-34.4	-35.6	-37.5	-44.3	-45.7	-36.3	3.7
80°	2	-34.2						-41.2	-37.7	4.9

L Band, VV Polarization

Angle	N	σ°_{max}	σ°_{5}	σ°_{25}	Median	σ°_{75}	σ°_{95}	σ°_{min}	Mean	Std. Dev.
0°	93	15.0	12.3	9.7	7.2	4.1	-2.4	-6.8	6.7	4.4
10°	56	4.6	2.0	-3.3	-7.7	-10.7	-17.5	-18.3	-7.2	5.6
20°	78	-3.6	-6.6	-10.0	-14.6	-17.8	-24.4	-27.5	-14.2	5.4
30°	58	-6.8	-13.0	-18.1	-20.4	-22.2	-26.1	-29.6	-19.9	4.0
50°	104	-14.4	-18.1	-19.4	-21.3	-24.9	-30.1	-36.5	-22.4	4.0
60°	3	-22.4						-28.2	-24.6	3.2
70°	81	-20.5	-22.7	-26.4	-28.1	-30.1	-33.5	-38.7	-28.3	3.4
80°	2	-25.1						-31.3	-28.2	4.4

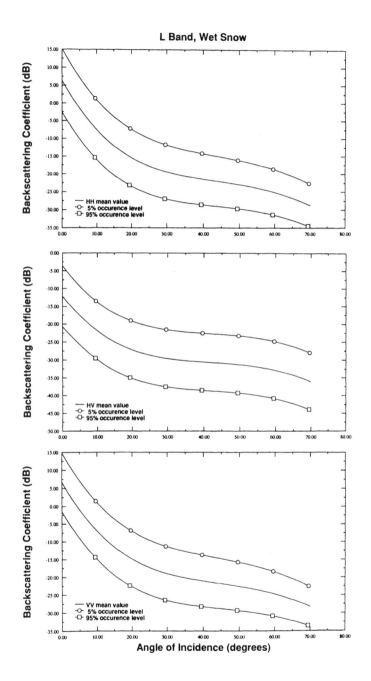

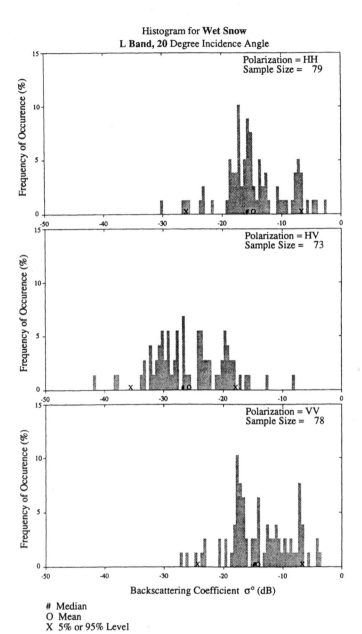

Histogram for **Wet Snow**
L Band, 20 Degree Incidence Angle

Median
O Mean
X 5% or 95% Level

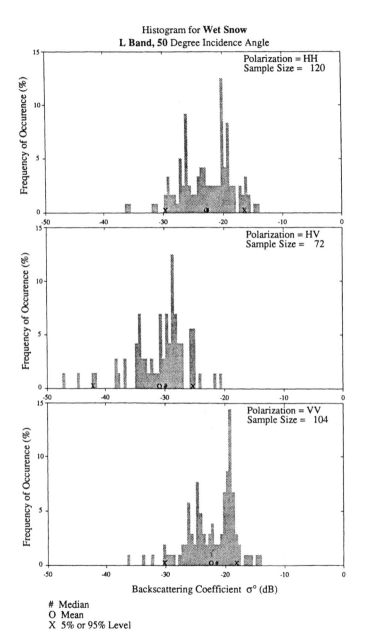

Histogram for **Wet Snow**
L Band, 50 Degree Incidence Angle

Median
O Mean
X 5% or 95% Level

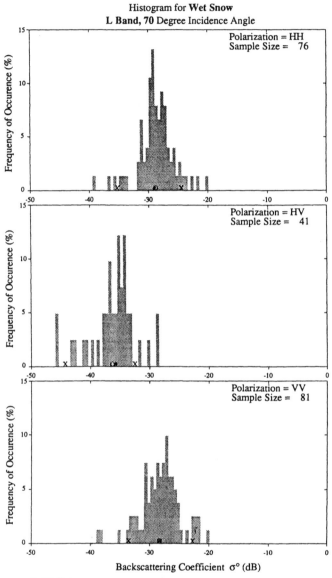

Histogram for **Wet Snow**
L Band, 70 Degree Incidence Angle

Median
O Mean
X 5% or 95% Level

I-3 S-Band Data:

Statistical Distribution Table for **Wet Snow**

S Band, HH Polarization

Angle	N	σ^o_{max}	σ^o_5	σ^o_{25}	Median	σ^o_{75}	σ^o_{95}	σ^o_{min}	Mean	Std. Dev.
0°	99	18.7	15.0	10.8	7.5	5.9	1.9	-1.3	8.2	4.0
10°	56	2.2	-1.4	-5.3	-7.7	-10.2	-13.2	-16.1	-7.3	3.7
20°	100	-3.1	-6.9	-9.4	-11.9	-15.3	-18.0	-20.6	-12.3	3.6
30°	60	-9.6	-11.4	-12.9	-14.6	-17.4	-24.0	-28.1	-15.8	3.9
50°	147	-11.9	-14.6	-17.8	-19.4	-21.5	-24.7	-30.3	-19.6	3.2
70°	97	-15.0	-22.1	-24.3	-25.6	-27.6	-32.9	-35.2	-26.0	3.1
80°	2	-28.6						-28.8	-28.7	0.1

S Band, HV Polarization

Angle	N	σ^o_{max}	σ^o_5	σ^o_{25}	Median	σ^o_{75}	σ^o_{95}	σ^o_{min}	Mean	Std. Dev.
0°	63	-2.2	-4.4	-7.6	-9.9	-12.2	-18.0	-22.4	-10.3	4.3
10°	52	-16.1	-17.3	-18.4	-20.5	-22.8	-28.8	-31.7	-21.1	3.5
20°	88	-15.5	-18.2	-20.5	-23.6	-25.1	-29.9	-32.9	-23.2	3.4
30°	57	-21.1	-22.3	-24.0	-25.9	-27.5	-32.0	-34.7	-25.9	2.8
50°	119	-22.2	-24.0	-26.3	-28.8	-31.3	-35.9	-38.6	-29.0	3.5
70°	82	-26.9	-29.0	-31.5	-33.2	-35.3	-39.1	-42.4	-33.5	3.2
80°	2	-35.9						-38.1	-37.0	1.6

S Band, VV Polarization

Angle	N	σ^o_{max}	σ^o_5	σ^o_{25}	Median	σ^o_{75}	σ^o_{95}	σ^o_{min}	Mean	Std. Dev.
0°	91	19.2	14.4	10.8	7.6	5.6	2.1	-1.4.	8.2	4.0
10°	56	2.0	-1.1	-4.6	-7.2	-9.4	-12.9	-17.1	-7.1	3.8
20°	91	-2.7	-6.4	-8.8	-11.4	-14.4	-17.7	-20.9	-11.6	3.7
30°	59	-9.3	-11.1	-13.1	-15.0	-16.6	-22.4	-22.8	-15.1	3.1
50°	125	-13.6	-16.0	-17.6	-19.0	-21.0	-24.8	-28.5	-19.4	2.8
70°	98	-16.2	-22.0	-24.1	-25.2	-26.3	-29.3	-35.1	-25.3	2.6
80°	3	-27.0						-32.0	-29.2	2.6

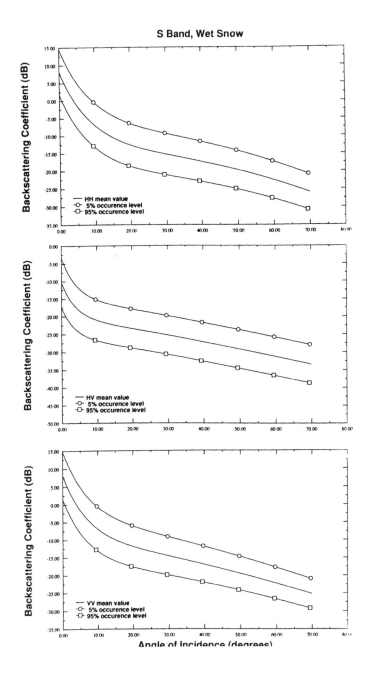

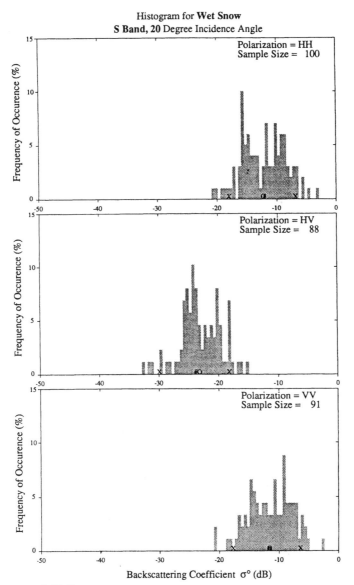

Histogram for **Wet Snow**
S Band, 20 Degree Incidence Angle

Median
O Mean
X 5% or 95% Level

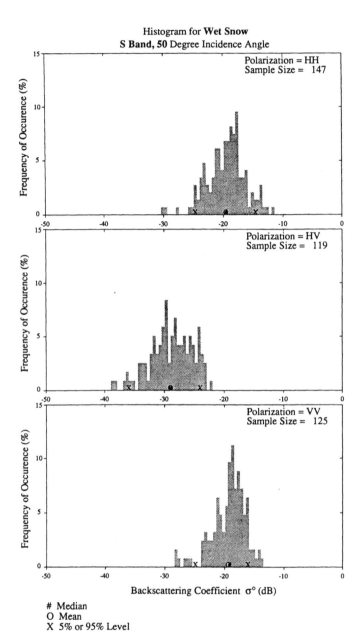

Histogram for **Wet Snow**
S Band, 50 Degree Incidence Angle

Median
O Mean
X 5% or 95% Level

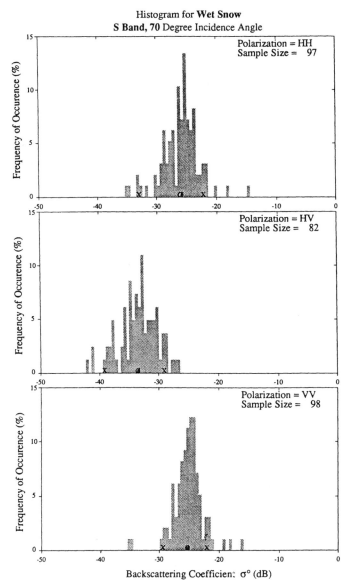

Histogram for **Wet Snow**
S Band, 70 Degree Incidence Angle

Polarization = HH
Sample Size = 97

Polarization = HV
Sample Size = 82

Polarization = VV
Sample Size = 98

Backscattering Coefficient: σ° (dB)

\# Median
O Mean
X 5% or 95% Level

I-4   C-Band Data:

Statistical Distribution Table for **Wet Snow**

C Band, **HH** Polarization

Angle	N	σ^o_{max}	σ^o_5	σ^o_{25}	Median	σ^o_{75}	σ^o_{95}	σ^o_{min}	Mean	Std. Dev.
0°	141	26.5	13.2	10.3	8.4	5.4	1.8	-0.8	8.0	3.9
10°	97	2.3	-0.2	-7.2	-9.3	-11.2	-16.4	-21.6	-9.1	4.4
20°	148	-5.0	-6.8	-9.9	-12.2	-14.2	-21.4	-27.9	-12.5	4.0
30°	102	-5.4	-10.0	-12.2	-14.7	-16.9	-23.2	-26.6	-15.0	4.1
50°	214	-10.4	-12.9	-15.7	-17.9	-20.2	-28.9	-35.7	-18.4	4.5
70°	138	-8.9	-16.0	-20.9	-22.6	-25.3	-30.3	-32.6	-22.6	4.4
80°	2	-27.1						-27.2	-27.2	0.1

C Band, **HV** Polarization

Angle	N	σ^o_{max}	σ^o_5	σ^o_{25}	Median	σ^o_{75}	σ^o_{95}	σ^o_{min}	Mean	Std. Dev.
0°	85	5.0	2.2	-2.3	-5.8	-10.9	-14.7	-17.6	-6.1	5.4
10°	92	-8.5	-15.3	-18.2	-20.2	-22.9	-30.0	-31.4	-20.7	4.1
20°	123	-12.7	-15.5	-19.2	-21.4	-24.4	-29.4	-32.4	-21.8	4.1
30°	100	-13.6	-18.3	-21.5	-23.7	-26.8	-30.7	-33.1	-24.1	3.9
50°	137	-16.2	-19.3	-23.1	-26.2	-28.5	-36.5	-39.8	-26.3	4.7
70°	115	-18.7	-21.7	-24.4	-28.7	-30.9	-35.2	-36.5	-28.0	4.1
80°	2	-35.9						-41.1	-38.5	3.7

C Band, **VV** Polarization

Angle	N	σ^o_{max}	σ^o_5	σ^o_{25}	Median	σ^o_{75}	σ^o_{95}	σ^o_{min}	Mean	Std. Dev.
0°	130	26.9	13.3	10.5	8.1	4.3	0.3	-1.3	7.6	4.5
10°	97	2.0	-1.3	-8.2	-9.6	-12.2	-16.6	-21.1	-9.5	4.2
20°	124	-1.6	-6.8	-9.4	-12.2	-14.7	-20.2	-25.1	-12.5	4.2
30°	104	-7.6	-11.0	-13.4	-14.9	-17.7	-23.2	-26.4	-15.6	3.7
50°	171	-11.3	-13.1	-15.9	-18.1	-20.7	-30.3	-33.7	-18.9	4.6
70°	103	-11.7	-15.1	-19.2	-22.3	-25.5	-29.7	-33.8	-22.2	4.6
80°	2	-25.2						-25.9	-25.5	0.5

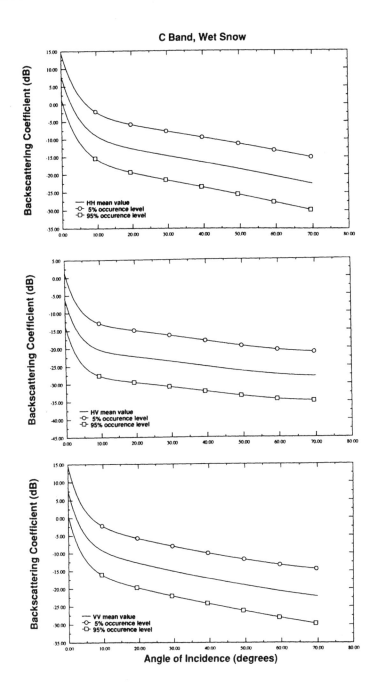

C Band, Wet Snow

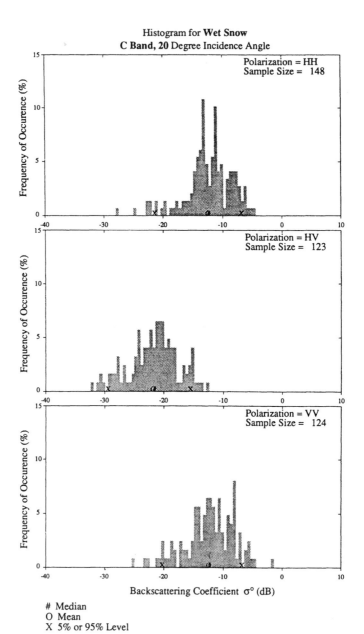

Histogram for **Wet Snow**
C Band, 20 Degree Incidence Angle

Median
O Mean
X 5% or 95% Level

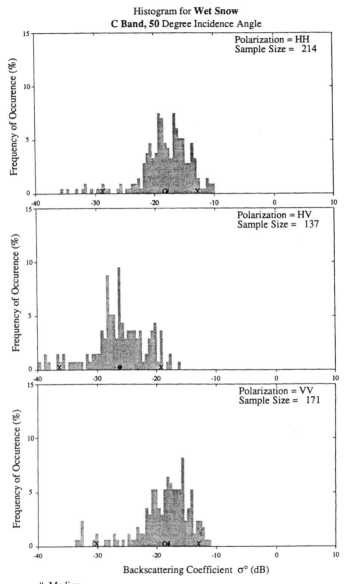

Histogram for **Wet Snow**
C Band, 50 Degree Incidence Angle

Polarization = HH
Sample Size = 214

Polarization = HV
Sample Size = 137

Polarization = VV
Sample Size = 171

Backscattering Coefficient σ° (dB)

Median
O Mean
X 5% or 95% Level

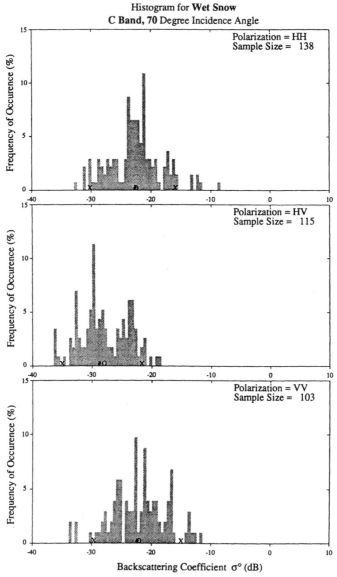

Histogram for **Wet Snow**
C Band, 70 Degree Incidence Angle

Median
O Mean
X 5% or 95% Level

I-5 X-Band Data:

Statistical Distribution Table for Wet Snow

X Band, HH Polarization

Angle	N	σ^o_{max}	σ^o_5	σ^o_{25}	Median	σ^o_{75}	σ^o_{95}	σ^o_{min}	Mean	Std. Dev.
0°	101	10.0	9.5	6.8	3.0	0.0	0.0	-5.0	3.5	3.9
5°	54	7.8	-0.2	-2.1	-3.5	-5.2	-13.1	-16.3	-4.2	4.1
10°	37	4.9	-2.5	-6.5	-8.7	-9.1	-13.6	-19.1	-7.6	3.9
20°	111	-4.0	-5.2	-8.1	-10.6	-12.9	-15.8	-21.2	-10.6	3.4
25°	48	-9.0	-9.7	-11.8	-13.5	-16.2	-18.4	-18.7	-13.8	2.7
30°	36	-10.4	-12.1	-12.8	-14.6	-16.3	-24.0	-29.8	-15.2	3.4
35°	13	-12.0						-19.5	-16.0	2.7
40°	38	-10.8	-12.3	-14.9	-16.3	-19.7	-25.0	-28.1	-17.0	3.6
45°	3	-16.7						-19.0	-18.2	1.3
50°	236	-8.9	-12.1	-15.6	-18.3	-20.2	-23.0	-28.9	-18.0	3.4
55°	54	-12.0	-14.9	-16.3	-18.1	-20.3	-21.9	-22.2	-18.1	2.5
60°	16	-9.8	-14.5	-19.8	-22.3	-24.5	-29.7	-30.0	-21.8	5.4
70°	87	-6.8	-15.2	-17.1	-19.4	-22.1	-28.2	-33.0	-20.0	4.3
75°	11	-17.6						-28.6	-22.2	4.2
80°	12	-3.8						-38.0	-26.5	11.3

X Band, HV Polarization

Angle	N	σ^o_{max}	σ^o_5	σ^o_{25}	Median	σ^o_{75}	σ^o_{95}	σ^o_{min}	Mean	Std. Dev.
0°	54	9.4	5.3	2.3	0.8	-0.9	-3.8	-4.6	0.9	2.8
5°	10	-11.5						-18.8	-14.6	2.6
10°	20	-6.9	-10.4	-11.4	-12.9	-14.0	-15.0	-15.0	-12.6	2.0
20°	79	-8.4	-10.4	-13.6	-15.6	-18.2	-20.6	-22.2	-15.5	3.3
25°	13	-14.8						-22.3	-18.4	2.5
30°	27	-14.6	-15.7	-16.3	-18.1	-19.8	-22.4	-22.4	-18.2	2.1
35°	13	-15.8						-28.0	-20.3	3.4
40°	29	-18.4	-19.4	-22.5	-23.4	-24.1	-27.8	-27.8	-23.3	2.5
45°	3	-23.1						-24.5	-23.8	0.7
50°	121	-14.8	-17.3	-19.2	-23.2	-24.9	-28.7	-29.3	-22.5	3.5
55°	8	-18.1						-25.9	-22.2	3.1
60°	8	-17.6						-30.7	-26.6	4.8
70°	76	-16.8	-18.9	-21.3	-23.8	-26.0	-31.8	-36.6	-24.2	4.1
75°	26	-24.1	-24.5	-26.1	-26.9	-27.8	-34.8	-35.6	-27.4	2.9
80°	3	-28.9						-34.6	-32.3	3.0

X Band, VV Polarization

Angle	N	σ^o_{max}	σ^o_5	σ^o_{25}	Median	σ^o_{75}	σ^o_{95}	σ^o_{min}	Mean	Std. Dev.
0°	71	9.8	9.1	6.2	0.0	0.0	0.0	-5.0	2.8	3.7
5°	19	9.6	5.5	1.7	-6.9	-11.7	-14.1	-16.0	-4.0	7.5
10°	35	6.3	-0.1	-4.9	-8.9	-9.5	-10.7	-11.1	-6.9	4.1
20°	73	-0.8	-2.3	-6.4	-10.5	-12.3	-15.8	-21.2	-9.4	4.4
25°	13	-8.9						-16.8	-13.6	2.5
30°	33	-8.7	-11.2	-12.5	-14.8	-16.3	-20.9	-29.8	-14.8	3.6
35°	13	-11.2						-20.7	-16.0	3.1
40°	13	-10.5						-28.1	-18.7	4.4
45°	3	-16.0						-17.9	-16.9	1.0
50°	101	-10.3	-11.1	-13.7	-15.9	-18.6	-23.6	-28.9	-16.4	3.9
55°	11	-13.5						-22.2	-17.5	2.9
60°	15	-9.8	-12.7	-14.1	-21.3	-24.5	-29.7	-30.0	-20.1	6.1
70°	77	-6.8	-14.1	-16.0	-18.3	-21.2	-27.2	-33.0	-18.8	4.2
80°	10	-3.8						-37.5	-25.0	11.7

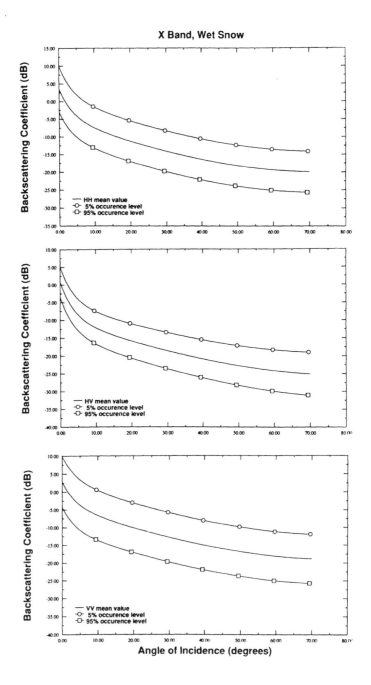

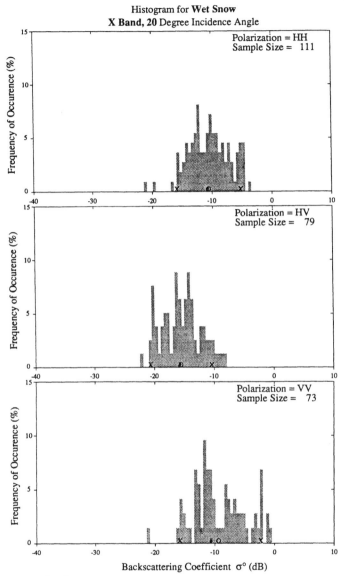

Histogram for **Wet Snow**
X Band, 20 Degree Incidence Angle

Median
O Mean
X 5% or 95% Level

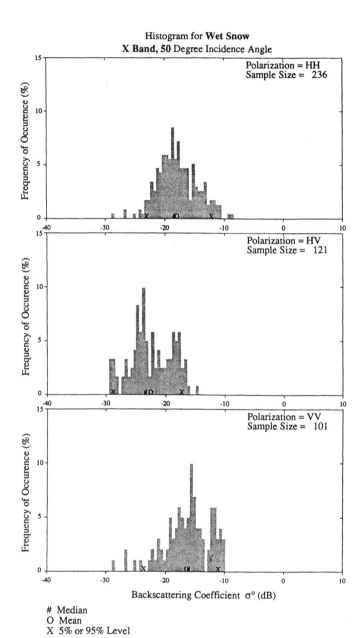

Histogram for **Wet Snow**
X Band, 50 Degree Incidence Angle

Median
O Mean
X 5% or 95% Level

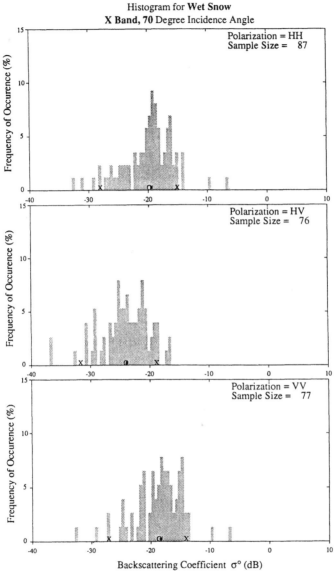

Histogram for **Wet Snow**
X Band, 70 Degree Incidence Angle

Median
O Mean
X 5% or 95% Level

I-6 Ku-Band Data:

Statistical Distribution Table for **Wet Snow**

Ku Band, **HH** Polarization

Angle	N	σ^o_{max}	σ^o_5	σ^o_{25}	Median	σ^o_{75}	σ^o_{95}	σ^o_{min}	Mean	Std. Dev.
0°	212	15.0	13.5	11.2	9.0	5.4	0.0	-3.7	8.1	4.1
5°	63	4.6	-1.6	-3.2	-4.5	-6.2	-13.4	-15.5	-5.0	3.7
10°	91	3.0	2.0	-0.2	-4.2	-7.3	-9.7	-10.4	-3.8	3.9
20°	225	1.5	-1.7	-4.6	-8.3	-11.0	-15.1	-16.7	-8.0	4.2
25°	57	-4.4	-5.4	-9.1	-10.9	-13.2	-17.3	-17.5	-10.9	3.3
30°	90	0.4	-1.5	-3.2	-7.9	-12.6	-17.0	-17.5	-8.2	5.4
35°	15	-8.8	-9.1	-10.1	-17.1	-18.6	-19.0	-19.1	-14.5	4.3
40°	52	-12.6	-14.1	-17.9	-19.1	-20.0	-21.7	-21.9	-18.6	2.3
45°	3	-14.7						-21.8	-18.4	3.6
50°	395	-1.6	-4.8	-10.5	-14.9	-18.0	-22.1	-24.7	-14.3	5.3
55°	52	-7.5	-9.4	-12.5	-14.7	-16.8	-20.5	-20.9	-14.6	3.3
60°	10	-8.3						-26.1	-18.9	7.4
70°	197	-6.8	-8.3	-11.7	-15.0	-18.5	-24.6	-29.7	-15.4	4.8
75°	49	-18.1	-19.2	-21.2	-22.8	-24.8	-28.3	-30.6	-22.9	2.7
80°	3	-18.4						-24.6	-21.7	3.1

Ku Band, **HV** Polarization

Angle	N	σ^o_{max}	σ^o_5	σ^o_{25}	Median	σ^o_{75}	σ^o_{95}	σ^o_{min}	Mean	Std. Dev.
0°	176	13.6	10.2	6.5	1.3	-1.3	-4.9	-7.5	2.4	4.8
5°	11	-8.7						-18.9	-13.8	4.1
10°	84	-0.5	-1.3	-3.3	-7.2	-10.4	-13.0	-13.9	-6.9	4.0
20°	197	-2.1	-4.1	-9.5	-13.4	-17.0	-21.5	-23.1	-12.9	5.2
25°	15	-11.9	-12.2	-12.9	-18.0	-19.5	-20.3	-20.4	-16.2	3.3
30°	90	-2.9	-4.4	-6.3	-11.2	-15.2	-20.0	-20.7	-11.0	5.1
35°	13	-12.1						-21.5	-16.8	4.0
40°	52	-16.6	-19.2	-23.3	-24.2	-25.0	-27.3	-27.8	-23.8	2.4
45°	3	-19.6						-25.4	-22.4	2.9
50°	278	-4.4	-7.5	-14.9	-19.8	-24.9	-27.5	-30.1	-19.2	6.4
55°	5	-16.1						-24.5	-19.0	3.7
60°	10	-16.1						-28.0	-23.3	4.8
70°	173	-3.9	-11.6	-15.7	-20.9	-25.0	-29.3	-32.7	-20.4	5.8
75°	49	-22.9	-23.5	-25.3	-28.3	-29.4	-31.4	-31.9	-27.6	2.5
80°	3	-26.4						-28.8	-27.7	1.2

Ku Band, **VV** Polarization

Angle	N	σ^o_{max}	σ^o_5	σ^o_{25}	Median	σ^o_{75}	σ^o_{95}	σ^o_{min}	Mean	Std. Dev.
0°	168	14.8	14.2	12.7	10.2	3.3	0.0	-1.2	8.5	5.0
5°	18	8.2	4.9	-1.8	-5.3	-10.9	-12.8	-13.2	-4.9	6.3
10°	91	3.7	2.8	0.4	-2.9	-6.8	-10.4	-11.2	-3.2	4.2
20°	172	1.3	-0.3	-2.7	-6.0	-10.2	-14.3	-17.4	-6.6	4.7
25°	15	-7.4	-7.6	-8.8	-12.6	-14.9	-16.0	-17.1	-11.6	3.3
30°	90	0.2	-0.8	-2.7	-6.5	-12.9	-16.6	-17.5	-7.5	5.2
35°	15	-7.8	-8.4	-9.1	-15.0	-16.2	-17.0	-17.6	-12.9	3.7
40°	6	-11.4						-20.7	-17.2	3.4
45°	3	-13.7						-19.5	-17.0	3.0
50°	209	-1.9	-3.4	-7.3	-11.3	-15.3	-18.2	-23.0	-11.2	4.9
55°	12	-8.6						-19.2	-14.2	4.2
60°	10	-7.4						-25.1	-16.8	7.0
70°	166	-5.7	-6.8	-10.3	-13.6	-15.7	-19.2	-24.5	-13.2	3.9
75°	3	-22.3						-28.6	-25.8	3.2
80°	3	-18.1						-25.3	-21.2	3.7

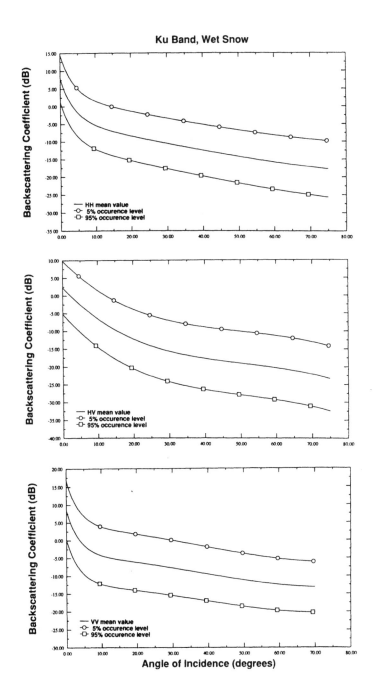

Ku Band, Wet Snow

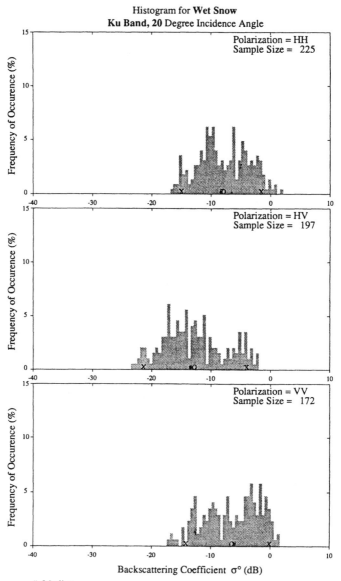

Histogram for **Wet Snow**
Ku Band, 20 Degree Incidence Angle

Median
O Mean
X 5% or 95% Level

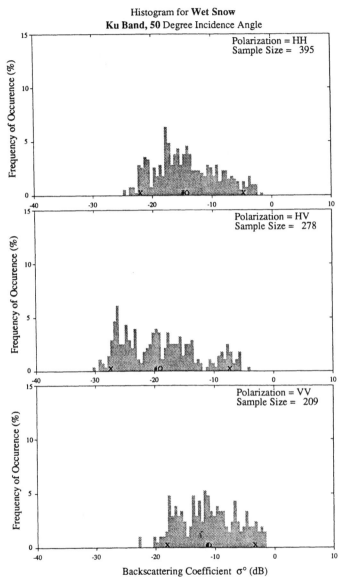

Histogram for **Wet Snow**
Ku Band, 50 Degree Incidence Angle

Median
O Mean
X 5% or 95% Level

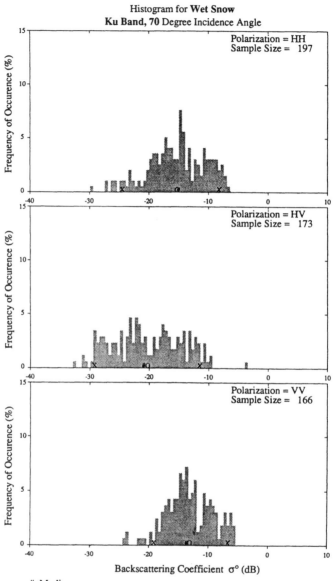

Histogram for **Wet Snow**
Ku Band, 70 Degree Incidence Angle

Median
O Mean
X 5% or 95% Level

I-7 Ka-Band Data:

Statistical Distribution Table for Wet Snow

Ka Band, HH Polarization

Angle	N	σ^o_{max}	σ^o_5	σ^o_{25}	Median	σ^o_{75}	σ^o_{95}	σ^o_{min}	Mean	Std. Dev.
0°	48	15.6	11.7	8.1	3.1	1.1	-12.6	-14.1	3.6	6.4
5°	13	5.9						-18.5	-2.6	8.5
10°	20	7.3	6.1	3.8	-1.2	-6.4	-19.3	-19.3	-2.7	8.6
20°	48	7.7	4.2	1.7	-1.2	-4.1	-17.5	-19.3	-2.3	6.2
25°	10	4.6						-25.6	-5.0	8.7
30°	20	6.5	5.1	4.0	-1.1	-12.7	-23.2	-23.2	-4.4	9.7
50°	114	4.8	2.4	-1.9	-7.9	-15.6	-20.3	-40.6	-8.5	8.2
55°	18	3.4	3.2	0.9	-4.9	-9.2	-12.5	-19.5	-4.6	6.1
70°	50	-0.3	-2.2	-4.6	-7.2	-12.2	-32.1	-32.5	-9.8	8.3
75°	5	-5.8						-28.2	-18.2	8.1
80°	4	-11.2						-39.9	-28.8	12.5

Ka Band, HV Polarization

Angle	N	σ^o_{max}	σ^o_5	σ^o_{25}	Median	σ^o_{75}	σ^o_{95}	σ^o_{min}	Mean	Std. Dev.
0°	9	-3.8						-27.4	-17.5	10.0
5°	6	-19.1						-30.5	-23.5	4.4
10°	5	-20.5						-31.4	-26.4	5.4
20°	6	-20.9						-28.7	-25.5	3.7
30°	6	-23.1						-26.2	-23.9	1.2
40°	47	-5.1	-6.8	-9.7	-24.9	-27.2	-30.0	-30.2	-19.6	9.1
50°	29	-22.4	-25.0	-27.2	-27.9	-28.7	-29.8	-29.8	-27.6	1.8
60°	6	-29.0						-33.6	-31.5	1.8
65°	2	-27.8						-33.8	-30.8	4.2
70°	3	-32.9						-38.4	-34.7	3.2
75°	27	-22.0	-25.8	-27.0	-29.0	-30.6	-35.5	-39.3	-29.1	3.3
80°	3	-31.9						-45.2	-39.4	6.8

Ka Band, VV Polarization

Angle	N	σ^o_{max}	σ^o_5	σ^o_{25}	Median	σ^o_{75}	σ^o_{95}	σ^o_{min}	Mean	Std. Dev.
0°	52	14.0	10.8	6.9	1.9	-0.3	-14.0	-14.6	2.1	6.4
5°	18	3.5	2.0	0.3	-4.2	-11.6	-18.2	-18.4	-5.7	7.1
10°	22	7.6	5.2	4.5	-2.7	-10.6	-19.3	-19.3	-3.8	8.7
20°	46	7.7	4.7	0.0	-1.4	-5.8	-15.9	-18.0	-2.8	6.1
30°	20	6.6	4.9	3.5	-4.0	-14.3	-15.1	-15.2	-4.6	8.2
40°	25	-0.5	-0.8	-1.5	-3.1	-3.8	-17.3	-19.7	-4.5	5.2
50°	76	9.4	2.8	-0.5	-4.3	-10.0	-19.1	-19.6	-5.4	6.5
55°	19	0.0	-0.6	-2.1	-8.0	-11.2	-14.1	-17.9	-7.1	5.0
70°	43	-2.1	-2.6	-3.6	-5.4	-12.0	-23.0	-27.4	-8.4	6.6
75°	7	-9.2						-27.8	-19.6	6.2
80°	4	-9.1						-33.3	-23.9	11.0

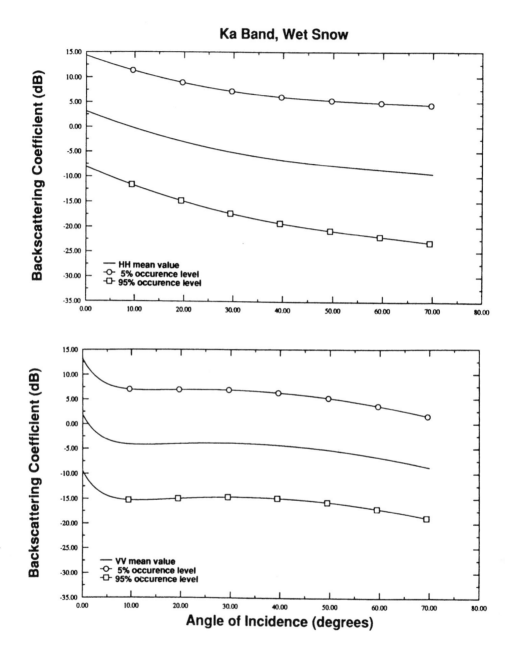

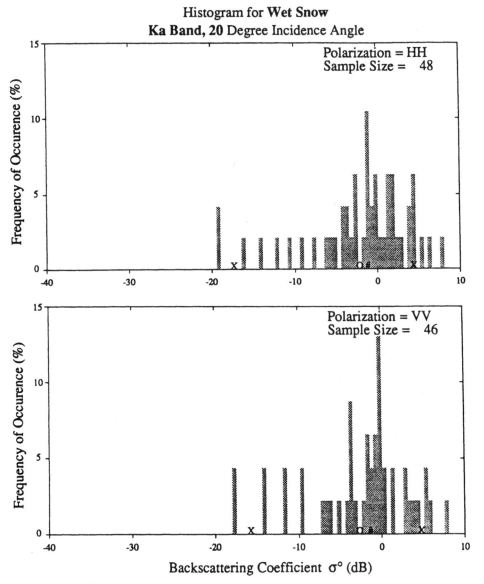

Histogram for **Wet Snow**
Ka Band, 20 Degree Incidence Angle

Polarization = HH
Sample Size = 48

Polarization = VV
Sample Size = 46

Frequency of Occurence (%)

Backscattering Coefficient σ° (dB)

\# Median
O Mean
X 5% or 95% Level

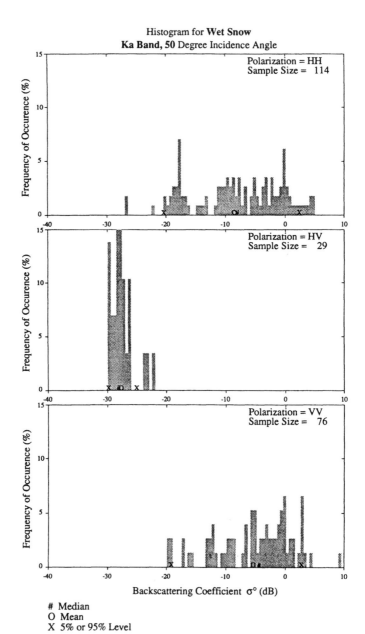

Histogram for **Wet Snow**
Ka Band, 50 Degree Incidence Angle

Median
O Mean
X 5% or 95% Level

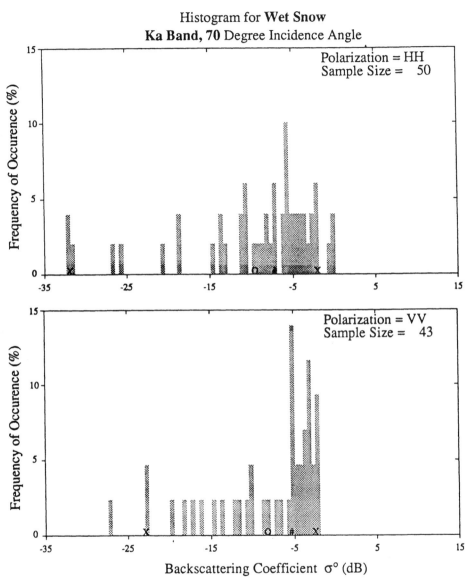

Histogram for **Wet Snow**
Ka Band, 70 Degree Incidence Angle

Median
O Mean
X 5% or 95% Level

I-8 W-Band Data:

Statistical Distribution Table for Wet Snow

W Band, HH Polarization

Angle	N	σ^o_{max}	σ^o_5	σ^o_{25}	Median	σ^o_{75}	σ^o_{95}	σ^o_{min}	Mean	Std. Dev.
0°	6	7.2						0.0	4.3	2.5
40°	16	2.8	2.6	1.9	1.1	0.7	-0.2	-0.3	1.2	0.9
45°	2	0.3						-5.0	-2.3	3.7
75°	2	-8.0						-16.0	-12.0	5.7

W Band, HV Polarization

Angle	N	σ^o_{max}	σ^o_5	σ^o_{25}	Median	σ^o_{75}	σ^o_{95}	σ^o_{min}	Mean	Std. Dev.
0°	9	2.2						-11.5	-3.2	5.5
5°	6	-6.0						-12.0	-9.3	2.1
10°	9	-6.0						-12.0	-9.3	2.1
20°	9	-5.5						-12.5	-9.8	2.4
30°	9	-6.0						-13.5	-10.1	2.5
35°	10	-12.0						-21.0	-17.8	2.7
40°	31	-0.9	-1.2	-1.8	-2.9	-12.0	-18.8	-22.0	-6.2	5.9
45°	12	-4.0						-23.0	-16.4	5.2
50°	12	-8.0						-23.0	-13.5	4.3
55°	14	-11.0						-26.0	-20.1	4.1
60°	12	-8.0						-26.0	-15.4	5.4
65°	10	-17.5						-27.0	-23.2	3.1
75°	13	-18.0						-27.5	-23.4	3.0

W Band, VV Polarization

Angle	N	σ^o_{max}	σ^o_5	σ^o_{25}	Median	σ^o_{75}	σ^o_{95}	σ^o_{min}	Mean	Std. Dev.
0°	10	5.4						-10.7	0.2	5.4
5°	10	-2.5						-10.5	-7.3	2.5
10°	11	-1.1						-11.6	-6.9	3.0
20°	14	-0.4						-12.0	-7.2	3.3
30°	14	-1.3						-13.0	-8.7	3.7
35°	13	-4.5						-13.5	-10.2	2.5
40°	31	3.5	2.6	1.6	0.6	-9.7	-12.5	-12.5	-3.1	5.9
45°	19	1.0	-5.8	-9.7	-11.5	-13.5	-15.2	-15.5	-10.7	3.8
50°	13	-5.1						-15.0	-11.0	3.2
55°	20	-7.0	-7.5	-11.2	-12.5	-14.5	-17.0	-17.0	-12.4	2.9
60°	13	-7.0						-19.0	-14.2	3.4
65°	14	-12.5						-18.5	-15.4	1.8
70°	4	-11.0						-19.0	-14.4	3.4
75°	22	-3.0	-13.9	-16.7	-19.5	-21.0	-23.2	-23.5	-18.0	5.3
80°	3	-14.0						-21.5	-17.8	3.8

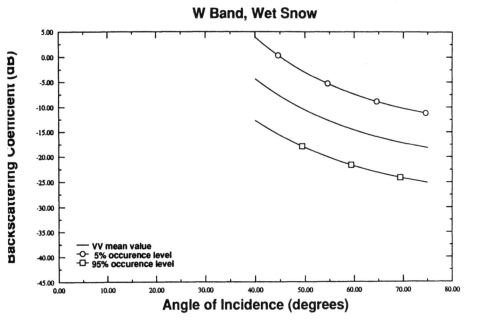

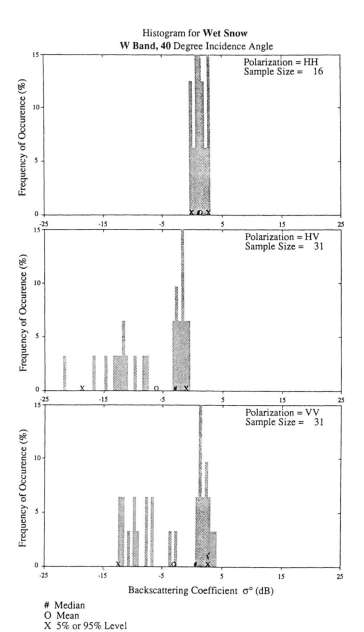

Histogram for **Wet Snow**
W Band, 40 Degree Incidence Angle

Median
O Mean
X 5% or 95% Level

PART III
COMPANION SOFTWARE BY
JOSÉ LUIS ÁLVAREZ-PÉREZ

APPENDIX J
PYTHON LIBRARY ACCOMPANYING THIS BOOK

J-1   Introduction

This appendix describes the functionality of the Python library developed to facilitate the extraction of data information from the database associated with this book. A similar set of MATLAB® routines are available in Appendix K, and instructions for using a user-friendly graphical user interface (GUI) are included in Appendix L.

The Python software is intended to provide three classes of output products, all pertaining to the radar backscattering coefficient $\sigma°$ (dB):

- Data in tabular form;
- Angular plots of $\sigma°$ (dB);
- Histograms of $\sigma°$ (dB).

The software includes thorough protection against errors. For example, if an object of the *histData* class is intended to be created with an argument, the code informs the user that no arguments were included in the initialization of the object:

```
>> dataexample = rsst.histData(5)
```

ERROR: root : Class must be instantiated with no arguments

Another useful feature of the software is the *__doc__* description for each class:

```
>> print(rsst.histData.__doc__)
```

Creates an object for the histogram data.
Usage:
>>> data3 = rsst.histData()
Similar __*doc*__ calls are available for the other classes and also for their applications:

```
>> print (rsst.modelData.plotdata.__doc__)
```

Plots the sigma–nought vs the incidence angle in different formats.
Usage (example):

```
>>> <object_name>.plotdata('f','L','hv')
```

plots a single with one plot for the mean value
and a shaded strip for the low and occurrence limit (5%, 95%)

```
>>> <object_name>.plotdata('f','L','hv',style = 'fill')
# plots a graph with a shaded area for the range between
the 5% and # the 95% occurrence limits
>>> <object_name>.plotdata('f','L','hv',style = 'line')
# plots a single graph with the mean value
```

J-2 Scope and Objectives

The software described herein is a Python package called Radar Scattering Statistics for Terrain, or *rsst*, that enables the user to read, plot, and examine the database of the radar backscattering coefficient $\sigma°$ (dB) for the various types of terrain provided in this book, at different radar bands and for different incidence angles. The use of this library requires familiarity with Python.

The user should be able to use the library or module, as it is called in Python context, to write his/her own programs so as to further expand the use of the database. It is ideally suited for the construction of analysis tools beyond those provided in this book.

J-3 System Requirements

The Python version recommended for the use of this library is Python 3.7.1. The following libraries also are required and should be installed:

- *numpy*
- *matplotlib*
- *pandas*

- *sqlalchemy*
- *scipy*

Python 3.7.1 can be installed by downloading it from the site https://www. python.org/downloads/release/python-371/. After installing Python 3.7.1, all of the above packages can be installed with the command *pip*. This command is included by default with all Python binary installers for Python versions 3.4 and higher.

To install a Python package, use

```
>> pip install <package-name>
```

with the "package-name" replaced by the name of the desired package. Note that if you are installing as a user, you may need to add the "--user" option at the end of this command. If the Python is already installed but requires updating, use

```
>> pip install --upgrade <package-name>
```

Whenever possible, *rsst* detects the versions in use and adjusts to them. For example, the *density* keyword argument used for histogram plots in matplotlib does not exist in older matplotlib versions, but it is necessary for versions 3.1 and higher. Thus, *rsst* uses this option internally if the version is 3.x, with x equal or higher than 1.

J-4 Data Formats

Data in this book is stored in four different formats to make it possible for the user to work with them under different environments. They include

1. Data files in Structured Query Language (SQL). These files have the extension .sqlite. The relational database management system SQLite is the SQL paradigm that was chosen for exercising the data files to allow easy interaction with C/C++ programming libraries, as well as Python, Java, MATLAB, and R.
2. Data files in Hierarchical Data Format (HDF5), which is easily accessible from C/C++, FORTRAN, Python, MATLAB, R, and IDL. These files have the extension .h5.
3. Data files in Python binary format (*pickle* format). This is the most effective format for Python and it is the format used by *rsst*.
4. Data files in MATLAB binary format (*mat* format). This is the preferred format for MATLAB and is the format with which the MATLAB routines are used, as described in Appendix K.

J-5 Data Parameters

For each terrain type, as listed in Table J.1, $\sigma°$ (dB) database is organized in terms of the following parameters:
Frequency band:

L = L-Band

S = S-Band

C= C-Band

X = X-Band

Ku = Ku-Band

Ka = Ka-Band

W = W-Band

Radar polarization:

hh = hh polarization

vv = vv polarization

hv = hv polarization

Incidence angle:

0, 5, 10, ..., 80 degrees

J-6 Importing rsst

Once Python and associated packages have been installed, start the python shell by entering "python" at the command line. The user may then import *rsst* by typing[1]

```
In[1]: import rsst
```

The module *rsst* provides three data classes:

- *sdtData*, which works with the statistical distribution tables of the data provided in this book.
- *modelData*, which facilitates the generation of plots of $\sigma°$ (dB) versus incidence angle for all types of terrains, frequency bands, and polarizations.
- *histData*, which deals with the data mining and plotting of the histograms. All three classes will be described in more detail in the sections that follow.

---

1. All the boxes with "In[m]:" or "Out[m]:" refer to commands typed or output in the *ipython* shell and would correspond, in a normal *python* shell, to the usual prompt ">>>".

Table J.1   Terrain Type

General Types						
a: Soil and rock surfaces	b: Trees	c: Short vegetation c1: Grasses c2: Flooded crops c3: Shrubs		d: Road surfaces	e: Urban areas	f: Dry snow cover	g: Wet snow cover
Specific Types						
a1: Smooth soil a2: Medium rough soil a3: Rough soil a4: Crop residues	b1: Coniferous trees b2: Deciduous trees (with leaves) b3: Deciduous trees (with no leaves)	c1a: Short grass c1b: Tall grass c1c: Small grains c1c1: Barley c1c2: Oats c1c3: Wheat	c3a: Large grains c3a1: Corn c3a2: Sorghum c3b: Root crops c3c: Legumes c3c1: Rapeseed c3c2: Soybeans c3d: Forage crops	d1: Asphalt d2: Concrete d3: Gravel d4: Dirt		

All of the aforementioned data classes are imported automatically by *rsst*. If a particular package is missing, the system will generate an error message, in which case the user should reinstall it using the *pip* command.

J-7   The sdtData Class

The *sdtData* Class is the simplest class and has two applications, one for reading the database and another for extracting data from it. In particular, we can define the name of the data file to be read and then read it:

```
In[2]: data1=rsst.sdtData()
In[3]: filename='RSST_SDT.pickle'
In[4]: data1.readfile(filename)
```

Now, the whole database consisting of the statistical distribution tables, which is stored in 'RSST_sdt.pickle', is available by querying the object *data1*. For example, if the user types

```
In[5]: data1.querydata('f','L','hh','sn5')
```

the response is

```
Out [5]:
Incidence angle
0.0 11.1
5.0 NaN
10.0 -3.0
15.0 NaN
20.0 -7.6
25.0 NaN
30.0 -17.5
```

```
35.0 NaN
40.0 NaN
45.0 NaN
50.0 -17.6
55.0 NaN
60.0 NaN
65.0 NaN
70.0 -26.2
75.0 NaN
80.0 NaN
Name: (f, L, hh, sn5), dtype : object
```

The book database is based on measured radar data. In some cases, there are missing measurements for certain angles of incidence for certain types of terrain. Hence, items indicated with "NaN" correspond to data that have not been measured. These missing data points appear as "NULL" in the database.

Using the definitions listed in Section J.5, let us now examine the contents of "In [5]" in the above example. The parameters ('f', 'L', 'hh', 'sn5') refer to the following:

- 'f' specifies the terrain type as 'Dry Snow'
- 'L' specifies the frequency band as L-Band
- 'hh' specifies the polarization as hh
- 'sn5' specifies the statistical attribute from among the following:
  - 'N' = number of data points available in the data distribution at a specified angle of incidence
  - 'snmax' = maximum value of $\sigma°$ (dB) contained in the data distribution at the specified angle
  - 'sn5' = the value of $\sigma°$ (dB) in the data distribution exceeded only 5% of the time
  - 'sn25' = the value of $\sigma°$ (dB) in the data distribution exceeded only 25% of the time
  - 'sn50' = the value of $\sigma°$ (dB) in the data distribution exceeded only 50% of the time
  - 'sn75' = the value of $\sigma°$ (dB) in the data distribution exceeded only 75% of the time
  - 'sn95' = the value of $\sigma°$ (dB) in the data distribution exceeded only 95% of the time
  - 'snmin' = minimum value of $\sigma°$ (dB) contained in the data distribution, at the specified angle
  - 'mean' = mean value of $\sigma°$ (dB) contained in the data distribution
  - 'stdev' = standard deviation of $\sigma°$ (dB) relative to the mean

The method *querydata* of the object *data1* requires this full list of (code, band, polarization, statistic); otherwise it returns an error message. Alternatively, it is possible to access the database using the so-called DataFrame in *pandas* format, which is stored as an *attribute* of *data1*, namely, *dataTable*. This *dataTable* is hierarchically organized in columns according to the following categories: code, band, polarization, and statistic, just like the *querydata*-method list of arguments. Each element of the column corresponds to an incidence angle value. With *data1.dataTable*, the user can access a more complex subset of data by deleting as many arguments as desired starting from the end. The outcome is a larger subset of the database. Here are some examples:

```
In[6]: data1.dataTable['f']['L']['hv']['mean']
# equivalent to
# data1.querydata('f','L','hv','mean')
In[7]: data1.dataTable['f']['L']['hv']
# returns a Data Frame
# (pandas' object) with all the statistical attributes;
# data1.querydata('f','L','hv') would not work
In[8]: data1.dataTable['f']['L']
# returns a Data Frame (pandas' object) with
# all the statistical attributes and polarizations;
# data1.querydata('f','L') would not work
In[9]: data1.dataTable['f']
# returns a Data Frame (pandas' object)
# with all the statistical attributes, polarizations and
bands;
# data1.querydata('f') would not work
In[10]: data1.dataTable
# returns a Data Frame (pandas' object) with
# all the data; data1.querydata() would not wok
```

If the objective is to access a single piece of data, such as the mean value for dry snow at L-Band, HH polarization, and an incidence angle of zero degrees, the following query can be used:

```
In[11]: data1.querydata('f','L','hv','mean',0)
Out[11]: -12.6
```

or

```
In[12]: data1.dataTable['f']['L']['hv']['mean'].loc[0]
Out[12]: -12.6
```

which returns a float type. Summing up, *data1.dataTable* is a pandas' DataFrame with the structure of a multidimensional matrix of the shape *data1.dataTable[$t] [$B][$pq][$a].loc[$wN]* where the first four indices correspond to hierarchical columns in pandas' nomenclature and the index inside *loc* represents the row. Indices can take the following values:

- *$t* specifies the code of any terrain type of Table J-1;
- *$B* is the name of the band in the list given in Section J-5;
- *$pq* is any of the three polarization configurations listed in Section J-5;
- *$a* is any of the statistical attributes listed in this section after Out[5];
- *$wN* is the combination of letter *w* and any incidence angle *N* listed in Section J-5.

We conclude this section by noting that the code includes a built-in dictionary called *terrainDict*, accessible directly as *rsst.terrainDict*, once the module has been imported. Let us suppose we want to know the code for all of the available snow types. Simply use the word 'snow' with the function *rsst.whichCode*, as in

```
In[13]: rsst.whichCode('snow')
Dry Snow: "f"
Wet Snow: "g"
```

The search term is not case-sensitive, and only a part of the word would suffice ('sno', for instance). If no matching words are found, an error indicating so is shown. Inversely, the user can look up the full name of a terrain code with

```
In[14]: rsst.terrainDict['f']
'Dry Snow'
```

J-8   The modelData Class

The creation of an object of the class *modelData* allows access to the set of modeling parameters described in the book, with which $\sigma°$ (dB) can be computed and represented against incidence angle for a choice of terrain type, band, and polarization. In a similar fashion to the methodology used to create an object of the class *sdtData* described in the previous section, we should start as follows:

```
In[15]: data2=rsst.modelData()
In[16]: filename='RSST_dat.pickle'
In[19]: data2.readfile(filename)
```

As in the case of *sdtData*, an attribute of the object called *dataTable* contains all the data of this second database. Again, there are two ways of having access to it: a dedicated one, which in this case is the method *sigmanoughtmodel*, and a direct one, through the pandas' Data Frame attribute named *dataTable* that we just mentioned.

Regarding the first option, we can write

```
In[18]: data2.sigmanoughtmodel('f','L','hv',0)
[-12.418605119198546, -4.512735119198545,
-20.324475119198546]
```

which returns three values: the mean (dB), the 5% occurence level (dB) and the 95% occurence level (dB) for the dry snow ('f') terrain type, at L-Band, HV polarization, and zero incidence angle. As for the modeling parameters; namely, 'thetamin','thetamax', 'P1', 'P2', 'P3', 'P4', 'P5', 'P6', 'M1', 'M2', and 'M3' (see listing below), they become accessible attributes of *data2* after the call to the *sigmanoughtmodel* method in *In*[18]. Thus, for example:

```
In [19]: data2.thetamax
70
```

and

```
In[20]: data2.p1
-91.341
```

output the values of the *thetamax* and *P1* parameters. The reason why the mean (dB) or the occurrence levels are given as outputs and yet the internal parameters of the database are internal attributes is that the former are values computed during the run-time of the class method according to the model described in Section 5-4.3, whereas the parameters are not computed during run-time but stored in the database (and then copied as selected parameters as attributes).

The second option is to access the parameters of the model via the *dataTable* attribute as

```
In[21]: data2.dataTable['f','L','hv']['thetamax']
#returns the value of thetamax just like in In[19]

In[22]: data2.dataTable['f','L','hv']
#returns a Data Frame
#(pandas' object) with  all the parameters for the
combination
#('f','L','hv')

In[23]: data2.dataTable['f','L']
# returns a Data Frame
```

```
    # (pandas' object) with     all the parameters for
the combination
    # ('f','L')(all polarizations)

In[24]: data2.dataTable['f']
    # returns a Data Frame
    # (pandas' object) with     all the parameters for the
terrain type 'f'
    # (all bands and polarizations)

In[25]: data2.dataTable
    # returns a Data Frame (pandas' object )
    # with all the model parameters for all terrain
    # types, bands and polarizations
```

The matrix-like notation of the pandas' DataFrame *data2.dataTable* is thus in general given by *data2.dataTable[$t,$B,$qp][$a]* or *data2.dataTable[$t][$B,'hv'].loc['thetamax']* (both are equivalent) with

- *$t* specifies the code of any X terrain type of Table J.1,
- *$B* is the name of the band in the list given in Section J-5,
- *$pq* is any of the three polarization configurations listed in Section J-5,
- *$a* is any of the following modeling parameters:
    - 'thetamin' = minimum incidence angle satisfying the $N \geq 16$ criteria for sufficient sample size;
    - 'thetamax' = maximum incidence angle satisfying the $N \geq 16$ criteria for sufficient sample size;
    - 'P1' to 'P6' = values for modeling parameters P1 to P6 ([5.1]. in Section 5-4.2)
- 'M1' to 'M3' = values for modeling parameters M1 to M3 ([5.2]. in Section 5-4.2)

The class *modelData* includes plotting options for three different output formats. The first plot type displays the mean (dB) value as a line and the 90% confidence level as a shaded strip and it is produced by calling the method *plotdata* with the option *style='fill'*,

```
In[26]: data2.plotdata('f','L','hv',style = 'fill')
```

An example is shown in Figure J.1. Alternatively, we might be interested in only the mean value. The option for that is *style='line'*

```
In[27]: data2.plotdata('f','L','hv',style = 'line')
```

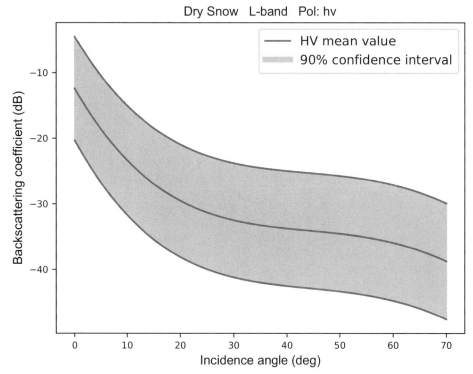

Figure J.1 Angular plot of mean value and 95% interval for dry snow at L-Band and hv polarization.

The default value for the argument *style*, is 'fill', so that if no argument is given, the output is the shaded angular plot in Figure J.1. A plot of the mean value alone is shown in Figure J.2.

J-9 The histData Class

The *histData* class is the third Python class available in *rsst*. It reads the histogram databases of the book, plots the histograms, and then computes probability distribution functions that best fit the histograms.

The first few steps follow the procedures outlined earlier for the other two classes. To create an object and read the associated data, we use

```
In[28]: data3=rsst.histData()
...  filename = 'RSST_his.pickle'
...  data3.readfile(filename)
```

Consequently, the whole database is now stored in the attribute *dataTable*,

```
In[29]: data3.dataTable
```

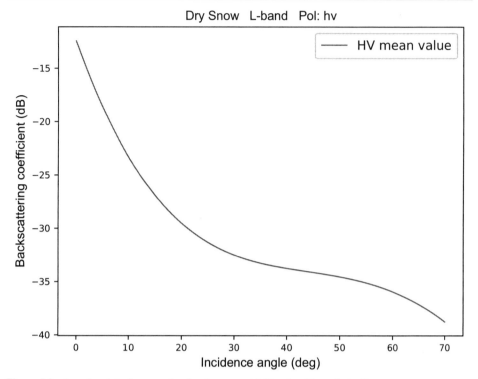

Figure J.2 Angular plot of mean value for dry snow at L-Band and hv polarization.

If we wish to inspect the first ten lines, we can type

```
In[30]: data3.dataTable.head()
```

This database allows us to generate a histogram for $\sigma°$ (dB) for any combination of terrain type, frequency band, radar polarization, and incidence angle. Additionally, the histogram data can then be fitted to any of five different types of pdfs (probability density functions):

1. The kernel density estimation (KDE): This is a so-called nonparametric fit; instead of fitting the data to a specific probability density function (pdf) for which the best parameters are chosen, usually by maximum likelihood estimation methods, we search for a fit to a more general pdf resulting from the sum of pdfs belonging to a functional basis, which in this case consists of Gaussian functions.

2. The Gaussian pdf.

3. A parametric fit to a Weibull pdf for the amplitude of the field in linear units. When plotted together with the histogram in dB, such pdf values are correspondingly transformed to power logarithmic units.

4. A parametric fit to a Gamma pdf for the intensity of the field in linear units. When plotted together with the histogram in dB, such pdf values are correspondingly transformed to power logarithmic units.

5. A parametric fit to a K-distribution pdf, in which the algorithm includes two steps. First, the measured data contained in the book is assimilated to a Gamma distribution pdf to represent the $\sigma°$ texture component. Then, a multiplicative speckle contribution is incorporated as a second Gamma distribution pdf whose order is the number of looks. The resulting two-component histogram tends to the original texture histogram as the number of looks increases. The use of this fitting technique requires only the input of the number of looks and all the intermediary steps are performed internally.

Specification of the data segment whose histogram is to be fitted includes the terrain type, band, polarization, and incidence angle, and it must follow the sequence in the following example:

```
In[31]: data3.createitem('c1','L','hh',80)
```

where the specification corresponds to "Grasses" ('c1'), L-Band, HH polarization, and 80-deg incidence angle. The data can be accessed via the subobject *hist*,

```
In[32]: data3.hist.terraincode
'c1'
In[33]: data3.hist.band
'L'
In[34]: data3.hist.polarization
'hh'
In[35]: data3.hist.inc_angle
80
In[36]: data3.hist.dataItem

Statistic
N          86
snmin    -38
snmax    -21
countData  [1.0, 0.0, 1.0, ...] , dtype : object
```

The member *data3.hist.dataItem* contains, as seen above, the following variables:

- 'N' = total number of observations available in the source data for a given combination of terrain type, band, polarization and incidence angle;
- 'snmin' = minimum value of the N $\sigma°$ (dB) observations rounded to the nearest half-dB;

- 'snmax' = maximum value of the N $\sigma°$ (dB) observations rounded to the nearest half-dB;
- 'countData' = array of number or occurrences of $\sigma°$ (dB) values in all the half-dB bins between 'snmin' and 'snmax'.

The occurrence counts contained in the database are plotted with the command *plotmat*:

```
In[37]: data3.plotmat()
```

Next, we consider how to compute the five aforementioned pdfs.

J-9.1 KDE

We run the KDE fitting method with

```
In[38]: data3.kdefit()
```

Now, we have a Python object of the type *scipy.stats* stored in *data3.hist. kde*. One of the useful features of this object is that we can use it to estimate the probability of occurrence of any specified value of $\sigma°$ (dB). In the example below

```
In[39]: data3.hist.kde.pdf(-29.3)
array([0.12485128])
```

we obtain the result that the probability density of having a $\sigma°$ value of −29.3 dB is 0.12485128. Therefore, the probability of finding such a value within a range of ±0.5 dB would be 0.12485128 × 0.5 = 0.06242564; that is, 6.24%.

We can also generate the plots for the normalized occurrence counts together with the KDE fit.

```
In[40]: data3.plotsns(fitType = 'kde')
```

We note that the histograms in Figures J.3 and J.4 do not look identical. This is because in Figure J.4 the width of the bins was optimized to suit the pdf fit. Additionally, the histogram is now normalized so that the cumulative probability is 1. However, it is still possible to generate the original (unnormalized) histogram by using the keyword argument *nbins* = 'original':

```
In[41]: data3.plotsns(fitType = 'kde', nbins =
'original')
```

The quality of the fit can be assessed by using the *kdechi* attribute of the object

```
In[42]: data3.kdechi
```

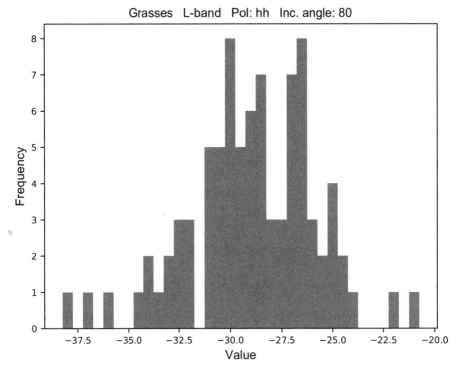

Figure J.3   Histogram of σ° (dB) for grasses at L-Band, HH polarization, and 80-deg incidence.

```
Power_divergence
Result (statistic 0.5552780019253866, pvalue = 1.0)
```

which contains the result of the chi-square test for binned data. The binning information is stored in the attributes *data3.kdeb.n* (counts) and *data3.kdeb.bins* (bin borders).

J-9.2   Gaussian Probability Distribution

We now run the fitting method with

```
In[43]: data3.gaussfit()
```

and obtain its corresponding parameters with the attribute *data3.gausspar*

```
In[44]: data3.gausspar
(-28.924418604651162, 3.076086914976733)
```

which are the mean and the standard deviation, respectively.

We also have a *scipy.stats* object stored in *data3.hist.gauss* that produces the pdf value of this fit:

```
In[45]: data3.hist.gauss.pdf(-29.3)
```

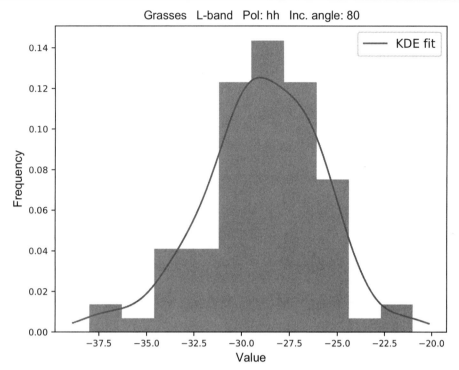

Figure J.4 Histogram and KDE fit.

```
0.12872837749575058
```

Again, we can get the plots for the normalized occurrence counts together with the Gaussian fit, as in the example below:

```
In[46]: data3.plotsns(fitType = 'norm')
```

The result of the chi-square test is available in the *gausschi* attribute:

```
In[47]: data3.gausschi
Power_divergence
Result (statistic = 0.07394990053650263,
pvalue = 0.9999999933381345)
```

The keyword argument *nbins*='original' is again an available option for this and for all the fits below. The Gaussian fit for the grasses example is shown in Figure J.5.

J-9.3 Weibull Probability Distribution

The Weibull fit is slightly different inside the code because it is performed in amplitude but is computed by the user in a way similar to the previous cases:

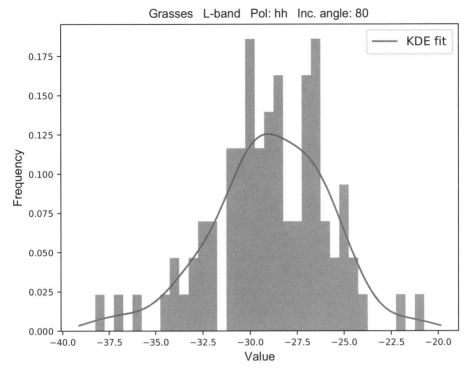

Figure J.5 Histogram and Gaussian fit.

```
In[48]: data3.weibullfit()
```

The fitting parameters are stored in the attribute *data3.weibpar*:

```
In[49]: data3.weibpar
(2.9719289016164985 ,0 ,0.042500704122617486)
```

where the first argument is the so-called shape parameter of the Weibull distribution and the third one is the scale. The second parameter is always zero and would represent a displacement in the function that does not apply in the fitting strategy applicable here. To compute the pdf for a specific σ° (dB) value, we apply the *pdf* method as in this example:

```
In[50]: data3.hist.weibull.pdf(-29.3)
0.10652605716131937
```

Finally, the Weibull pdf can be plotted with

```
In[51]: data3.plotsns(fitType = 'weibull')
```

The result of the chi-square test is available in the *weibullchi* attribute:

```
In[52]: data3.weibullchi
```

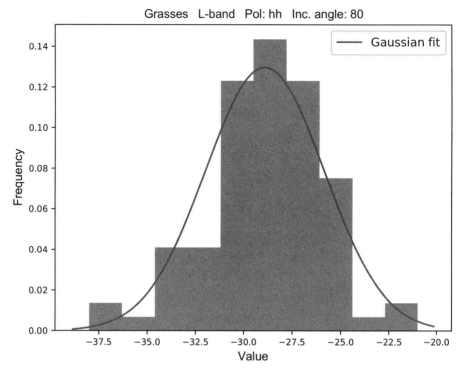

Figure J.6 Histogram with Weibull pdf.

```
Power_divergence
Result (statistic = 0.10013295706594322,
pvalue = 0.9999999742173699)
```

The fit of the Weibull distribution is shown in Figure J.6.

J-9.4 Gamma Probability Distribution

The Gamma fit is performed by

```
In[53]: data3.gammafit()
```

The fitting parameters are stored in the attribute *data3.gammapar*:

```
In[54]: data3.gammapar
(2.2782147251826284, 0, 0.0007114217195762417)
```

where the first parameter is the shape parameter of the Gamma distribution and the third one is the scale. As in the case of the Weibull distribution, the Gamma function is not displaced with respect to its origin and therefore the second parameter is set to zero. Again, to compute the pdf of a specific $\sigma°$ (dB) value, we use the *pdf* method:

```
In[55]: data3.hist.gamma.pdf(-29.3)

0.12022819267247607
```

The Gamma pdf can be plotted in a similar fashion to the previous cases:

```
In[56]: data3.plotsns(fitType = 'gamma')
```

The result of the chi-square test is available in the *gammachi* attribute. The plot of the Gamma distribution is shown in Figure J.7.

J-9.5 K-Distribution

In the database contained in this book, each individual value of $\sigma°$ (dB) represents the average value of a large number of spatial measurements referred to here as the number of Looks L. In a radar image, if L is smaller than about 10 looks, the image looks speckly, as discussed in Section 3-5. To simulate what a distribution would look like had the individual values of $\sigma°$ (dB) been an average of a limited number of looks, we can use the formulation in Section 3-5.2 to incorporate speckle effects into the pdf through the use of a second Gamma distribution that depends on the number of looks L:

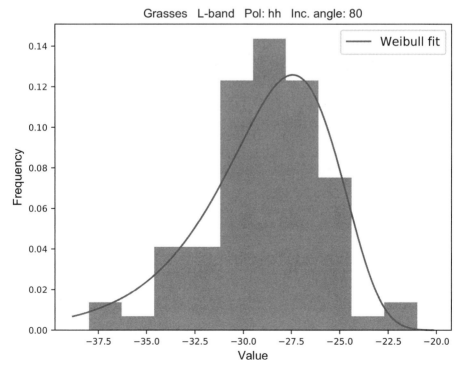

Figure J.7 Histogram and Gamma pdf.

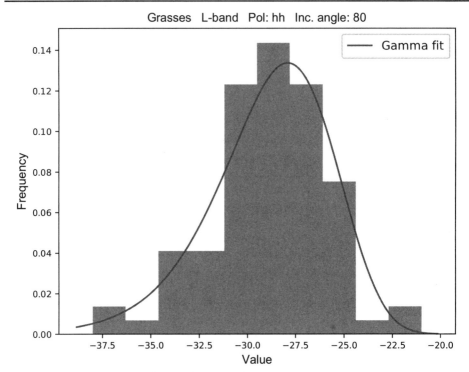

Figure J.8 Histogram and K-distribution with *L*=5 looks.

```
In[57]: data3.kdistfit(L=5)
```

We should note that in Chapter 5 we use the symbol N to denote the number of looks, but in the software, the symbol L is used instead.

To compute the pdf $\sigma°$ (dB) according to the K-distribution fit, we use the normalized pdf value

```
In[58]: data3.hist.kdist.pdf(-29.3)
 0.10659739698889292
```

The multilook K-distribution pdf can be plotted with

```
In[59]: data3.plotsns(fitType = 'kdist', L = 5)
```

An example is shown below for *L*=5 looks in Figure J.8.

APPENDIX K
MATLAB® LIBRARY ACCOMPANYING THIS BOOK

K-1 Introduction

The goal of this appendix is to describe the software tools accessible from MATLAB to extract and manipulate the database associated with this book. It mostly replicates the functionalities available from the Python code described in Appendix J, but with an orientation for MATLAB users. Therefore, it produces three classes of output products, all pertaining to the radar backscattering coefficient $\sigma°$ (dB), in the same manner as the Python library does:

- Data in tabular form;
- Angular plots of $\sigma°$ (dB);
- Histograms of $\sigma°$ (dB).

This MATLAB package is based on functions or *methods* accessible via the creation of class objects. As will be demonstrated in the forthcoming sections, no previous knowledge of object-oriented programming (OOP) is required and this package can be used either in non-OPP MATLAB code or in programs written by expert users of such a paradigm.

The usual *help* functionality of AMTLAB routines is also available in this package both for classes and methods.

K-2 Scope and Objectives

The software described herein is a MATLAB package called Radar Scattering Statistics for Terrain, or *rsst*, which mirrors the purposes of its Python counterpart. Thus, it enables the user to read, plot, and examine the database of the radar backscattering coefficient $\sigma°$ (dB) for the various types of terrain provided in this book, at different radar bands and for different incidence angles. The use of this library requires basic familiarity with MATLAB.

The user should be able to use the included functions to write his/her own programs so as to further expand the use of the database. It is ideally suited for the construction of analysis tools beyond those provided in this book.

K-3   System Requirements

This piece of software has been written with MATLAB R2018b, but none of the implemented functions are posterior to R2014b.

K-4   Data Formats

Data in this book is stored in four different formats as described in Section J-4. The user should refer to it for a full description of them.

K-5   Data Parameters

As in the previous section, the user should refer to the corresponding description included in Appendix J for the Python version of the software package, which in this case is Section J-5.

K-6   Databases And MATLAB® Classes

From the MATLAB terminal the three databases are easy to load:

```
>> load RSST_SDT.mat
>> load RSST_dat.mat
>> load RSST_his.mat
```

After being loaded, each one of these databases is stored in the *Workspace* of MATLAB with the following variable names:

- *scattdata* (from RSST_SDT.mat), which contains the statistical distribution tables of the data provided with in this book (see Section 5-4.1):
- *modelpar* (data from RSST_dat.mat), which stores the model parameters necessary to construct the angular plots of $\sigma°$ (dB) versus incidence angle (see Section 5-4.2);
- *scatthis* (data from RSST_his.mat), which contains the $\sigma°$ (dB) occurrence distributions or histograms of the measurements reported in this book.

Each variable is a structure of data with the format described in the next three sections. It is not necessary to simultaneously load the three databases, as it has been done in the lines above. If only one database is required, it can be loaded independently from the others.

To work with these three databases, the package *rsst* includes three data classes:

1. *sdtData*, which works with the statistical distribution tables database;
2. *modelData*, which facilitates the computation of the $\sigma°$ (dB) versus incidence angle for all types of terrains, frequency bands, and polarizations with the modeling parameters of *modelpar*;
3. *histData*, which deals with the data mining and plotting of the histograms.

These classes are complemented with *histItem*, which is used internally by *histData*, and the functions *whichCode* and *whichType*, defined inside files of the same name. The main three classes are described in more detail in the following sections.

K-7   The scattdata Structure

After loading RSST_SDT.mat as in Section K-6, the variable *scattdata* is a MAT-LAB structure that includes as *members* all the data blocks of the statistical distribution tables in this book. *Scattdata* has two levels:

1. It contains as members a number of substructures for which the nomenclature is *scattdata.$t_$B_$qp_$a*

where

- *$t* specifies the code of any terrain type of Table J.1
- *$B* is the name of the band in the list given in Section J.5
- *$pq* is any of the three polarization configurations listed in Section J.5

  $a is any of the following statistical attributes:

  - 'N' = number of data points available in the data distribution at a specified angle of incidence;
  - 'snmax' = maximum value of $\sigma°$ (dB) contained in the data distribution at the specified angle;
  - 'sn5' = the value of $\sigma°$ (dB) in the data distribution exceeded only 5% of the time;
  - 'sn25' = the value of $\sigma°$ (dB) in the data distribution exceeded only 25% of the time;
  - 'sn50' = the value of $\sigma°$ (dB) in the data distribution exceeded only 50% of the time;
  - 'sn75' = the value of $\sigma°$ (dB) in the data distribution exceeded only 75% of the time;
  - 'sn95' = the value of $\sigma°$ (dB) in the data distribution exceeded only 95% of the time;
  - 'snmin' = minimum value of $\sigma°$ (dB) contained in the data distribution at the specified angle;

- 'mean' = mean value of $\sigma°$ (dB) contained in the data distribution;
- 'stdev' = standard deviation of $\sigma°$ (dB) relative to the mean.

2. Each one of these substructures *scattdata.$t_$B_$qp_$a* contains seventeen member variables, each of which represents an incidence angle, indicated as *w$A* where *$A* is the number of degrees of such an angle (again see Section J-5 for all incidence angle values).

Thus, *scattdata.f_L_vv_sn5.w0* is the statistical attribute 'sn5'; that is, the value of $\sigma°$ (dB) in the data distribution exceeded only 5% of the time, for a dry-snow scene with L-band data measured in a VV-polarization configuration. The example *scattdata.f_L_vv_sn5* produces

```
>> scattdata.f_L_vv_sn5

ans = struct with fields:

     w0:  11.2000
     w5:  NaN
    w10:  -1.5000
    w15:  NaN
    w20:  -8.6000
    w25:  NaN
    w30:  -18.2000
    w35:  NaN
    w40:  NaN
    w45:  NaN
    w50:  -19
    w55:  NaN
    w60:  NaN
    w65:  NaN
    w70:  -26.3000
    w75:  NaN
    w80:  NaN
```

where NaN means that the corresponding data item has not been measured nor included in the database. Each field or member is accessible with its name, as in *scattdata.f_L_vv_sn5.w0*:

```
>> scattdata.f_L_vv_sn5.w0

ans =

    11.2000
```

K-8   The sdtData Class

The *sdtData* class is the simplest class and has two methods: one constructor that reads all the data and one method to query a specific subset of data. Instead of loading RSST_SDT.mat, the database can be read by creating an object, which stores the corresponding data:

```
>> data1=sdtData('RSST_SDT.mat');
```

The structure of data described in Section K-7 is now stored in the object *data1*. Thus, we have

```
>> data1.scattdata.f_L_vv_sn5
```

```
ans = struct with fields:

     w0:  11.2000
     w5:  NaN
    w10:  -1.5000
    w15:  NaN
    w20:  -8.6000
    w25:  NaN
    w30:  -18.2000
    w35:  NaN
    w40:  NaN
    w45:  NaN
    w50:  -19
    w55:  NaN
    w60:  NaN
    w65:  NaN
    w70:  -26.3000
    w75:  NaN
    w80:  NaN
```

In addition to this, the object can be queried in a more formal manner if you type

```
>> data1.query('f','L','vv','sn5');
```

```
ans = struct with fields:

     w0:  11.2000
     w5:  NaN
    w10:  -1.5000
    w15:  NaN
    w20:  -8.6000
    w25:  NaN
    w30:  -18.2000
    w35:  NaN
    w40:  NaN
```

```
w45:  NaN
w50:  -19
w55:  NaN
w60:  NaN
w65:  NaN
w70:  -26.3000
w75:  NaN
w80:  NaN
```

The method *query* of the object data1 requires this full list of arguments: code, band, polarization, and statistic; otherwise it returns an error:

```
>> data1.query('a','L','hh');
Four or five arguments are mandatory.
```

One of the most important reasons for using this class is its potentiality for using it as an inherited class if the user intends to elaborate more complex applications suited to his/her needs.

We conclude this section with the functions that allow the identification of the code letter for a specific terrain type, and conversely to identify terrain type from the code letter. For the former functionality we use *whichCode*, and for the latter we use *whichType*

```
>> whichCode('snow')
Dry Snow | 'f'
Wet Snow | 'g'

>> whichType('c1')
c1   | 'Grasses'
c1a  | 'Short Grass'
c1b  | 'Tall Grass'
c1c  | 'Small Grains'
c1c1 | 'Barley'
c1c2 | 'Oats'
c1c3 | 'Wheat'
c3c1 | 'Rapeseed'
```

The search term is not case-sensitive and a part of the word would work as well ('sno', for instance). If no matching words are found, an error indicating so is shown.

K-9   The modelPar Structure

After loading RSST_dat.mat as in Section K-6, the variable *modelpar* is a MAT-LAB structure that includes as *members* all the modeling parameters required to produce plots of $\sigma°$ (dB) versus incidence angle for all types of terrains, frequency bands and polarizations. Like *scattdata*, *modelpar* has two levels:

1. It contains as members a number of substructures for which the nomenclature is *modelpar.$t_$B_$qp*, where *$t* is the type of terrain and it consists of one to four characters (see Table J.1), *$B* is the name of the band, and *$qp* is a two-character indicator of the polarization receive-transmit pair (see Section J-5 for a listing of all the possibilities of *$t*, *$B*, and *$qp*).

2. Each one of these substructures *modelpar.$t_$B_$qp* contains 11 member variables: 'thetamin','thetamax', 'P1', 'P2', 'P3', 'P4', 'P5', 'P6', 'M1', 'M2' and 'M3', described in Section 5-4.2:

   ● 'thetamin' = minimum incidence angle satisfying the $N > 16$ criteria for sufficient sample size;
   ● 'thetamax' = maximum incidence angle satisfying the $N > 16$ criteria for sufficient sample size;
   ● 'P1' to 'P6' = values for modeling parameters P1 to P6 (Equation (5.1));
   ● 'M1' to 'M3' = values for modeling parameters M1 to M3 (Equation (5.2)).

The example *modelpar.f_L_vv* produces

```
>> modelpar.f_L_vv

ans = struct with fields:

    thetamin: 0
    thetamax: 70
          P1: -77.0320
          P2: 99
          P3: 1.4150

          P4: -30
          P5: 1.7200
          P6: 0.9970
          M1: 4.4870
          M2: -1.0000e-03
          M3: -5.7250
```

and *modelpar.f_L_vv.P1* outputs

```
>> modelpar.f_L_vv.P1

ans =

  -77.0320
          P6: 0.9970
          M1: 4.4870
```

```
M2: -1.0000e-03
M3: -5.7250
```

K-10   The modelData Class

In a similar fashion to the methodology described in Section K-8 when creating an object of the class *sdtData*, we can create an object of the *modelData* class:

```
>> data2=modelData('RSST_dat.mat');
```

An attribute of *data2* called *data2.modelpar* contains all the data as the structure *modelpar* described in Section K-9. Now, this class provides the functionality of computing the mean (dB), the 5% occurence-level (dB) and the 95% occurence-level (dB) with the method *sigmanoughtmodel*, as in the following example:

```
>> [snmean,sn5,sn95]=data2.sigmanoughtmodel('f','L',
'hh',0)

snmean =

    6.4504

sn5 =

    14.1359

sn95 =

    -1.2350
```

where the choice for the scene has been soil and rock surfaces ('a') at L-Band, HH polarization, and zero incidence angle. The relevant modeling parameters 'thetamin', 'thetamax', 'P1', 'P2', 'P3', 'P4', 'P5', 'P6', 'M1', 'M2', and 'M3' become accessible via the attribute *modelpar* of *data2* after the previous call to the *sigmanoughtmodel* method:

```
>> data2.modelpar

ans =

  struct with fields:

      thetamin: 0
      thetamax: 70
            P1: -74.0190
            P2: 99
            P3: 1.5920
```

```
P4:  -30
P5:  1.9280
P6:  0.9050
M1:  -9
M2:  13.6720
M3:  0.0640
```

Modeling parameters are stored in the database as members of structures, whereas the values of *snmean*, *sn5*, *sn95* are computed during run-time.

The class *modelData* allows for plotting the models of σ° (dB) versus incidence angle in two different output formats. The first plot type displays the mean (dB) value as a line and the confidence levels as a shaded strip. It is produced by calling the method *plotdata* with the argument '*fill*' after the arguments corresponding to terrain type, band, polarization and plot number:

```
>> data2.plotdata('a','L','hh',1,'fill');
```

An exmple is shown in Figure K.1 for solid and rock surfaces. Alternatively, we might be interested only in the mean vale. The option for that is

```
>> data2.plotdata('a','L','hh',2,'single');
```

In that case the outcome will be as shown in Figure K.2.

K-11   The scatthis Structure

The loading of RSST_his.mat creates the variable *scatthis* in the Workspace of MATLAB as a structure. It includes as *members* all the data required to reproduce the probability distributions in the shape of histograms σ° (dB) that are reported in this book. Like *scattdata* and *modelpar*, *scatthis* has two levels:

1. It contains as members a number of substructures for which the nomenclature is *scatthis.$t_$B_$qp_$a*, where *$t* is the type of terrain and it consists of one to four characters (see Table J.1), *$B* is the name of the band, *$qp* is a two-character indicator of the polarization receive-transmit pair, and *$a* is a string containing a short name for a statistical attribute relevant for the construction of an histogram. Sections J-5 and J-7 provide the listing of all the possibilities of *$t*, *$B*, and *$qp*. As for *$a*, it can take the following values:

   • 'N' = total number of observations available in the source data for a given combination of terrain type, band, polarization, and incidence angle;

   • 'snmin' = minimum value of the N σ° (dB) observations rounded to the nearest half-dB;

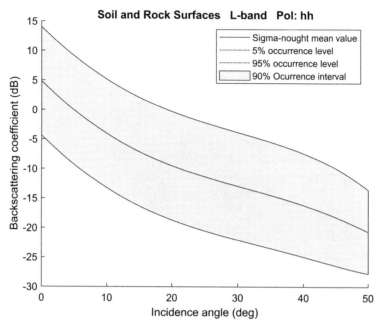

Figure K.1 Angular plot of mean value and 95% interval for dry snow at L-Band and hv polarization.

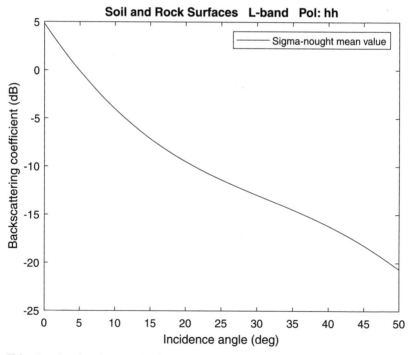

Figure K.2 Angular plot of mean value for dry snow at L-Band and hv polarization.

- 'snmax' = maximum value of the N $\sigma°$ (dB) observations rounded to the nearest half-dB;
- 'countData' = array of number or occurrences of $\sigma°$ (dB) values in all the half-dB bins between 'snmin' and 'snmax'.

2. Each one of these substructures *scatthis.$t_$B_$qp_$a* contains 17 member variables, each of which represents an incidence angle, indicated as *w$A* where *$A* is the number of degrees of such an angle (again see Section J-5 for all incidence angle values).

Thus, *scatthis.f_L_vv_snmin.w0* is the statistical attribute 'snmin'; that is, the smallest recorded value of $\sigma°$ (dB) in the data distribution for a dry-snow scene with L-Band data measured in a VV-polarization configuration. The example *scatthis.f_L_vv_snmin* produces

```
>> scatthis.f_L_vv_snmin

ans =

  struct with fields:

     w0: -7
     w5: NaN
    w10: -22
    w15: NaN
    w20: -28
    w25: NaN
    w30: -31
    w35: NaN
    w40: NaN
    w45: NaN
    w50: -37.5000
    w55: -22.5000
    w60: -24
    w65: -26.5000
    w70: -41
    w75: -31
    w80: -34.5000
```

Each field or member is accessible with its name, like in *scatthis.f_L_vv_snmin.w0*:

```
>> scatthis.f_L_vv_snmin.w0

ans =

  -7
```

K-12   The histData Class

The *histData* class is the third MATLAB class available in *rsst* designed to work with the databases of this book. It reads the database in RSST_his.mat, plots the histograms, and then fit probability distribution functions to the histograms.

To create an object of this class and read the database, we use

```
>> data3=histData('RSST_his.mat');
```

Now the whole database is now stored in the *data3* member variable *scatthis*, which is the same structure described in Section K-11.

As in the case of the Python code, this database allows us to generate a histogram for σ° (dB) for any combination of terrain type, frequency band, radar polarization, and incidence angle. Additionally, the histogram data can then be fitted to any of five different types of pdfs:

1. The kernel density estimation (KDE): This is a so-called nonparametric fit. Instead of fitting the data to a specific pdf for which the best parameters are chosen, usually by maximum likelihood estimation methods, we search for a fit to a more general pdf resulting from the sum of pdfs belonging to a functional basis, which in this case consists of Gaussian functions.

2. The Gaussian pdf.

3. A parametric fit to a Weibull pdf for the amplitude of the field in linear units. When plotted together with the histogram in dB, such pdf values are correspondingly transformed to power logarithmic units.

4. A parametric fit to a Gamma pdf for the intensity of the field in linear units. When plotted together with the histogram in dB, such pdf values are correspondingly transformed to power logarithmic units.

5. A parametric fit to a K-distribution pdf, in which the algorithm includes two steps. First, the measured data contained in this book are assimilated to a Gamma distribution pdf to represent the σ° texture component. Then, a multiplicative speckle contribution is incorporated as a second Gamma distribution pdf, whose order is the number of looks. The resulting two-component histogram tends to the original texture histogram as the number of looks increases. The use of this fitting technique requires only the input of the number of looks and all the intermediary steps are performed internally.

Specification of the data segment whose histogram is to be fitted includes the terrain type, band, polarization, and incidence angle, and it must follow the sequence in the following example:

```
>> data3.createitem('c1','L','hh',80);
```

where the selection corresponds to "Grasses" ('c1'), L-Band, HH polarization, and 80-deg incidence angle. This data now remains stored in the subobject *scatthisItem*:

```
>> data3.scatthisItem

ans =

    struct with fields:

                    N: 86
                snmin: -38
                snmax: -21
            countData: [1 0 1 0 1 0 0 1 2 1 2 3 3 0 5 5 8 5
6 7 3 3 7 8 3 2 4 2 1 0 0 0 1 0 1]
              histSel: [1×1 histItem]
       dataItemRepeat: [1×86 double]
                angle: 80
          terraincode: 'c1'
                 band: 'L'
         polarization: 'hh'
```

The occurrence counts contained in the database are plotted with the command *plotmat*:

```
>> data3.plotmat(1,'original');
```

which produces the original histogram in this book on the MATLAB figure canvas number 1, as shown in Figure K.3. By original, it is meant that the same binning as in this book is selected. By choosing a number of bins, this original binning can be altered. Examples of this will be shown in the following sections.

Next, we consider how to compute the five aforementioned pdfs.

K-12.1 KDE

After creating a data item with *data3.createitem*, we can also obtain the plots of the normalized histogram together with the KDE fit with

```
>> data3.plotsns(2,'kde','original');
```

where 2 is the number of the figure and 'original' implies that the original histogram binning from the Handbook is used. However, this affects only the representation of the histogram and not the fit, which is performed using an optimal binning scheme for fitting purposes. The same applies to the other fits of subsequent sections.

The KDE fitting method can be used to compute individual values of the probability density function, as in

```
>> data3.kdefitdB(-30)
```

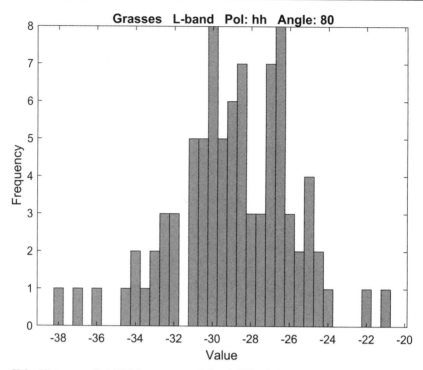

Figure K.3 Histogram of $\sigma°$ (dB) for grasses at L-Band, HH polarization, and 80 incidence.

```
ans =

   0.1172
```

The resultant KDE plot is shown in Figure K.4. A different binning is obtained if instead of 'original' we select a different number of bins by inputting the number of them instead, as in

```
>> data3.plotsns(3,'kde',10);
```

An example displaying the KDE fit when using 10 bins is shown in Figure K.5.

K-12.2   Gaussian Probability Distribution

In a similar fashion, we can obtain the plots for the normalized histograms together with the Gaussian fit with

```
>> data3.plotsns(4,'norm',10);

pd =

   NormalDistribution
```

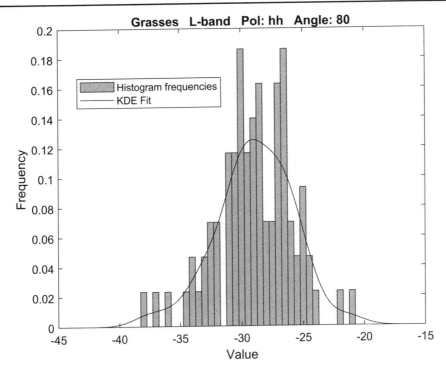

Figure K.4 Histogram and KDE fit (original binning).

```
Normal distribution
      mu = -28.9244   [-29.5878, -28.261]
   sigma =  3.09413   [2.69078, 3.64085]
```

where the quality of the fitting is provided through confidence intervals as explained in the MATLAB documentation for the command 'fitdist', whose output is *pd* in the lines above.

We can also use the fitting method to compute individual values of the probability density function:

```
>> data3.gaussfitdB(-30)

ans =

    0.1214
```

The Gaussian fit is shown in Figure K.6.

K-12.3   Weibull Probability Distribution

The Weibull fit is slightly different because it is performed in amplitude. Nevertheless, this is opaque to the user, as the input is again similar to the other cases:

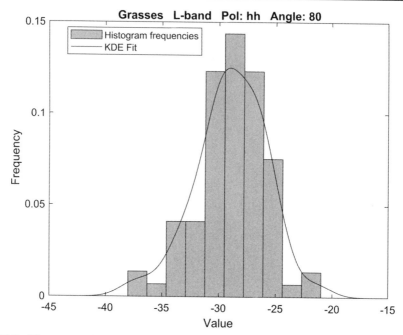

Figure K.5   Histogram and KDE fit (10 bins).

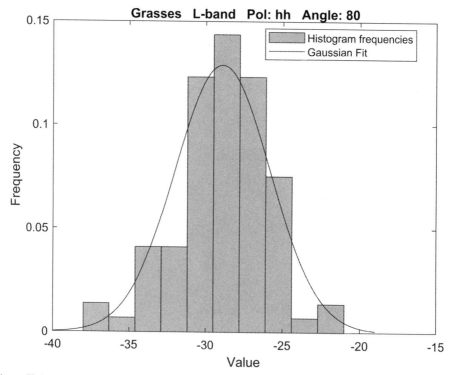

Figure K.6   Histogram and Gaussian fit (10 bins).

```
>> data3.plotsns(5,'weibull',10);

pd =

  WeibullDistribution

  Weibull distribution
    A = 0.0425007    [0.039417, 0.0458257]
    B =   2.97191    [2.55601, 3.45548]
```

where again the quality of the fitting is provided through confidence intervals as explained in the MATLAB documentation for the command 'fitdist', whose output is *pd* in the lines above.

Once again we can use the fitting method to compute individual values of the probability density function:

```
>> data3.weibullfitdB(-30)

ans =

   0.0938
```

The Weibull PDF is shown in Figure K.7.

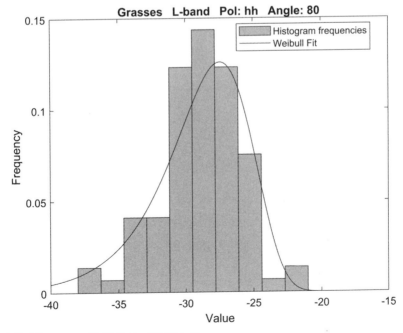

Figure K.7   Histogram with Weibull pdf (10 bins).

K-12.4   Gamma Probability Distribution

The Gamma fit is performed in the intensity:

```
>> data3.plotsns(6,'gamma',10);

pd =

    GammaDistribution

    Gamma distribution
        a =       2.27821    [1.72222, 3.01371]
        b = 0.000711422    [0.000520299, 0.00097275]
```

where the quality of the fitting is, like before, described in the documentation of the MATLAB 'fitdist' function.

Individual values of the probability density function are obtained with

```
>> data3.gammafitdB(-30)

ans =

    0.1065
```

The Gamma PDF is shown in Figure K.8.

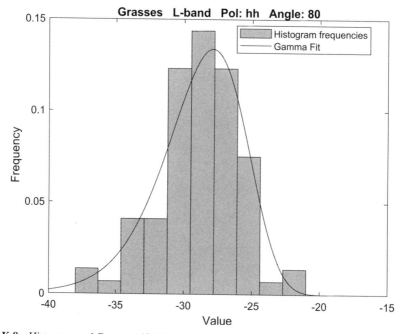

Figure K.8   Histogram and Gamma pdf (10 bins).

K-12.5   K-Distribution

In the database contained in this book, each individual value of $\sigma°$ (dB) represents the average value of a large number of spatial measurements referred to here as the number of looks L. In a radar image, if L is smaller than about 10 looks, the image looks speckly, as discussed in Section 3-5. To simulate what a distribution would look like had the individual values of $\sigma°$ (dB) been an average of a limited number of looks, we can use the formulation in Section 3-5.2 to incorporate speckle effects into the pdf through the use of a second Gamma distribution that depends on the number of looks L:

```
>> data3.plotsns(7,'kdist',10,5);

pd =

    GammaDistribution

    Gamma distribution
       a =      2.27821    [1.72222, 3.01371]
       b = 0.000711422    [0.000520299, 0.00097275]
```

The output of *pd* corresponds to the underlying, nonspeckle texture $\sigma°$ (dB). We should also note that in Chapter 5 we use the symbol N to denote the number of looks, but in the software, the symbol L is used instead.

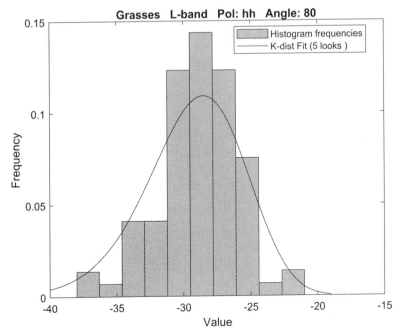

Figure K.9   Histogram and K-distribution with L = 5 looks.

To compute the pdf $\sigma°$ (dB) according to the K-distribution fit, we use the normalized pdf value

```
>> data3.kdistfitdB(-30,5)

ans = 0.1006
```

The plot of the K-distribution is shown in Figure K.9.

APPENDIX L
GRAPHICAL USER INTERFACE

L-1   Introduction

The database in this book is organized in terms of four parameters: (1) terrain type, (2) frequency band, (3) radar polarization, and (4) incidence angle.

1. Terrain type

    a = Soil and rock surfaces

    b = Trees

    c1 = Grasses

    c2 = Shrubs

    c3 = Short vegetation

    d = Road surfaces

    e = Urban areas

    f = Dry snow cover

    g = Wet snow cover

2. Frequency band

    L = L-Band

    S = S-Band

    C= C-Band

    X = X-Band

    Ku = Ku-Band

    Ka = Ka-Band

    W = W-Band

3. Radar polarization

    hh = hh polarization

    vv = vv polarization

    hv = hv polarization

4. Incidence angle

    0, 5, 10, ..., 80 degrees

L-2   Data Products

The graphical user interface (GUI) can be used to generate the following types of products.

L-2.1   Statistical Distribution Table

For a specified set of parameters (terrain type, frequency band, and radar polarization), the statistical distribution table displays the following attributes of $\sigma°$(dB):

'N' = number of data points available in the data distribution at a specified angle of incidence;

'snmax' = maximum value of $\sigma°$(dB) contained in the data distribution at the specified angle;

'sn5' = the value of $\sigma°$(dB) in the data distribution exceeded only 5% of the time;

'sn25' = the value of $\sigma°$(dB) in the data distribution exceeded only 25% of the time;

'sn50' = the value of $\sigma°$(dB) in the data distribution exceeded only 50% of the time;

'sn75' = the value of $\sigma°$(dB) in the data distribution exceeded only 75% of the time;

'sn95' = the value of $\sigma°$(dB) in the data distribution exceeded only 95% of the time;

'snmin' = minimum value of $\sigma°$(dB) contained in the data distribution, at the specified angle;

'mean' = mean value of $\sigma°$(dB) contained in the data distribution;

'stdev' = standard deviation of $\sigma°$(dB) relative to the mean.

    Examples of soil and rock surfaces at X-Band and HH polarization are shown in Table L.1.

L-2.2   Angular Plots

Upon specifying the terrain type and frequency band of polarization, the GUI generates plots of $\sigma°$(dB) versus incidence angle for the mean value, the 5% level (value of $\sigma°$(dB) exceeded only 5% of the time), and the 95% level (value of $\sigma°$(dB) exceeded 95% of the time). An example is shown in Figure L.1 for dry snow.

Table L.1 Statistical Distribution Table for Soil and Rock Suraces

X Band, HH Polarization

Angle	N	σ°_{max}	σ°_5	σ°_{25}	Median	σ°_{75}	σ°_{95}	σ°_{min}	Mean	Std. Dev.
0°	25	15.5	14.3	11.4	5.1	-3.2	-5.7	-6.3	5.0	7.4
5°	8	6.0						-7.5	1.5	4.5
10°	43	10.0	4.3	1.3	-1.6	-4.7	-12.9	-15.1	-1.6	5.1
15°	309	6.5	4.5	2.0	-1.0	-3.0	-5.0	-9.1	-0.5	3.1
20°	55	3.0	1.3	-2.0	-4.5	-6.9	-14.4	-17.0	-4.8	4.2
30°	94	3.5	-1.7	-5.6	-8.0	-10.4	-15.9	-19.1	-8.1	4.3
40°	58	3.0	0.0	-6.1	-9.2	-12.2	-20.4	-21.2	-9.1	5.5
45°	16	-7.3	-9.4	-10.3	-11.1	-12.3	-18.7	-19.2	-12.0	3.2
50°	94	3.0	-2.0	-8.4	-11.5	-13.9	-19.6	-24.9	-10.8	5.2
60°	126	1.0	-6.0	-12.2	-15.0	-21.0	-27.0	-31.0	-16.3	6.4
70°	57	-3.5	-6.4	-12.2	-14.5	-18.8	-29.1	-37.0	-15.2	6.3
80°	27	-7.5	-11.5	-13.2	-16.7	-20.5	-28.3	-29.5	-17.4	5.6

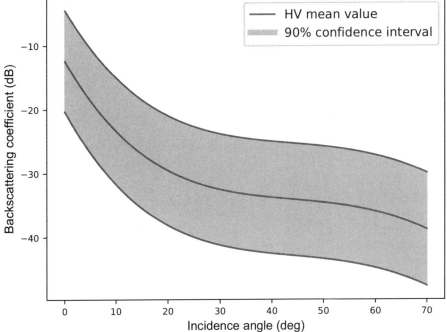

Figure L.1 Angular variation of the mean value and 90% interval for σ° (dB).

L-2.3 Histograms

For the following set of specified parameters, the GUI generates a data histogram of σ° (dB), plus a best-fit continuous histogram of the type specified in the input list:

- Terrain type
- Frequency band

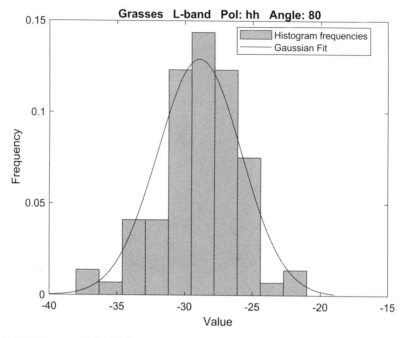

Figure L.2   Histogram of σ° (dB) for grasses at L-Band, HH polarization, and $\theta = 80^\circ$.

- Radar polarization
- Incidence angle
- Type of histogram (KDE, Gaussian, or Gamma)

An example of a Gaussian histogram is shown in Figure L.2.

About the Authors

Fawwaz T. Ulaby is the Emmett Leith Distinguished University Professor of Electrical Engineering and Computer Science at the University of Michigan. Previously, he served the university as vice president for research from 1999 to 2005. His research involves the use of radar to map terrestrial geophysical features from satellite platforms. In 2006 and 2007 he served as the chair of the Radar Review Team for the Phoenix spacecraft that landed on Mars in May 2008. Prof. Ulaby is a member of the National Academy of Engineering, and recipient of the 2006 Thomas Edison Medal and the 2012 IEEE Education Medal. He has authored/co-authored 17 books, including *Microwave Radar and Radiometric Remote Sensing*, Artech House, 2014.

M. Craig Dobson served as a NASA program manager from 2006 until 2018, with responsibility for the Space Archaeology Program and Earth Surface and Interior focus area, which includes Space Geodesy, Geodetic Imaging, and Earth Surface and Interior Programs. Craig is also program scientist for the NI-SAR mission, non-U.S. radar missions, the UAVSAR Airborne Program, as well as the Distributed Active Archive Centers (DAAC) at the Alaska Satellite Facility (ASF), the Socio-Economic Data and Applications Center (SEDAC), and the Crustal Dynamics Data Information System (CDDIS). He joined NASA HQ after spending a long research career at the university focused on understanding the microwave interactions with terrestrial materials such as (rock, soil, and vegetation), including their dielectric properties, scattering/emission behaviors, and associated applications of airborne and spaceborne synthetic aperture radar. Craig received B.A. degrees in geology and anthropology from the University of Pennsylvania and an M.A. in geography from the University of Kansas.

José Luis Álvarez-Pérez is associate professor of signal theory and communications at the University of Alcala in Madrid, Spain. His research interests include the theoretical study of scattering processes that occur in random media and their relevance for quantitative radar and radiometric remote sensing, as well as image enhancement techniques and antenna design and optimization. Dr. Álvarez-Pérez received an M.Sc. in physics from the University Autonoma of Madrid (UAM), Spain, in 1993, and a Ph.D. from the University of Nottingham in the United Kingdom in 2001. From 2000 to 2005, he worked for the Microwaves and Radar Institute (HR) of the German Aerospace Center (DLR) at Oberpfaffenhofen, Germany. His responsibilities included the SAR antenna subsystems of the Envisat and TerraSAR-X satellites. In 2005, he moved to the Technical University of Catalonia (UPC) in Barcelona, Spain, and two years later he joined the Department of Signal Theory and Communications at the University of Alcala in Madrid.

Index

Autocorrelation function, 22, 70
Autocovariance function, 71
Background distribution, 60
Backscattering coefficient, 2, 14
Backscattering data, see Appendices A to I
Calibration, 79
Correlation coefficient, 71
Clutter, see Radar clutter statistics
Confidence interval, 49
Cumulative distribution, 42
Data quality, 79
Decorrelation bandwidth, 70, 72
Decorrelation distance, 67
Fading random variable for
    power, 43
    voltage, 41
Frequency agility, 69
Frequency averaging, 72
Frequency bands, 88
Grazing angle, 5, 14
Illumination integral, 16
Imaging radar, 4, 16
Incidence angle, 5, 14
Independent samples, 44-48, 66, 72
Monostatic radar, 11
PDF, see Probability density function
Point target, 11
Polarimetric radar, 7
Polarization, 12, 14
Precision, 79
Probability density function (PDF), 41
    exponential, 43
    Gamma, 63
    log-normal, 36, 50-53, 63
    Rayleigh, 36, 38, 42
    Rice, 36, 56-58
    Weibull, 36, 54-55
Radar clutter statistics, 1, 4
    model for nonuniform terrain, 59-64
    model for point target in background, 36, 56-58
    model for uniform targets, 38, 42
    and frequency averaging, 69-78
    and spatial averaging, 67-69

Radar cross section, 13
Radar equation, 14, 16, 17
Real aperture radar (RAR), 4, 17
Signal fading, 2, 38
Speckle, 2, 3
Statistical distribution table (SDT), 90
Synthetic aperture radar (SAR), 4, 17
Target description, 79
Terrain classification, 80-87
    grasses, 86
    shrubs and bushy plants, 86
    snow, 87
    soil and rock surfaces, 83
    trees, 86
    urban land, 86
    vegetated land, 85
Texture, 2

The Artech House Remote Sensing Series

Fawwaz T. Ulaby, Series Editor

Backscattering from Multiscale Rough Surfaces with Application to Wind Scatterometry, Adrian K. Fung

Digital Processing of Synthetic Aperture Radar Data: Algorithms and Implementation, Ian G. Cumming and Frank H. Wong

Digital Terrain Modeling: Acquisitions, Manipulation, and Applications, Naser El-Sheimy, Caterina Valeo, and Ayman Habib

Handbook of Radar Scattering Statistics for Terrain, F. T. Ulaby, M. C. Dobson, and José Luis Álvarez-Pérez

Low-Power and High-Sensitivity Magnetic Sensors and Systems, Eyal Weiss and Roger Alimi

Magnetic Sensors and Magnetometers, Pavel Ripka, editor

Measurement Systems and Sensors, Second Edition, Waldemar Nawrocki

Microwave Radiometer Systems Design and Analysis, Second Edition Niels Skou and David Le Vine

Microwave Remote Sensing: Fundamentals and Radiometry, Volume I, F. T. Ulaby, R. K. Moore, and A. K. Fung

Microwave Remote Sensing: Radar Remote Sensing and Surface Scattering and Emission Theory, Volume II, F. T. Ulaby, R. K. Moore, and A. K. Fung

Microwave Remote Sensing: From Theory to Applications, Volume III, F. T. Ulaby, R. K. Moore, and A. K. Fung

Microwave Scattering and Emission Models for Users, Adrian K. Fung and K. S. Chen

Neural Networks in Atmospheric Remote Sensing, William J. Blackwell and Frederick W. Chen

Radargrammetric Image Processing, F. W. Leberl

Radar Polarimetry for Geoscience Applications, C. Elachi and F. T. Ulaby

Spectral-Spatial Classification of Hyperspectral Remote Sensing Images, Jón Atli Benediktsson and Pedram Ghamisi

Understanding Synthetic Aperture Radar Images, Chris Oliver and Shaun Quegan

Wavelets for Sensing Technologies, Andrew K. Chan and Cheng Peng

For further information on these and other Artech House titles, including previously considered out-of-print books now available through our In-Print-Forever® (IPF®) program, contact:

Artech House
685 Canton Street
Norwood, MA 02062
Phone: 781-769-9750
Fax: 781-769-6334
e-mail: artech@artechhouse.com

Artech House
16 Sussex Street
London SW1V 4RW UK
Phone: +44 (0)20-7596-8750
Fax: +44 (0)20-7630-0166
e-mail: artech-uk@artechhouse.com

Find us on the World Wide Web at:
www.artechhouse.com